Elaeis guineensis

Edited by Hesam Kamyab

Published in London, United Kingdom

IntechOpen

Supporting open minds since 2005

Elaeis guineensis
http://dx.doi.org/10.5772/intechopen.92931
Edited by Hesam Kamyab

Contributors
Razak Wahab, Mohd Sukhairi Mat Rasat, Mohammad Abdus Salam, Muhammad Haziq Razak, Nur Hafizah Ramle, Mohamad Saiful Sulaiman, Norashikin Mohd Fauzi, Hashim W. Samsi, Sitti Fatimah Mhd. Ramle, Zubaidah Aimi Abdul Hamid, Zaiton Haron, Suhaida Galip, Khairulzan Yahya, Roslli Noor Mohamed, Nadirah Darus, Herni Halim, Manoel Teixeira Souza Junior, Cleiton Barroso Bittencourt, Philippe de Castro Lins, Alessandra de Jesus Boari, Betania Ferraz Quirino, Wenceslau Geraldes Teixeira, José Sebastião Cidreira Vieira, Wilame Silva Castro, Makson Rangel de Melo Rodrigues, Neemias da Silva Nascimento, Farhan M. Said, Nor Farhana Hamid, Nur Fathin Shamirah Daud, Mohamad Al-Aamin Razali, Siti Khadijah A. Karim, Tchindjang Mesmin, Fideline Mboringong, Ludovic Miaro Iii, Gilles Etoga, Eric Voundi, Emmanuel Pierre Jonathan Ngom, Hasan Fauzi, Susilaningtyas B. Kurniawati, Akpan Noah, Dickson Osei Darkwah, Meilina Ong-Abdulllah, Famous Baa Adade, Samuel O. Agele, Friday E. Charles, Appolonia E. Obi, Ademola I. Agbona, Azman Abdullah, Ahmed Atia, Nadia Salem Alrawaiq, Mohd Kamil Md Yusof, Muhammad Fadzli Rusli, Alexis Gonzalez-Diaz, Jesus-Alberto Garcia-Nuñez, Folorunso A. Adewale Olabiyi, Yapo G. Aboua, Thomas K. Monsees, Ahmad Hussaini Jagaba, Shamsul Rahman Mohamed Kutty, Azmatullah Noor, Mohammad Fakhuma Ubaidillah Bin Md Hafiz, Nura Shehu Aliyu Yaro, Ibrahim Mohammed Lawal, Abdullahi Haruna Birniwa, Abdullahi Usman Kilaco, Gasim Hayder Ahmed Salih, Anwar Ameen Hezam Saeed

First published in London, United Kingdom, 2022 by IntechOpen
IntechOpen is the global imprint of INTECHOPEN LIMITED, registered in England and Wales, registration number: 11086078, 5 Princes Gate Court, London, SW7 2QJ, United Kingdom
Printed in Croatia

British Library Cataloguing-in-Publication Data
A catalogue record for this book is available from the British Library

Additional hard and PDF copies can be obtained from orders@intechopen.com

Elaeis guineensis
Edited by Hesam Kamyab
p. cm.
Print ISBN 978-1-83962-755-2
Online ISBN 978-1-83962-762-0
eBook (PDF) ISBN 978-1-83962-769-9

For EU product safety concerns:
IN TECH d.o.o., Prolaz Marije Krucifikse Kozulić 3, 51000 Rijeka, Croatia, info@intechopen.com or visit our website at intechopen.com.

Meet the editor

Dr. Hesam Kamyab is currently a postdoctoral research and teaching fellow at Malaysia–Japan International Institute of Technology, Universiti Teknologi Malaysia. He was appointed as the junior board member in the Journal of Cleaner Production, Elsevier (IF: 9.297) following his active paper review role. Also, he was appointed as managing guest editor in Energy Journal, Elsevier (IF: 7.147), guest editor in Environmental Science and Pollution Research, Springer (IF: 4.223), and guest editor in Sustainability Journal, MDPI (IF: 3.251). He has a chair in IWA Emerging Water Leader (EWL). He has published more than 110 papers in reputed journals and has been serving as an editorial board member of repute. In recent years, he has reviewed more than 1,000 papers.

Contents

Preface

Palm oil (*Elaeis guineensis*) originated in West Africa as a native plant useful for a variety of purposes, and several plantations have since spread widely, particularly in the South American and Asian continents. Until the mid-1960s, Nigeria was thought to be the world's largest producer of palm oil. As a result of rapid growth in the palm oil industry, Malaysia's palm oil production surpassed that of Nigeria. Malaysia was the world's largest producer and exporter of palm oil and its associated products until 2007 when Indonesia surpassed Malaysia. Palm oil tree grows well in tropical climates with a lot of rain, responsible for the successful cultivation of the plant in Indonesia and Malaysia. These two countries produce approximately 85% of the world's palm oil, with Indonesia and Malaysia contributing 44 and 41% of yields, respectively.

Annually, large amounts of palm oil biomass are generated, with only a small portion being converted into value-added products and a large portion remaining unutilized. Several researchers have published reports on the various technologies available for converting palm oil biomass into useful bioproducts such as biofuel, biogas, biofertilizers, biocomposite, and briquettes. In general, these technologies are either underutilized or insufficient for full conversion of this abundantly available biomass; thus, there is an urgent need for such technologies to be upgraded. Palm oil biomass produced on the plantation includes oil palm trunk (OPT) and oil palm fronds (OPF); biomass produced at the palm oil processing mills includes empty fruit bunch (EFB), palm kernel shell (PKS), mesocarp fiber (MF), and palm oil mill effluent (POME).

Nowadays, the palm oil enterprise is developing quickly and turning into a noteworthy agriculture-based industry in the world. While the palm oil industry has been recognized strongly for its contribution toward economic growth and rapid development, it has also contributed to environmental pollution due to the production of large quantities of by-products during the process of oil extraction.

Palm oil is the most productive oil-producing plant in the world, with 1 hectare of palm tree producing between 10 and 35 tons of fresh fruit bunch (FFB) per year. During the processing of palm oil, more than 70% of the weight of the FFB was left over as waste. Usually, the harvested part is the fruit where oil is obtained from its fleshy mesocarp. Despite the importance of edible oil and fats extracted from palm fruits, the POME contains residual oil that affects the environment and thus cannot be ignored. Treatment and disposal of oily wastewater such as POME are presently one of the serious environmental problems. Palm oil mill wastes have existed for years, but their effects on the environment are at present more noticeable. The oily waste has to be removed to prevent problems concerning hazardous pollutants particularly in the aquatic environments because they are highly toxic to the organisms. Discharging the effluents or by-products to lands or releasing them to rivers may lead to pollution and might deteriorate the surrounding environment. In order to conserve the environment, an efficient management system in the treatment of these by-products is needed.

Dr. Hesam Kamyab
Malaysia–Japan International Institute of Technology,
Universiti Teknologi Malaysia,
Jalan Sultan Yahya Petra, Kuala Lumpur, Malaysia

Section 1

Palm Oil

Chapter 1

Minor Compounds of Palm Oil: Properties and Potential Applications

Alexis Gonzalez-Diaz and Jesús Alberto García-Núñez

Abstract

The oil contained in ripe fruits produced by cultivars of African oil palm *Elaeis guineensis Jacq*., as well as that obtained from fresh fruit bunches of certain interspecific hybrid cultivars derived from crossbreeding between *Elaeis oleifera* (*Kunth*) *Cortés* and *E. guineensis* Jacq., have shown to be lipid substrates rich in valuable phytochemicals with exceptional biological properties and functional applications for multiple human health tasks. Eight isoforms of vitamin E (four tocopherols and four tocotrienols), α- and β-carotene, squalene, and various phenolic structures, make up the largest group of minor compounds in palm oil and are essential nutrients with physiological functions that include, but are not limited to their antioxidant properties. Vitamin E regulates the redox (oxidation-reduction) balance in the body, and compounds such as squalene and carotenoids are ubiquitously distributed throughout the body, including cell membranes and lipoproteins. Several studies suggest that regular intake of foods rich in this group of phytonutrients minimizes the reactivity of oxidative chemical species at the cellular level and serves as an effective adjunct in the treatment of oxidative stress.

Keywords: vitamin E, provitamin A, carotenes, phenolic compounds, phytochemicals, nutraceuticals

1. Introduction

In palm oil mills, crude palm oil (CPO) is obtained by mechanical pressing ripe fruits produced by commercial cultivars of African palm *Elaeis guineensis* Jacq. D × P type (i.e., Dura × Pisifera breed) (D × P CPO) or by *Elaeis oleifera (Kunth) Cortés* × *E. guineensis Jacq.* Breeds are commonly known as O × G interspecific hybrids (O × G CPO) under specific pressure and temperature conditions. In its natural unprocessed state, CPO is dark red, a distinctive feature that is attributed to the carotenoid fraction contained in its lipid structure, which includes α- and β-carotene (the precursor to vitamin A that gives carrots their characteristic color), and lycopene (which gives fruits and vegetables their red color) to a lesser extent [1, 2].

CPO is a fatty compound comprising an important fraction of biologically active molecules with varied physiological properties that, in appropriate amounts, stimulate the proper functioning of the immune, digestive, and reproductive systems [3–5]; facilitate the recovery of connective tissue [6]; promote the correct development of vision [5, 7]; have positive effects on the cardiovascular health of adults and the elderly [3]; limit the action of free radicals, provide protection against other reactive oxygen species, and fight oxidative stress [8–14]. A high concentration of tocopherols and tocotrienols, carotenoids, squalene, and phenolic compounds gives CPO its antioxidant power.

Antioxidants are compounds that have the ability to prevent or delay the oxidation of other molecules by inhibiting the initiation or spread of chemical reactions [15]. This allows them to protect the body against the possible effects attributed to the action of free radicals and other reactive oxygen species—ROS—(organic and inorganic oxygen ions and peroxides) [4, 16]. Depending on their source, antioxidants can be classified into two groups, one made up of those synthesized by the body (endogenous) and the other made up of those derived from food intake (exogenous) [17]. Over the last decade, the role of antioxidants in the diet and their impact on human health and the treatment of different diseases have gained significant scientific interest [18–20]. Different studies suggest that antioxidants supplied to the body *via* food intake play a key role in slowing the development of chronic diseases with the greatest impact worldwide, such as neoplastic [21, 22], neurodegenerative [23, 24], and cardiovascular [25, 26] diseases.

Furthermore, CPO is refined and fractioned by physical or chemical processes to obtain refined, bleached, and deodorized (RBD) palm olein (liquid fraction: 65–70% of unsaturated fatty acids) and RBD palm stearin (solid fraction: 30–35% of saturated fatty acids). Refining is the most effective mechanism to remove the natural color, odor, taste, and impurities of CPO [27]. However, about 99% of carotenoids are removed during the bleaching stage of palm oil refining [28], while approximately 36% of vitamin E is degraded during its refining and fractioning [29]. For a few decades now, has minimally processed and refined red palm oil been introduced into Western markets, with varying results in the consumers' perception of the product. In some cases, the natural color of red palm oil proved to be unattractive to some buyers, while for others, this property represented a high nutritional value and the richness in carotenoids of this vegetable oil [30].

In recent decades, several studies have revealed much of the biological functions of the micronutrients found to some extent in palm oil, such as phenols and tocotrienols, β-carotene, squalene, and phytosterols, which make this fatty constituent a unique and ideal raw material for various food applications given its versatility. This chapter highlights the most relevant properties of the most abundant group of minor compounds in palm oil of different sources while proposing it as a suitable material to formulate and develop functional foods enriched with palm phytochemicals.

2. Palm phytochemicals

In addition to triglycerides (>95%) [31], diacylglycerols, and free fatty acids, CPO contains a significant amount of minor compounds representing at least 1% of their lipid composition by weight (**Table 1**). These compounds can be of two types—glycerolipids such as monoglycerides, diglycerides, and phospholipid; and non-glycerolipids, which include tocopherols, tocotrienols, phytosterols, carotenoids, and other vitamins, proteins and amino acids, phenolic and polyphenolic compounds, and free fatty acids [42, 43]. Hence, the content of biologically active phytochemicals in CPO cannot be overlooked, given the attractive biological properties and the nutritional value attributed to this type of substances, as well as the marked preference of the pharmaceutical and nutraceutical industries for natural raw materials to exploit these phytonutrients [44–47].

Minor compounds	D × P CPO ($mg \cdot kg^{-1}$)	Coari La × Mé O × G CPO ($mg \cdot kg^{-1}$)	References
Tocopherols and tocotrienols (vitamin E)	500–800	876–1843	[32, 33]
Total carotenoids	988	514–1042 1172.1–1449.6	[33–35]
Total phytosterols	~300	735–1135	[32, 33]
Squalene	200–500	253.86 247.4 ± 3.3	[36–38]
Total phenolic compounds	~61–91	215–224*, †	[28]
Aliphatic alcohols	100–200	N. D	[36]
Phospholipids	20–80 5–130	N. D	[36, 39]
Isoprenoid alcohols	40–80	160.7–251.3 269.3 ± 60.0	[36, 38, 40]
Methyl sterols	40–80	6.9–14.9 12.7 ± 1.5	[36, 38, 40]
Ubiquinones	18–25 10–80	N. D	[36, 41]
Aliphatic hydrocarbons	50	N. D	[36]

N.D: no data.
**Data from Colombian Oil Palm Research Center—Cenipalma.*
†Expressed in gallic acid equivalent milligrams.

Table 1.
Minor compounds in crude palm oil from different sources.

However, palm oil refining is the most widely implemented conventional process to remove unwanted compounds such as free fatty acids, residual phospholipids, remaining metals, soap traces, volatile oxidation products, and other contaminants [48–50]. To date, there is no known refining technology on an industrial scale that is effective and selective to remove components considered as harmful or that cause adverse effects on the organoleptic qualities of the final product and that to preserve most of the original phytochemicals of CPO. This brings new opportunities for the palm sector worldwide, considering the current trends in the food market with "functional" characteristics and their influence on consumer behavior. Furthermore, this situation translates into new challenges for the oils and fats industry.

Some of the most relevant properties of the minor compounds group in the CPO of different sources are described below.

2.1 Tocopherols and tocotrienols

Tocopherols and tocotrienols are well-known isoforms of vitamin E (**Figure 1**), which greatly improve the oxidative stability of vegetable oils, thanks to their antioxidant properties [52]. In nature, tocopherols are freely found as alcohols, while tocotrienols are found in esterified forms [53]. The term vitamin E refers to eight isoforms of fat-soluble vitamins that can be classified in four tocopherol isoforms (α-, β-, γ-, and δ-Tocopherol) and in four tocotrienol isoforms (α-, β-, γ-, and δ-Tocotrienol) [54, 55], in which the position and number of methyl groups ($-CH_3$) in the chromanol ring of their structures are unequal (**Figure 1**).

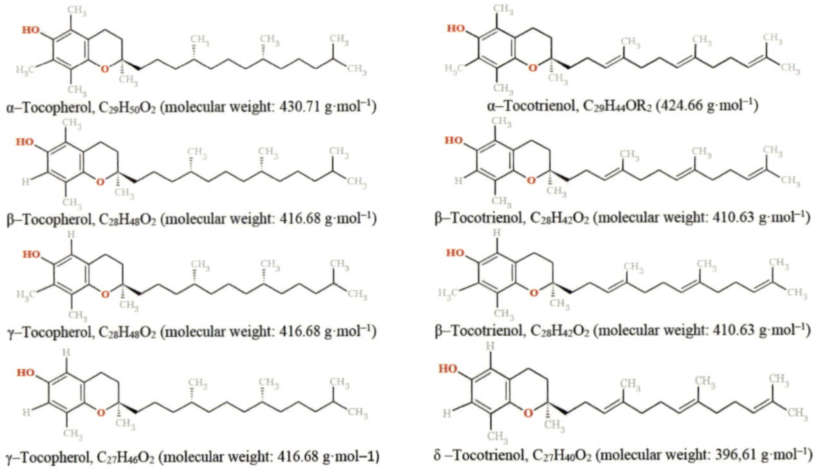

Figure 1.
Tocopherols and tocotrienols in palm oil. Chemical structured developed in ACD/CHEMSktech software [51].

In virgin vegetable oils, the concentration of the different isomeric forms of tocopherols and tocotrienols may depend on the type and quality of the raw material. In some vegetable oils, part of the original vitamin E content is removed involuntarily during refining, especially during the deodorization stage [56]. The main food sources of tocopherols and tocopherols are O × G CPO extracted from the Coari × La Mé cultivar (1316 mg·kg^{-1}) [33]; D × P CPO (914 mg·kg^{-1}) [57]; olive oil (10.4 mg·kg^{-1}) [57]; and barley germ, canola, corn germ, cottonseed, oat bran, peanut, rapeseed, rice bran, rice bran, sesame, soy, sunflower, and wheat germ oils [58, 59].

2.2 Provitamin A: carotenoids

Carotenes are pigments with an organic structure found in plants and other photosynthetic organisms [60]. The α- and β-carotenes (**Figure 2**) are tetraterpenes biochemically synthesized from eight isoprene units (methyl-1,3-butadiene) [61] and are part of more than 600 liposoluble carotenoids identified in natural sources around the world [62].

β-Carotene is a biological precursor (inactive form) of vitamin A or retinol (**Figure 2**), also responsible for the biosynthesis of other retinoids (retinol ester, retinaldehyde or retinal, retinoic acids and its analogs) [63]. β-Carotene is

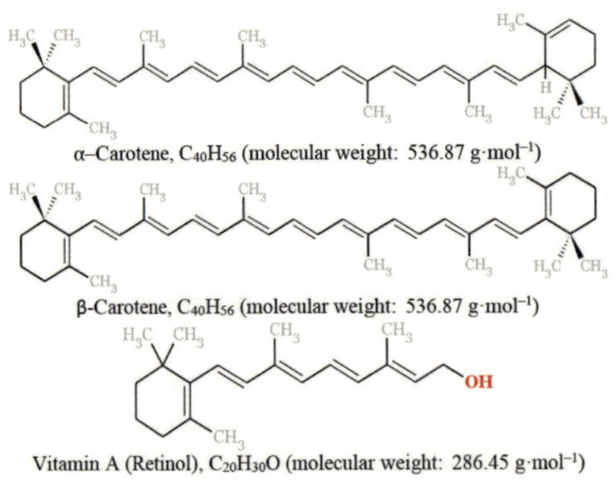

Figure 2.
Molecular structure of the most predominant carotenoids in crude palm oil (α- and β-) and vitamin A (retinol). Chemical structured developed in ACD/CHEMSktech software [51].

considered an indispensable compound for life, which must be obtained from the diet. This substance is capable of producing two retinol molecules thanks to the enzymatic action of β,β-Carotene-15,15'monooxygenase [13, 64].

Structurally, α- and β-carotene consist of 40 carbon atoms and two rings of β-ionone located at each end of the chain (**Figure 2**) [60, 65]. D × P CPO contains between 500 and 700 mg·kg^{-1} of carotenoids, with α-carotene (~ 35%) and β-carotene (~ 56%) being the most prevalent in the matrix [66]. In addition, concentrations between 514 and 1042 mg·kg^{-1} of these compounds have been found in O × G CPO extracted from the Coari × La Mé hybrid cultivar, with β–carotene accounting for approximately 73% of the total carotenoids [33]. The group of foods with high carotenoid content includes vegetables, milk and dairy products, meat and meat products, fish and seafood, eggs and derivatives, fruits, D × P CPO and O × G CPO (**Table 1**), and other vegetable fats, sauces, herbs, and spices [67].

2.3 Squalene

Squalene is a polyunsaturated triterpene made up of six isoprene units, resulting in a compound with six double bonds between carbon atoms in its structure. As a result, squalene is classified as the molecule with the highest degree of unsaturation among lipids, which makes it highly sensitive to oxidation [68]. Squalene belongs to the group of natural antioxidants known as isoprenoids, classified as a bioactive compound with the ability to prevent or minimize the negative effects of free radicals on cells in the human body [69, 70]. Some studies suggest that the squalene secreted in the fatty mantle of human skin provides protection against ultraviolet radiation [71]. D × P CPO has been found to contain between 200 and 500 mg·kg^{-1} of squalene [36], whereas O × G CPO of the Coari × La Mei hybrid cultivar has been found to contain 253.86 mg·kg^{-1} of squalene on average [37]. Currently, squalene is classified as a component with nutritional and medicinal properties with vast expectations for application in the pharmaceutical industry. Some of these properties include cardioprotective, antioxidant, antibacterial, antifungal, anticancer, and detoxifying effects [72].

2.4 Phenolic compounds: phenols and polyphenols

CPO contains significant amounts of phenolic phytohormones (e.g., p-salicylic acid), phenolic aldehydes (e.g., protocatechuic aldehyde), and phenolic acids (e.g., vanillic acid, protocatechuic acid, gallic acid, and ferulic acid) (**Figure 3**) which together make up the largest proportion of phenolic compounds in this type of oil [73]. In plants, phenolic compounds are secondary natural metabolites that are biologically synthesized by the shikimic acid (shikimate-phenylpropanoid) pathway, resulting in phenylpropanoids [74], or by the acetate-malonate pathway (polyketide route), in which monomeric and polymeric phenols and polyphenols are produced [75].

These compounds have important physiological functions in plants and play a key role as defense compounds when environmental stress, pathogen attack, herbivory, and nutrient deficiency lead to a systematic increase in the production of free radicals and other oxidative chemical species [75]. Furthermore, phenolic compounds are regularly described as bioactive substances with antioxidant properties at the cellular level, partly attributed to their ability to act as chelators of metal ions [76–78].

In foods, phenolic compounds influence their appearance, quality, acceptability, and stability because they act as dyes [79], antioxidants [80], and flavorings [81]. Cereals and legumes (e.g., wheat flour, soy, and oats), as well as fruits (e.g., sweet orange, yellow raspberry, and apples) and vegetables (e.g., red cabbage, broccoli,

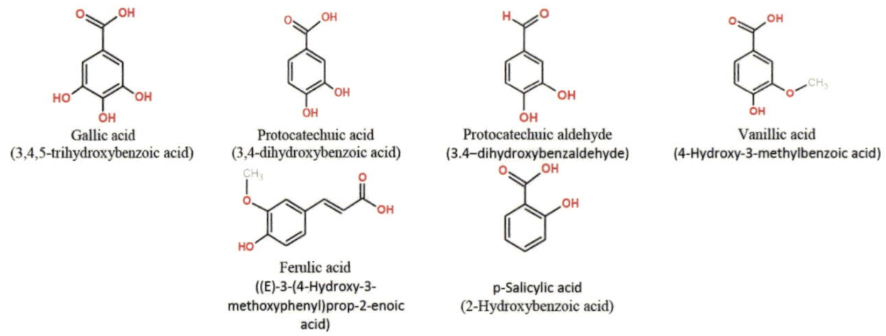

Figure 3.
Phenolic compounds of major relevance in palm oil. Chemical structured developed in ACD/CHEMSktech software [51].

carrots, tomatoes, and spinach) [82], and some vegetable oils (e.g., palm [28, 83], olive [84], soy and cotton [85], coconut [86], sesame and sunflower [87] oils), are part of the food sources of phenolic compounds.

2.5 Phytosterols and phytostanols

Phytosterols and phytostanols are biologically active compounds regularly found in plants and various foods of plant origin. Phytosterols differ from cholesterol in that they have a different elemental distribution in the side chain that forms their chemical structure (**Figure 4**), whereas phytostanols are 5α-saturated derivatives of phytosterols [88]. These structural changes, although minimal, make phytosterols, phytostanols, and cholesterol have particular physicochemical characteristics and differentiate them from each other metabolically and functionally.

Phytosterols and phytostanols are not synthesized by the human body [89, 90]; therefore, they must be supplied to the body through the intake of foods rich in these compounds. The main food sources of phytosterols are vegetable oils, fats, and edible fatty derivatives [91, 92], as well as nuts [93], cereals and derivatives [92], and vegetables [94]. The most abundant phytosterols and phytostanols in the human diet are β-sitosterol, campesterol, sitostanol, and campestanol [95].

D × P CPO contains a substantial amount of natural sterols (325–527 mg·kg^{-1}) [36], mainly constituted by β-sitosterol (46.55 ± 0.93%), campesterol (31.91 ± 0.5%), and stigmasterol (21.54 ± 0.86%) [59]. Similarly, concentrations between 735 and 1135 mg·kg^{-1} of these compounds have been found in O × G CPO extracted from the Coari × La Mé hybrid cultivar, with β-carotene accounting for approximately 63% of the total sterols [33].

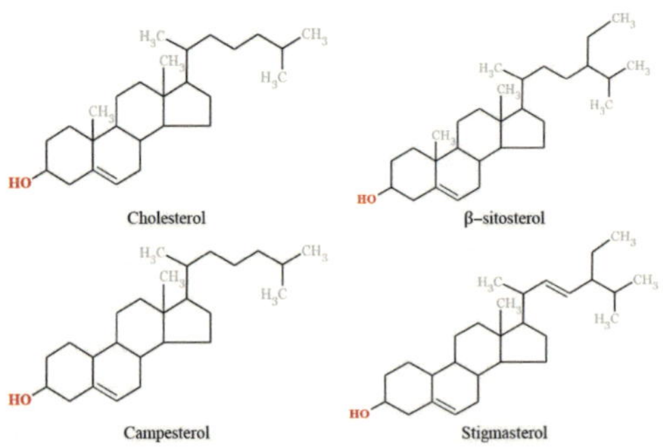

Figure 4.
Phytosterols most abundant in palm oil and cholesterol. Chemical structured developed in ACD/CHEMSktech software [51].

3. Oil palm as natural ingredient rich in biologically active constituents

Both D × P CPO and O × G CPO of the Coari × La Mé cultivar are naturally occurring lipid materials with important tocopherols and tocotrienols contents (**Table 1**), with a wide range of uses in various productive sectors. For the pharmaceutical sector, as for the food and nutraceutical industries, palm oil of different sources may be an active component to enrich various edible matrices or to formulate and develop new products. Given the high content of vitamin E in its lipid composition, palm oil can be incorporated into the formulation of products that may be useful to prevent or treat vitamin E deficiency, associated with health disorders such as peripheral neuropathy, retinopathy pigmentosa ataxia, and myopathy [96–98].

Carotenoids in palm oil can be biologically active primary components in the formulation of new products; furthermore, the amount of tocopherols, tocotrienols, and squalene naturally found in this oil could add even more value to products that may contain them [99–102]. Also, the β-carotene in palm oil can serve as an active component in food aggregates for human and animal diets [103–105] due to the properties that have been identified in this compound at the biological level and due to the reported benefits of its consumption for human [106, 107] and animal [108] health.

On the other hand, the phytosterols that make up the complex of minor compounds in palm oil (**Figure 4**) have industrial uses as essential elements required to manufacture various products [109–111]. β-Sitosterol, one of the natural sterols found in greater amounts in palm oil, could be included in food preparations aimed at reducing low-density lipoproteins (LDL), which are closely related to the development of cardiovascular diseases [112, 113].

To another extent, squalene is a constituent of high biological value used as an aggregate in different products. This compound is part of the raw materials used in cosmetics [114] and the formulation of pharmaceutical and food products [115, 116]. In the food and cosmetic industries, squalene is used as an additive due to the several benefits for human health that have been reported in various works [72, 117]. Squalene in palm oil could contribute to the enrichment of diverse food matrices and, together with tocopherols, tocotrienols and carotenoids, could collectively supplement much of the deficiency of these substances in some organisms. In addition to the aforesaid, some research has found that squalene is an effective chemotherapeutic agent for the treatment of colon carcinomas [118], breast cancer [119], and pancreatic tumors [120].

4. Available technologies to extract, concentrate, and/or purify palm phytochemicals

Several mechanisms have been developed to extract, fraction, and refine the phytochemicals, which are found in CPO. Such processes include, but are not limited to, extraction with supercritical fluids [121], molecular distillation and crystallization [122], and molecular distillation with prior esterification/transesterification of oil [123]. In general, these technologies are repeatedly used to produce oily extracts rich in squalene, carotenoids, tocopherols, and tocotrienols.

By means of solid-phase extraction and fractionation in polar phase mobile bed, a product right in α-tocotrienol free of other isomers of tocotrienols, tocopherols, and carotenoids was obtained from CPO [124]. Squalene, vitamin E, and phytosterols were fractioned from CPO through a process that included esterification, transesterification, vacuum distillation, saponification, crystallization, and exclusion of organic solvents stages [125]. On the other hand, the implementation of supercritical fluids was useful in the production of extracts rich in tocochromanols and carotenoids from CPO.

According to several authors, the vitamin complex composed of tocopherols and tocotrienols can be extracted, fractioned, and purified from various biomaterials by using technologies such as solvent extraction (direct extraction [126], Soxhlet extraction [127], and pressurized fluid extraction (PLE) [128]), supercritical fluid extraction (SFE) [129], enzyme extraction [130], extraction with prior chemical modification of the oil's lipidic matrix (saponification [127] and esterification [131]), absorption [132], sequential adsorption-desorption [133], molecular distillation [134], microwave-assisted extraction (MAE) [135], and membrane filtration [136].

On the other hand, some of the most widely implemented methodologies to extract carotenoids from different vegetable materials include processes such as liquid extraction at atmospheric pressure with maceration [137], Soxhlet extraction [138], MAE [139], ultrasound-assisted extraction (UAE) [140], accelerated solvent extraction (ASE) [141], pulsed electric field-assisted extraction (PEF) and moderate electric field extraction (MEF) [142], SFE [143], complex enzyme-assisted extraction [144], PLE [145], and extraction with green solvents [146]. After an additional separation stage, the analysis of carotenoids has been carried out using instrumental techniques such as high-efficiency liquid chromatography with diode array (HPLC-DAD), thin-layer chromatography, and gas chromatography coupled to mass spectrometry [147].

Currently, fractions rich in phenolic compounds from biomaterials are obtained by implementing technologies such as solid-liquid extraction (SLE) [148], PLE [149], ASE [150], SFE [151], UAE [152], MAE [153], ultrafiltration (UF) [154], and complex enzyme-assisted extraction [155]. These processes have guaranteed the extraction of phenolic compounds with good yields. Likewise, the purification of these compounds by implementing liquid and solid phases has guaranteed the acquisition of extracts with abundant phenol and polyphenol contents with high levels of purity [156].

Finally, squalene and phytosterols have been extracted from natural sources using supercritical carbon dioxide (CO2) (SFE) [157]. In addition, high levels of these same substances have been found in deodorization distillates during the refining of some vegetable oils, such as palm and olive [158, 159]. Furthermore, the most widely used techniques to quantify the compounds mentioned above include gas chromatography with flame ionization detector (FID) and gas chromatography coupled to mass spectrometry [72].

5. Conclusions

D × P CPO and O × G CPO Coari × La Mé are natural oils derived from ripe fruits of the African palm *Elaeis guineensis* Jacq., and from one of the interspecific hybrid cultivars between the species *Elaeis oleifera* (Kunth) Cortés and *E. guineensis* Jacq., respectively, with a significant content of tocopherols and tocotrienols, α- and β-carotene, phytosterols, squalene, and phenolic structures that, when incorporated into the human diet in appropriate doses, promote the correct physiological functioning of organisms. According to the various works mentioned in this chapter, the above components are considered biomolecules indispensable for life due to their biological functions and nutritional attributes.

At present, different processes have been developed to extract, recover, and purify the phytochemicals contained in palm oil with good yields and high concentration values in the extracts obtained as final products. Currently, there is a marked trend toward obtaining phytochemicals from various natural sources by means of green technologies. Furthermore, the number of companies engaged in this work is increasing.

This chapter aims to show the attributes and benefits of including D × P CPO and Coarí × Me O × G CPO in the human diet and seeks to propose them as raw materials to produce functional food rich in phytochemicals of nutritional value.

Acknowledgements

The authors thank the Palm Oil Promotion Fund, administered by the National Federation of Oil Palm Growers—Fedepalma, for funding this study. Moreover, the authors also thank the Colombian Oil Palm Research Center—Cenipalma for all the information support received.

Author details

Alexis Gonzalez-Diaz* and Jesús Alberto García-Núñez
Colombian Oil Palm Research Center—Cenipalma, Bogotá, Colombia

*Address all correspondence to: agonzalezd@cenipalma.org

References

[1] Ma T et al. Influence of technical processing units on the α-carotene, β-carotene and lutein contents of carrot (*Daucus carrot* L.) juice. Journal of Functional Foods. 2015;**16**:104-113. DOI: 10.1016/j.jff.2015.04.020

[2] Suwanaruang T. Analyzing lycopene content in fruits. Agriculture and Agricultural Science Procedia. 2016;**11**:46-48. DOI: 10.1016/j.aaspro.2016.12.008

[3] Lewis ED, Meydani SN, Wu D. Regulatory role of vitamin E in the immune system and inflammation. IUBMB Life. 2019;**71**(4):487-494. DOI: 10.1002/iub.1976

[4] Patlevič P, Vašková J, Švorc P, Vaško L, Švorc P. Reactive oxygen species and antioxidant defense in human gastrointestinal diseases. Integrative Medicine Research. 2016;**5**(4):250-258. DOI: 10.1016/j.imr.2016.07.004

[5] Amengual J. Bioactive properties of carotenoids in human health. Nutrients. 2019;**11**(10):1-6. DOI: 10.3390/nu111 02388

[6] Zerres S, Stahl W. Carotenoids in human skin. Biochimica et Biophysica Acta (BBA)—Molecular and Cell Biology of Lipids. 2020;**1865**(11):158588. DOI: 10.1016/j.bbalip.2019.158588

[7] Zielińska MA, Wesołowska A, Pawlus B, Hamułka J. Health effects of carotenoids during pregnancy and lactation. Nutrients. 2017;**9**(8):1-25. DOI: 10.3390/nu9080838

[8] Gul K, Singh AK, Jabeen R. Nutraceuticals and functional foods: The foods for the future world. Critical Reviews in Food Science and Nutrition. 2016;**56**(16):2617-2627. DOI: 10.1080/10408398.2014.903384

[9] Kumar PKP, Krishna AGG. Physico-chemical characteristics and nutraceutical distribution of crude palm oil and its fractions. Grasas y Aceites. 2014;**65**(June):1-2

[10] Ting-Fu K, Yih-Ming W, Chiou RY-Y. Squalene content and antioxidant activity of *Terminalia catappa* leaves and seeds. Journal of Agricultural and Food Chemistry. 2002;**50**(19):5343-5348

[11] May CY, Nesaretnam K. Research advancements in palm oil nutrition. European Journal of Lipid Science and Technology. 2014;**116**(10):1301-1315. DOI: 10.1002/ejlt.201400076

[12] Meléndez-Martínez AJ, Stinco CM, Mapelli-Brahm P. Skin carotenoids in public health and nutricosmetics: The emerging roles and applications of the UV radiation-absorbing colourless carotenoids phytoene and phytofluene. Nutrients. 2019;**11**(5). pp. 1-39. DOI: 10.3390/nu11051093

[13] Bohn T et al. β-Carotene in the human body: Metabolic bioactivation pathways - from digestion to tissue distribution and excretion. Proceedings of the Nutrition Society. 2019;**78**(1):68-87. DOI: 10.1017/S0029665118002641

[14] Kim SK, Karadeniz F. Biological Importance and Applications of Squalene and Squalane. 1st ed. Vol. 65. Busan, Republic of Korea: Elsevier Inc.; 2012. DOI: 10.1016/B978-0-12-416003-3.00014-7

[15] Lorenzo JM et al. Bioactive peptides as natural antioxidants in food products—A review. Trends in Food Science & Technology. 2018;**79**:136-147. DOI: 10.1016/j.tifs.2018.07.003

[16] Schumacker PT. Reactive oxygen species in cancer: A dance with the devil. Cancer Cell. 2015;**27**(2):156-157. DOI: 10.1016/j.ccell.2015.01.007

[17] Neha K, Haider MR, Pathak A, Yar MS. Medicinal prospects of

antioxidants: A review. European Journal of Medicinal Chemistry. 2019;**178**(June):687-704. DOI: 10.1016/j.ejmech.2019.06.010

[18] Singh B, Singh JP, Kaur A, Singh N. Phenolic composition, antioxidant potential and health benefits of citrus peel. Food Research International. 2020;**132**(February):109114. DOI: 10.1016/j.foodres.2020.109114

[19] Shahidi F, Ambigaipalan P. Phenolics and polyphenolics in foods, beverages and spices: Antioxidant activity and health effects—A review. Journal of Functional Foods. 2015;**18**:820-897. DOI: 10.1016/j.jff.2015.06.018

[20] Chang SK, Alasalvar C, Shahidi F. Review of dried fruits: Phytochemicals, antioxidant efficacies, and health benefits. Journal of Functional Foods. 2016;**21**(March):113-132. DOI: 10.1016/j.jff.2015.11.034

[21] Xie J et al. Apoptotic activities of brusatol in human non-small cell lung cancer cells: Involvement of ROS-mediated mitochondrial-dependent pathway and inhibition of Nrf2-mediated antioxidant response. Toxicology. 2021;**451**:152680. DOI: 10.1016/j.tox.2021.152680

[22] Chien JT, Chang R-H, Hsieh C-H, Hsu C-Y, Wang C-C. Antioxidant property of *Taraxacum formosanum* Kitam and its antitumor activity in non-small-cell lung cancer cells. Phytomedicine. 2018;**49**:1-10. DOI: 10.1016/j.phymed.2018.06.011

[23] Mazumder MK, Choudhury S, Borah A. An in silico investigation on the inhibitory potential of the constituents of Pomegranate juice on antioxidant defense mechanism: Relevance to neurodegenerative diseases. IBRO Reports. 2019;**6**(May):153-159. DOI: 10.1016/j.ibror.2019.05.003

[24] Losada-Barreiro S, Bravo-Díaz C. Free radicals and polyphenols: The redox chemistry of neurodegenerative diseases. European Journal of Medicinal Chemistry. 2017;**133**:379-402. DOI: 10.1016/j.ejmech.2017.03.061

[25] Kim JH. Pharmacological and medical applications of Panax ginseng and ginsenosides: A review for use in cardiovascular diseases. Journal of Ginseng Research. 2018;**42**(3):264-269. DOI: 10.1016/j.jgr.2017.10.004

[26] Gianazza E, Brioschi M, Fernandez AM, Banfi C. Lipoxidation in cardiovascular diseases. Redox Biology. 2019;**23**(January):101119. DOI: 10.1016/j.redox.2019.101119

[27] Gibon V. 12—Palm oil and palm kernel oil refining and fractionation technology. In: Lai O-M, Tan C-P, Akoh CCBT-PO, editors. Palm Oil—Production, Processing, Characterization, and Uses. Zaventem, Belgium: AOCS Press; 2012. pp. 329-375. DOI: 10.1016/B978-0-9818936-9-3.50015-0

[28] Szydłowska-Czerniak A, Trokowski K, Karlovits G, Szłyk E. Effect of refining processes on antioxidant capacity, total contents of phenolics and carotenoids in palm oils. Food Chemistry. 2011;**129**(3):1187-1192. DOI: 10.1016/j.foodchem.2011.05.101

[29] Nagendran B, Unnithan UR, Choo YM, Sundram K. Characteristics of red palm oil, a carotene- and vitamin E-rich refined oil for food uses. Food and Nutrition Bulletin. 2000;**21**(2):189-194. DOI: 10.1177/156482650002100213

[30] Andrianto M. Preliminary product acceptance and initial price of virgin red palm oil on African expatriate community in Jakarta. International Journal of Oil Palm. 2019;**2**(2):87-93. DOI: 10.35876/ijop.v2i2.27

[31] Pande G, Akoh CC, Lai OM. Food uses of palm oil and its components. In: Tan C-P, Akoh CC, Lai O-M, editors. Palm Oil: Production, Processing,

Characterization, and Uses. Athens, GA, USA and Sedang, Selangor, Malaysia: AOCS Press; 2012. pp. 561-586. DOI: 10.1016/B978-0-9818936-9-3.50022-8

[32] Sambanthamurthi R, Sundram K, Tan Y. Chemistry and biochemistry of palm oil. Progress in Lipid Research. 2000;**39**(6):507-558. DOI: 10.1016/s0163-7827(00)00015-1

[33] Rincón-Miranda SM et al. Use of phenological stages of the fruits and physicochemical characteristics of the oil to determine the optimal harvest time of oil palm interspecific OxG hybrid fruits. Industrial Crops and Products. 2013;**49**:204-210. DOI: 10.1016/j.indcrop.2013.04.035

[34] Ribeiro JAA, Almeida ES, Neto BAD, Abdelnur PV, Monteiro S. Identification of carotenoid isomers in crude and bleached palm oils by mass spectrometry. LWT - Food Science and Technology. 2018;**89**(November):631-637. DOI: 10.1016/j.lwt.2017.11.039

[35] Chaves G, Ligarreto-Moreno GA, Cayon-Salinas DG. Physicochemical characterization of bunches from American oil palm (*Elaeis oleifera* H.B.K. Cortes) and their hybrids with African oil palm (*Elaeis guineensis* Jacq.). Acta Agronomica. 2018;**67**(1):168-176. DOI: 10.15446/acag.v67n1.62028

[36] Zou Y, Jiang Y, Yang T, Hu P, Xu X. Minor constituents of palm oil: Characterization, processing, and application. In: Lai O-M, Tan C-P, Akoh CC, editors. Palm Oil: Palm Oil Production, Processing, Characterization, and Uses. Shanghai 200137, China and Aarhus C 8000, Denmark; 2012. pp. 471-526. DOI: 10.1016/B978-0-9818936-9-3.50019-8

[37] Gonzalez-Diaz A, Pataquiva-Mateus A, García-Núñez JA. Characterization and response surface optimization driven ultrasonic nanoemulsification of oil with high phytonutrient concentration recovered from palm oil biodiesel distillation. Colloids and Surfaces A: Physicochemical and Engineering Aspects, December. 2020;**612**:2021. DOI: 10.1016/j.colsurfa.2020.125961

[38] Mozzon M, Pacetti D, Frega NG, Lucci P. Crude Palm oil from interspecific hybrid *Elaeis oleifera* × *E. guineensis*: Alcoholic constituents of unsaponifiable matter. Journal of the American Oil Chemists' Society. 2015;**92**(5):717-724

[39] Panpipat W, Chaijan M. Palm phosp/holipids. In: Ahmad M, Xu X, editors. Polar Lipids: Biology, Chemistry, and Technology. 1st ed. Thasala, Nakhon Si Thammarat, Thailand: Elsevier Inc.; 2015. pp. 77-90. DOI: 10.1016/B978-1-63067-044-3.50008-X

[40] Lucci P, Pacetti D, Frega NG, Mozzon M. Phytonutrient concentration and unsaturation of glycerides predict optimal harvest time for *Elaeis oleifera* × *E. guineensis* palm oil hybrids. European Journal of Lipid Science and Technology. 2015;**117**(7):1027-1036. DOI: 10.1002/ejlt.201400599

[41] Mba OI, Dumont MJ, Ngadi M. Palm oil: Processing, characterization and utilization in the food industry—A review. Food Bioscience. 2015;**10**(1):26-41. DOI: 10.1016/j.fbio.2015.01.003

[42] Chen B, McClements DJ, Decker EA. Minor components in food oils: A critical review of their roles on lipid oxidation chemistry in bulk oils and emulsions. Critical Reviews in Food Science and Nutrition. 2011;**51**(10):901-916. DOI: 10.1080/10408398.2011.606379

[43] Aluyor EO, Ozigagu CE, Oboh OI, Aluyor P. Chromatographic analysis of vegetable oils: A review. Scientific Research and Essays. 2009;**4**(4):191-197

[44] Singh D, Chaudhuri PK. A review on phytochemical and pharmacological

properties of Holy basil (*Ocimum sanctum* L.). Industrial Crops and Products. 2018;**118**(August):367-382. DOI: 10.1016/j.indcrop.2018.03.048

[45] Bursać Kovačević D et al. Innovative technologies for the recovery of phytochemicals from *Stevia rebaudiana* Bertoni leaves: A review. Food Chemistry. 2018;**268**(June):513-521. DOI: 10.1016/j.foodchem.2018.06.091

[46] Mannino G, Perrone A, Campobenedetto C, Schittone A, Margherita Bertea C, Gentile C. Phytochemical profile and antioxidative properties of *Plinia trunciflora* fruits: A new source of nutraceuticals. Food Chemistry. 2020;**307**:125515. DOI: 10.1016/j.foodchem.2019.125515

[47] Ashraf MA. Phytochemicals as potential anticancer drugs: Time to Ponder Nature's Bounty. BioMed Research International. 2020;**2020**. pp. 1-7. DOI: 10.1155/2020/8602879

[48] Gibon V, De Greyt W, Kellens M. Palm oil refining. European Journal of Lipid Science and Technology. 2007; **109**(4):315-335. DOI: 10.1002/ejlt.200600307

[49] Rossi M, Gianazza M, Alamprese C, Stanga F. The role of bleaching clays and synthetic silica in palm oil physical refining. Food Chemistry. 2003;**82**(2): 291-296. DOI: 10.1016/S0308-8146(02)00551-4

[50] Choong CG, McKay A. Sustainability in the Malaysian palm oil industry. Journal of Cleaner Production. 2014;**85**: 258-264. DOI: 10.1016/j.jclepro.2013.12.009

[51] Advanced Chemistry Development Inc. (ACD/Labs). ACD/ChemSketch. Toronto; 2018

[52] Grosshagauer S, Steinschaden R, Pignitter M. Strategies to increase the oxidative stability of cold pressed oils. LWT. 2019;**106**:72-77. DOI: 10.1016/j.lwt.2019.02.046

[53] Combs GF, McClung JP. Chapter 8—Vitamin E. In: Combs GF, McClung JP, editors. The Vitamins Fundamental Aspects in Nutrition and Health. 5th ed. Westborough, MA: Academic Press; 2017. pp. 207-242. DOI: 10.1016/B978-0-12-802965-7.00008-3

[54] Peh HY, Tan WSD, Liao W, Wong WSF. Vitamin E therapy beyond cancer: Tocopherol versus tocotrienol. Pharmacology and Therapeutics. 2016;**162**:152-169. DOI: 10.1016/j.pharmthera.2015.12.003

[55] Bartella L, di Donna L, Napoli A, Sindona G, Mazzotti F. High-throughput determination of vitamin E in extra virgin olive oil by paper spray tandem mass spectrometry. Analytical and Bioanalytical Chemistry. 2019;**411**(13):2885-2890. DOI: 10.1007/s00216-019-01727-z

[56] W C on O and E O Processing. The proceedings of the world conference on oilseed and edible oils processing. In: Koseoglu SS, Rhee KC, Wilson RF, editors. Emerging Technologies, Current Practices, Quality Control, Technology Transfer, and Environmental Issues. World Conference on Oilseed and Edible Oils Processing. Vol. 1. 1998

[57] Azlan A et al. Comparison of fatty acids, vitamin E and physicochemical properties of Canarium odontophyllum Miq. (dabai), olive and palm oils. Journal of Food Composition and Analysis. 2010;**23**(8):772-776. DOI: 10.1016/j.jfca.2010.03.026

[58] Wanasundara PKJPD, Shahidi F, Shukla VKS. Endogenous antioxidants from oilseeds and edible oils. Food Reviews International. 1997;**13**(2):225-292. DOI: 10.1080/87559129709541106

[59] Li C et al. Comparison and analysis of fatty acids, sterols, and tocopherols in

eight vegetable oils. Journal of Agricultural and Food Chemistry. 2011;**59**(23):12493-12498. DOI: 10.1021/jf203760k

[60] Qian C, Decker EA, Xiao H, McClements DJ. Physical and chemical stability of β-carotene-enriched nanoemulsions: Influence of pH, ionic strength, temperature, and emulsifier type. Food Chemistry. 2012;**132**(3):1221-1229. DOI: 10.1016/j.foodchem.2011.11.091

[61] Khalil S et al. Retinoids: A journey from the molecular structures and mechanisms of action to clinical uses in dermatology and adverse effects. Journal of Dermatological Treatment. Nov. 2017;**28**(8):684-696. DOI: 10.1080/09546634.2017.1309349

[62] Gul K, Tak A, Singh AK, Singh P, Yousuf B, Wani AA. Chemistry, encapsulation, and health benefits of β-carotene—A review. Cogent Food & Agriculture. 2015;**1**(1):1-12. DOI: 10.1080/23311932.2015.1018696

[63] Saurat J-H, Sorg O. Retinoids. In: European Handbook of Dermatological Treatments. 3rd ed. Berlin Heidelberg, Germany: Springer-Verlag; 2015. pp. 1493-1511. DOI: 10.1007/978-3-662-45139-7

[64] Wu L et al. Molecular aspects of β, β-carotene-9′, 10′-oxygenase 2 in carotenoid metabolism and diseases. Experimental Biology and Medicine. 2016;**241**(17):1879-1887. DOI: 10.1177/1535370216657900

[65] Iftikhar HT, Zhao Y. Enrichment of β-carotene from palm oil using supercritical carbon dioxide pretreatment-solvent extraction technique. LWT—Food Science and Technology. 2017;**83**:262-266. DOI: 10.1016/j.lwt.2017.05.026

[66] Okogeri O, Uchenna-Onu U. Changes occurring in quality indices during storage of adulterated red palm oil. International Journal of Food Science and Nutrition. 2016;**1**(3):1-5

[67] Beltrán-de-Miguel B, Estévez-Santiago R, Olmedilla-Alonso B. Assessment of dietary vitamin A intake (retinol, α-carotene, β-carotene, β-cryptoxanthin) and its sources in the National Survey of Dietary Intake in Spain (2009-2010). International Journal of Food Sciences and Nutrition. 2015;**66**(6):706-712. DOI: 10.3109/09637486.2015.1077787

[68] Pham DM, Boussouira B, Moyal D, Nguyen QL. Oxidization of squalene, a human skin lipid: A new and reliable marker of environmental pollution studies. International Journal of Cosmetic Science. 2015;**37**(4):357-365. DOI: 10.1111/ics.12208

[69] Fernando IPS, Sanjeewa KKA, Samarakoon KW, Lee WW, Kim HS, Jeon YJ. Squalene isolated from marine macroalgae *Caulerpa racemosa* and its potent antioxidant and anti-inflammatory activities. Journal of Food Biochemistry. 2018;**42**(5). pp. 1-10. DOI: 10.1111/jfbc.12628

[70] Spanova M, Daum G. Squalene: Biochemistry, molecular biology, process biotechnology, and applications. European Journal of Lipid Science and Technology. 2011;**113**(11):1299-1320. DOI: 10.1002/ejlt.201100203

[71] J. J. Gaforio, C. Sánchez-Quesada, A. López-Biedma, M. del C. Ramírez-Tortose, and F. Warleta, “Molecular aspects of squalene and implications for olive oil and the mediterranean diet,” in The Mediterranean Diet: An Evidence-Based Approach, V. R. Preedy and R. Ross Watson, Eds. Jaén and Granada, Spain: Elsevier Inc, 2015, pp. 281-290. doi: 10.1016/B978-0-12-407849-9.00026-9.

[72] Lozano-Grande MA, Gorinstein S, Espitia-Rangel E, Dávila-Ortiz G, Martínez-Ayala AL. Plant sources,

extraction methods, and uses of squalene. International Journal of Agronomy. 2018;**2018**:. pp. 1-13. DOI: 10.1155/2018/1829160

[73] Rodríguez JC et al. Effects of the fruit ripening stage on antioxidant capacity, total phenolics, and polyphenolic composition of crude palm oil from interspecific hybrid *Elaeis oleifera* × *Elaeis guineensis*. Journal of Agricultural and Food Chemistry. 2016;**64**(4):852-859. DOI: 10.1021/acs.jafc.5b04990

[74] Vogt T. Phenylpropanoid biosynthesis. Molecular Plant. 2010;**3**(1):2-20. DOI: 10.1093/mp/ssp106

[75] Lattanzio V. Phenolic Compounds: Introduction. In: Ramawat KG, Mérillon J-M, editors. Natural Products: Phytochemistry, Botany and Metabolism of Alkaloids, Phenolics and Terpenes. New Delhi, India: Springer-Verlag Berlin Heidelberg; 2013. pp. 1-4242. DOI: 10.1007/978-3-642-22144-6

[76] Nimse SB, Pal D. Free radicals, natural antioxidants, and their reaction mechanisms. RSC Advances. 2015;**5**(35): 27986-28006. DOI: 10.1039/c4ra13315c

[77] Liochev SI. Reactive oxygen species and the free radical theory of aging. Free Radical Biology and Medicine. 2013;**60**: 1-4. DOI: 10.1016/j.freeradbiomed.2013.02.011

[78] Darvin ME, Fluhr JW, Meinke MC, Zastrow L, Sterry W, Lademann J. Topical beta-carotene protects against infra-red-light-induced free radicals. Experimental Dermatology. 2011;**20**(2):125-129. DOI: 10.1111/j.1600-0625.2010.01191.x

[79] Wojdyło A, Samoticha J, Chmielewska J. Effect of different pre-treatment maceration techniques on the content of phenolic compounds and color of Dornfelder wines elaborated in cold climate. Food Chemistry. 2021;**339**:127888. DOI: 10.1016/j.foodchem.2020.127888

[80] Vuolo MM, Lima VS, Maróstica Junior MR. In: Campos MRSBT-BC, editor. Phenolic Compounds: Structure, Classification, and Antioxidant Power. Campinas, Brazil and Coari, Brazil: Woodhead Publishing; 2019. pp. 33-50. DOI: 10.1016/B978-0-12-814774-0.00002-5

[81] Wang H, Chambers E, Kan J. Sensory characteristics of combinations of phenolic compounds potentially associated with smoked aroma in foods. Molecules. 2018;**23**(8). pp. 1-11. DOI: 10.3390/molecules23081867

[82] Reis Giada MdL, "Food phenolic compounds: Main classes, sources and their antioxidant power maria," Oxidative Stress and Chronic Degenerative Diseases A Role for Antioxidants, J. A. Morales-Gonzalez, Rio de Janeiro, Brazil: IntechOpen, 2013, 87-112.

[83] Ferreira CD et al. Physicochemical characterization and oxidative stability of microencapsulated crude palm oil by spray drying. Food and Bioprocess Technology. 2016;**9**(1):124-136. DOI: 10.1007/s11947-015-1603-z

[84] Gouvinhas I, Machado J, Gomes S, Lopes J, Martins-Lopes P, Barros AIRNA. Phenolic composition and antioxidant activity of monovarietal and commercial Portuguese olive oils. JAOCS, Journal of the American Oil Chemists' Society. 2014;**91**(7):1197-1203. DOI: 10.1007/s11746-014-2462-x

[85] Mohdaly AAE-R, El-Hameed Seliem KA, EL-Hassan AE-MMA, Mahmoud AAT. Effect of refining process on the quality characteristics of soybean and cotton seed oils. International Journal of Current Microbiology and Applied Sciences. 2017;**6**(1):207-222. DOI: 10.20546/ijcmas.2017.601.026

[86] Marina AM, Che Man YB, Nazimah SAH, Amin I. Antioxidant capacity and phenolic acids of virgin coconut oil. International Journal of

Food Sciences and Nutrition. 2009;**60**(SUPPL. 2):114-123. DOI: 10.1080/09637480802549127

[87] Janu C, Kumar DRS, Reshma MV, Jayamurthy P, Sundaresan A, Nisha P. Comparative study on the total phenolic content and radical scavenging activity of common edible vegetable oils. Journal of Food Biochemistry. 2014;**38**(1):38-49. DOI: 10.1111/jfbc.12023

[88] Gylling H, Simonen P. Phytosterols, phytostanols, and lipoprotein metabolism. Nutrients. 2015;7(9):7965-7977. DOI: 10.3390/nu7095374

[89] Piironen V, Lampi A-M. Occurrence and levels of phytosterols in foods. In: Dutta PC, editor. Phytosterols as Functional Food Components and Nutraceuticals. 1st ed. Helsinki, Finland: CRC Press; 2004. pp. 1-32. DOI: 10.1201/9780203913413

[90] Lagarda MJ, García-Llatas G, Farré R. Analysis of phytosterols in foods. Journal of Pharmaceutical and Biomedical Analysis. 2006;**41**(5):1486-1496. DOI: 10.1016/j.jpba.2006.02.052

[91] Yang R et al. Phytosterol contents of edible oils and their contributions to estimated phytosterol intake in the Chinese diet. Foods. 2019;**8**(8). pp. 1-12. DOI: 10.3390/foods8080334

[92] Pokkanta P, Sookwong P, Tanang M, Setchaiyan S, Boontakham P, Mahatheeranont S. Simultaneous determination of tocols, γ-oryzanols, phytosterols, squalene, cholecalciferol and phylloquinone in rice bran and vegetable oil samples. Food Chemistry. 2019;**271**:630-638. DOI: 10.1016/j.foodchem.2018.07.225

[93] Islam MA, Jeong B-G, Jung J, Shin E-C, Choi S-G, Chun J. Phytosterol determination and method validation for selected nuts and seeds. Food Analytical Methods. 2017;**10**(10):3225-3234. DOI: 10.1007/s12161-017-0877-3

[94] Naumoska K, Vovk I. Analysis of triterpenoids and phytosterols in vegetables by thin-layer chromatography coupled to tandem mass spectrometry. Journal of Chromatography A. 2015;**1381**:229-238. DOI: 10.1016/j.chroma.2015.01.001

[95] Srigley CT, Haile EA. Quantification of plant sterols/stanols in foods and dietary supplements containing added phytosterols. Journal of Food Composition and Analysis. 2015;**40**:163-176. DOI: 10.1016/j.jfca.2015.01.008

[96] Raederstorff D, Péter S, Weber P. Vitamin E in human health. In: Weber P, Birringer M, Blumberg J, Eggersdorfer M, Frank J, editors. Vitamin E in Human Health. 1st ed. Wellington: Humana Press; 2019. pp. 163-174. DOI: 10.1007/978-3-030-05315-4

[97] Sokol RJ. A new old treatment for vitamin E deficiency in cholestasis. Journal of Pediatric Gastroenterology and Nutrition. 2016;**63**(6):577-578. DOI: 10.1097/MPG.0000000000001330

[98] Rahmoune H, Boutrid N, Amrane M, Chekkour MC, Bioud B. Ataxia in children: Think about Vitamin E deficiency ! (comment on: Ataxia in children: Early recognition and clinical evaluation). Italian Journal of Pediatrics. 2017;**43**(1):1-2. DOI: 10.1186/s13052-017-0378-4

[99] Phoon KY, Ng HS, Zakaria R, Yim HS, Mokhtar MN. Enrichment of minor components from crude palm oil and palm-pressed mesocarp fibre oil via sequential adsorption-desorption strategy. Industrial Crops and Products. 2018;**113**(September 2017):187-195. DOI: 10.1016/j.indcrop.2018.01.039

[100] Nur Sulihatimarsyila AW, Lau HLN, Nabilah KM, Nur Azreena I. Refining process for production of refined palm-pressed fibre oil. Industrial Crops and Products.

2019;**129**(November 2018):488-494. DOI: 10.1016/j.indcrop.2018.12.034

[101] Sangkharak K, Pichid N, Yunu T, Kingman P. Separation of carotenes and vitamin e from palm oil mill effluent using silica from agricultural waste as an adsorbent. Walailak Journal of Science and Technology. 2016;**13**(11):939-947. DOI: 10.14456/vol13iss12pp%p

[102] Kua YL, Gan S, Morris A, Ng HK. Simultaneous recovery of carotenes and tocols from crude palm olein using ethyl lactate and ethanol. Journal of Physics: Conference Series. 2018;**989**(1). DOI: 10.1088/1742-6596/989/1/012005

[103] Álvarez R, Meléndez-Martínez AJ, Vicario IM, Alcalde MJ. Carotenoid and vitamin A contents in biological fluids and tissues of animals as an effect of the diet: A review. Food Reviews International. 2015;**31**(4):319-340. DOI: 10.1080/87559129.2015.1015139

[104] Berman J et al. Nutritionally important carotenoids as consumer products. Phytochemistry Reviews. 2015;**14**(5):727-743. DOI: 10.1007/s11101-014-9373-1

[105] Rubin LP, Ross AC, Stephensen CB, Bohn T, Tanumihardjo SA. Metabolic effects of inflammation on vitamin A and carotenoids in humans and animal models. Advances in Nutrition: An International Review Journal. 2017;**8**(2):197-212. DOI: 10.3945/an.116.014167

[106] Rodriguez-Concepcion M et al. A global perspective on carotenoids: Metabolism, biotechnology, and benefits for nutrition and health. Progress in Lipid Research. 2018;**70**:62-93. DOI: 10.1016/j.plipres.2018.04.004

[107] Woodside JV, McGrath AJ, Lyner N, McKinley MC. Carotenoids and health in older people. Maturitas. 2015;**80**(1):63-68. DOI: 10.1016/j.maturitas.2014.10.012

[108] Green AS, Fascetti AJ. Meeting the vitamin A requirement: The efficacy and importance of β-carotene in animal species. Scientific World Journal. 2016;**2016**. pp. 1-22. DOI: 10.1155/2016/7393620

[109] Ras RT et al. Increases in plasma plant sterols stabilize within four weeks of plant sterol intake and are independent of cholesterol metabolism. Nutrition, Metabolism and Cardio-vascular Diseases. 2016;**26**(4):302-309. DOI: 10.1016/j.numecd.2015.11.007

[110] Weingärtner O et al. Plant sterol ester diet supplementation increases serum plant sterols and markers of cholesterol synthesis, but has no effect on total cholesterol levels. Journal of Steroid Biochemistry and Molecular Biology. 2017;**169**(July 2016):219-225. DOI: 10.1016/j.jsbmb.2016.07.016

[111] Baumgartner S et al. Effects of plant stanol ester consumption on fasting plasma oxy(phyto)sterol concentrations as related to fecal microbiota characteristics. Journal of Steroid Biochemistry and Molecular Biology. 2017;**169**:46-53. DOI: 10.1016/j.jsbmb.2016.02.029

[112] Johnston TP, Korolenko TA, Pirro M, Sahebkar A. Preventing cardiovascular heart disease: Promising nutraceutical and non-nutraceutical treatments for cholesterol management. Pharmaco-logical Research. 2017;**120**:219-225. DOI: 10.1016/j.phrs.2017.04.008

[113] Vazquez-Vidal I, Jones PJH. Nutrigenetics and blood cholesterol levels in response to plant sterols. In: Principles of Nutrigenetics and Nutrigenomics. Vol. 2. Innipeg, MB, Canada and Vaudreuil-Dorion, QC, Canada: Elsevier Inc.; 2020. pp. 227-230. DOI: 10.1016/b978-0-12-804572-5.00029-x

[114] Lacatusu I, Arsenie LV, Badea G, Popa O, Oprea O, Badea N. New cosmetic formulations with broad photoprotective

and antioxidative activities designed by amaranth and pumpkin seed oils nanocarriers. Industrial Crops and Products. 2018;**123**:424-433. DOI: 10.1016/j.indcrop.2018.06.083

[115] Narayan Bhilwade H, Tatewaki N, Nishida H, Konishi T. Squalene as novel food factor. Current Pharmaceutical Biotechnology. 2010;**11**(8):875-880. DOI: 10.2174/138920110793262088

[116] Kim S-K, Karadeniz F. Chapter 14—Biological importance and applications of squalene and squalane. In: S-KBT-AF, Kim NR, editors. Marine Medicinal Foods. Vol. 65. Busan, Republic of Korea: Academic Press; 2012. pp. 223-233. DOI: 10.1016/B978-0-12-416003-3.00014-7.

[117] Jin R, Yin H, Wang H, Zhang D, Cao K, Yuan C. Preparation and characterization of squalene microcapsules by complex coacervation. Journal of Food Process Engineering. 2018;**41**(6):1-6. DOI: 10.1111/jfpe.12847

[118] Kim JH, Kim CN, Kang DW. Squalene epoxidase correlates E-cadherin expression and overall survival in colorectal cancer patients: The impact on prognosis and correlation to clinicopathologic features. Journal of Clinical Medicine. 2019;**8**(5):632. DOI: 10.3390/jcm8050632

[119] Cirmena G et al. Squalene epoxidase as a promising metabolic target in cancer treatment. Cancer Letters. 2018;**425**:13-20. DOI: 10.1016/j.canlet.2018.03.034

[120] Birhanu G, Javar HA, Seyedjafari E, Zandi-Karimi A. Nanotechnology for delivery of gemcitabine to treat pancreatic cancer. Biomedicine and Pharmacotherapy. 2017;**88**:635-643. DOI: 10.1016/j.biopha.2017.01.071

[121] Brunner G, Gast K, Chuang MH, Kumar S, Chan P, Chan WP. Process for the production of highly enriched fractions of natural compounds of palm oil with supercritical and almost critical fluids (Processo para a produção de frações altamente enriquecidas de compostos naturais de oleo de palma com fluidos supercrit). BRPI0706288B1. 2017

[122] Pathak PV, Charhate PS, Kurkarni MG. Process for the Preparation of Tocols & Squalene. PCT/IN2017/050557-WO2018/109780 A1. 2018

[123] May CY, Nang HLL, Wei PC, Ah Ngan M, Basiron Y. Recovery of phytonutrients from palm oil. EP 1398311 B1. 2007

[124] Oroskar AG, Sharma D, Oroskar A, Oroskar G. Recovery of highly pure alpha-tocotrienol from crude palm oil extract. US008937191B2. 2015

[125] Choo Yuen M, Lik Nang Lau H, Ngan MA, Basiron Y. Extraction of palm vitamin E, phytosterols and squalene from palm oil, US 2005/0250953 A1. 2005

[126] Peterson DM, Jensen CM, Hoffman DL, Mannerstedt-Fogelfors B. Oat tocols: Saponification vs. direct extraction and analysis in high-oil genotypes. Cereal Chemistry. 2007;**84**(1):56-60. DOI: 10.1094/CCHEM-84-1-0056

[127] Lee YY et al. Comparing extraction methods for the determination of tocopherols and tocotrienols in seeds and germinating seeds of soybean transformed with OsHGGT. Journal of Food Composition and Analysis. 2012;**27**(1):70-80. DOI: 10.1016/j.jfca.2012.03.010

[128] dos Santos Freitas L, Jacques RA, Richter MF, da Silva AL, Caramão EB. Pressurized liquid extraction of vitamin E from Brazilian grape seed oil. Journal of Chromatography A. 2008;**1200**(1):80-83. DOI: 10.1016/j.chroma.2008.02.067

[129] Lau HLN, Choo YM, Ma AN, Chuah CH. Selective extraction of palm carotene and vitamin E from fresh

palm-pressed mesocarp fiber (*Elaeis guineensis*) using supercritical CO2. Journal of Food Engineering. 2008;**84**(2):289-296. DOI: 10.1016/j.jfoodeng.2007.05.018

[130] Teixeira CB, Macedo GA, Macedo JA, da Silva LHM, Rodrigues AM d C. Simultaneous extraction of oil and antioxidant compounds from oil palm fruit (*Elaeis guineensis*) by an aqueous enzymatic process. Bioresource Technology. 2013;**129**:575-581. DOI: 10.1016/j.biortech.2012.11.057

[131] Jiang ST, Shao P, Pan LJ, Zhao YY. Molecular distillation for recovering tocopherol and fatty acid methyl esters from rapeseed oil deodoriser distillate. Biosystems Engineering. 2006;**93**(4):383-391. DOI: 10.1016/j.biosystemseng.2006.01.008

[132] Chandrasekaram K, Han NM, May CY, Hock CC. Concentration and isolation of individual vitamin e components in palm phytonutrients concentrate using high performance liquid chromatography with flourescence detection. Journal of Oil Palm Research. 2009;**21**(June):621-626

[133] Phoon KY, Ng HS, Zakaria R, Yim HS, Mokhtar MN. Enrichment of minor components from crude palm oil and palm-pressed mesocarp fibre oil via sequential adsorption-desorption strategy. Industrial Crops and Products. 2018;**113**:187-195. DOI: 10.1016/j.indcrop.2018.01.039

[134] Posada LR, Shi J, Kakuda Y, Xue SJ. Extraction of tocotrienols from palm fatty acid distillates using molecular distillation. Separation and Purification Technology. 2007;**57**(2):220-229. DOI: 10.1016/j.seppur.2007.04.016

[135] Zigoneanu IG, Williams L, Xu Z, Sabliov CM. Determination of antioxidant components in rice bran oil extracted by microwave-assisted method. Bioresource Technology. 2008;**99**(11):4910-4918. DOI: 10.1016/j.biortech.2007.09.067

[136] Aryanti N, Wardhani DH, Nafiunisa A. Ultrafiltration membrane for degumming of crude palm oil-isopropanol mixture. Chemical and Biochemical Engineering Quarterly. 2018;**32**(3):325-334. DOI: 10.15255/CABEQ.2017.1244

[137] Mezzomo N, Maestri B, dos Santos RL, Maraschin M, Ferreira SRS. Pink shrimp (*P. brasiliensis* and *P. paulensis*) residue: Influence of extraction method on carotenoid concentration. Talanta. 2011;**85**(3):1383-1391. DOI: 10.1016/j.talanta.2011.06.018

[138] Cardenas-Toro FP, Alcázar-Alay SC, Coutinho JP, Godoy HT, Forster-Carneiro T, Meireles MAA. Pressurized liquid extraction and low-pressure solvent extraction of carotenoids from pressed palm fiber: Experimental and economical evaluation. Food and Bioproducts Processing. 2015;**94**:90-100. DOI: 10.1016/j.fbp.2015.01.006

[139] Hiranvarachat B, Devahastin S. Enhancement of microwave-assisted extraction via intermittent radiation: Extraction of carotenoids from carrot peels. Journal of Food Engineering. 2014;**126**:17-26. DOI: 10.1016/j.jfoodeng.2013.10.024

[140] Tsiaka T, Zoumpoulakis P, Sinanoglou VJ, Makris C, Heropoulos GA, Calokerinos AC. Response surface methodology toward the optimization of high-energy carotenoid extraction from *Aristeus antennatus* shrimp. Analytica Chimica Acta. 2015;**877**:100-110. DOI: 10.1016/j.aca.2015.03.051

[141] Zaghdoudi K et al. Accelerated solvent extraction of carotenoids from: Tunisian Kaki (*Diospyros kaki* L.), peach (*Prunus persica* L.) and apricot (*Prunus armeniaca* L.). Food Chemistry. 2015;**184**:131-139. DOI: 10.1016/j.foodchem.2015.03.072

[142] Jaeschke DP, Menegol T, Rech R, Mercali GD, Marczak LDF. Carotenoid and lipid extraction from *Heterochlorella luteoviridis* using moderate electric field and ethanol. Process Biochemistry. 2016;**51**(10):1636-1643. DOI: 10.1016/j.procbio.2016.07.016

[143] Zaghdoudi K et al. Response surface methodology applied to Supercritical Fluid Extraction (SFE) of carotenoids from Persimmon (*Diospyros kaki* L.). Food Chemistry. 2016;**208**:209-219. DOI: 10.1016/j.foodchem.2016.03.104

[144] Sowbhagya HB, Chitra VN. Enzyme-assisted extraction of flavorings and colorants from plant materials. Critical Reviews in Food Science and Nutrition. 2010;**50**(2):146-161. DOI: 10.1080/10408390802248775

[145] Kwang HC, Lee HJ, Koo SY, Song DG, Lee DU, Pan CH. Optimization of pressurized liquid extraction of carotenoids and chlorophylls from chlorella vulgaris. Journal of Agricultural and Food Chemistry. 2010;**58**(2):793-797. DOI: 10.1021/jf902628j

[146] Yara-Varón E, Fabiano-Tixier AS, Balcells M, Canela-Garayoa R, Bily A, Chemat F. Is it possible to substitute hexane with green solvents for extraction of carotenoids? A theoretical versus experimental solubility study. RSC Advances. 2016;**6**(33):27750-27759. DOI: 10.1039/c6ra03016e

[147] Saini RK, Keum YS. Carotenoid extraction methods: A review of recent developments. Food Chemistry. 2018;**240**(April 2017):90-103. DOI: 10.1016/j.foodchem.2017.07.099

[148] Moreira BO et al. Application of response surface methodology for optimization of ultrasound-assisted solid-liquid extraction of phenolic compounds from *Cenostigma macrophyllum*. Journal of Chemometrics. 2020;**34**(10):1-12. DOI: 10.1002/cem.3290

[149] Tripodo G, Ibáñez E, Cifuentes A, Gilbert-López B, Fanali C. Optimization of pressurized liquid extraction by response surface methodology of Goji berry (*Lycium barbarum* L.) phenolic bioactive compounds. Electrophoresis. 2018;**39**(13):1673-1682. DOI: 10.1002/elps.201700448

[150] Gomes SVF et al. Accelerated solvent extraction of phenolic compounds exploiting a Box-Behnken design and quantification of five flavonoids by HPLC-DAD in Passiflora species. Microchemical Journal. 2017;**132**:28-35. DOI: 10.1016/j.microc.2016.12.021

[151] Rodríguez-Solana R, Salgado JM, Domínguez JM, Cortés-Diéguez S. Comparison of soxhlet, accelerated solvent and supercritical fluid extraction techniques for volatile (GC-MS and GC/FID) and phenolic compounds (HPLC-ESI/MS/MS) from lamiaceae species. Phytochemical Analysis. 2015;**26**(1):61-71. DOI: 10.1002/pca.2537

[152] Altemimi A, Watson DG, Choudhary R, Dasari MR, Lightfoot DA. Ultrasound assisted extraction of phenolic compounds from peaches and pumpkins. PLoS ONE. 2016;**11**(2):1-20. DOI: 10.1371/journal.pone.0148758

[153] Setyaningsih W, Saputro IE, Palma M, Barroso CG. Optimisation and validation of the microwave-assisted extraction of phenolic compounds from rice grains. Food Chemistry. 2015;**169**:141-149. DOI: 10.1016/j.foodchem.2014.07.128

[154] Conidi C, Cassano A, Caiazzo F, Drioli E. Separation and purification of phenolic compounds from pomegranate juice by ultrafiltration and nanofiltration membranes. Journal of Food Engineering. 2017;**195**:1-13. DOI: 10.1016/j.jfoodeng.2016.09.017

[155] Schroyen M, Vervaeren H, Vandepitte H, Van Hulle SWH, Raes K.

Effect of enzymatic pretreatment of various lignocellulosic substrates on production of phenolic compounds and biomethane potential. Bioresource Technology. 2015;**192**:696-702. DOI: 10.1016/j.biortech.2015.06.051

[156] Ajila CM, Brar SK, Verma M, Tyagi RD, Godbout S, Valéro JR. Extraction and analysis of polyphenols: Recent trends. Critical Reviews in Biotechnology. 2011;**31**(3):227-249. DOI: 10.3109/07388551.2010.513677

[157] Dąbrowski G, Czaplicki S, Konopka I. Fractionation of sterols, tocols and squalene in flaxseed oils under the impact of variable conditions of supercritical CO2 extraction. Journal of Food Composition and Analysis. 2019;**83**:103261. DOI: 10.1016/j.jfca.2019.103261

[158] Tarhan İ, Kara H. A new HPLC method for simultaneous analysis of sterols, tocopherols, tocotrienols, and squalene in olive oil deodorizer distillates using a monolithic column with chemometric techniques. Analytical Methods. 2019;**11**(36):4681-4692. DOI: 10.1039/c9ay01525f

[159] Bouriakova A, Mendes PSF, Elst K, De Clercq J, Thybaut JW. Techno-economic evaluation of squalene recovery from oil deodorizer distillates. Chemical Engineering Research and Design. 2020;**154**:122-134. DOI: 10.1016/j.cherd.2019.12.003

Chapter 2

CSR Practical Orientation in Small Medium Enterprises (SMEs): A Case Study in Solo City Indonesia

Susilaningtyas B. Kurniawati and Hasan Fauzi

Abstract

This study explores empirical data on the orientation of CSR practices to SMEs in City Surakarta seen from CSR practices that are employee, market, and environment-oriented as a whole. The sample in this study was 90 SMEs in the Surakarta area. The method used in this research is a survey with data analysis using Confirmatory Factor Analysis (CFA). Before the survey, a pilot study was conducted on 10 SMEs actors to test the instrument's validity and reliability. The results showed that the most dominant CSR practices at SMEs in City Surakarta were the community orientation indicated by the rotated component matrix table of 0.904 and the market orientation indicated by the rotated component matrix table of 0.876. Both orientations are more dominant than other orientations.

Keywords: Corporate Social Responsibility, CSR orientation, SMEs, MSMEs, Market orientation, Community orientation, Environmental orientation

1. Introduction

Corporate Social Responsibility (CSR) began to be adopted in accounting using a financial and management perspective. The financial perspective focuses on CSR disclosure or CSR reporting, while in the management perspective, it is related to measuring social performance [1]. This view is based on several main models in understanding the concept of CSR. First, Carroll's [2] view defines CSR in three dimensions: combining social responsibility at four different levels (economic, legal, ethical, and discretionary), the number of social problems facing companies, and the underlying philosophy. Second, the view of Wartick and Cochran [3], who adopted and perfected the CSR model of Carroll [2] by including the social problem management variable on CSR. Third, Wood, [4] sees CSR from the aspects of: configuration of CSR principles, social response processes, policies, programs, and outcomes that can be observed in the context of corporate and social relations. Fourth, the view of CSR from the perspective of stakeholders who feel satisfied with the company's performance is one of the successes of CSR [5]. This study uses CSR terminology from Wood's [4] model, which looks at CSR from responsive social processes, policies, programs, and observable outcomes (impacts) in the context of corporate and social relations.

CSR in Indonesia began to develop after Law no. 40 of 2007 [6] concerning Limited Liability Companies Article 74 concerning the obligations of companies engaged in natural resources to carry out social and environmental responsibilities. Along with the development of CSR in Indonesia, more and more companies are implementing social responsibility. CSR is an obligation and a strategy developed in large-scale companies but is expected to be further developed in Micro, Small, and Medium Enterprises (MSMEs) companies.

MSMEs in Indonesia have experienced significant development. Based on the Ministry of Cooperatives and Small and Medium Enterprises data, the number of MSMEs in Indonesia currently reaches 59.3 million [7]. That number of MSMEs in 2018 is known to contribute to Indonesia's GDP by 60% or Rp. 4.800 trillion [8]. The current condition of MSMEs needs better management further to increase their contribution to the country's economy.

The current phenomenon is that the MSME empowerment program is still not following the needs of MSMEs and is still routine and monotonous. Implementing the MSMEs empowerment program should be carried out by considering financial factors and developing environmentally friendly MSMEs businesses, especially production activities that can reduce industrial emissions. One example is the collaboration between Indonesia and South Korea at the 1st Indonesia-Korea Green Business Forum in Jakarta on December 6, 2016 [9]. So far, the concept of CSR is only practiced by large companies, so it has not touched the MSME sector. If CSR can be applied to MSMEs, it is hoped to provide integrity within the MSMEs organization, which thinks about profit and environmental and social. Therefore, all companies, both large and small, including MSMEs, can implement CSR as one of the things that can determine the sustainability of their business because, through CRS, companies can get support from stakeholders and the community around the company.

CSR activities by MSMEs can be carried out with different orientations. The CSR Practice Orientations include employee-oriented, community-oriented, market-oriented, environmental-oriented, and generally or comprehensively oriented [10]. However, CSR practices have always been associated with environmental orientation and have not led to other CSR practice orientations.

Based on data from the Central Java Province Cooperatives and MSMEs Service in the first quarter of 2019, the number of MSMEs in Central Java is 147,233 SMEs. One of the regencies/cities with the largest MSMEs is Surakarta City, which has 43,700 MSMEs in 2018 [11]. In other words, Surakarta City has 30% of MSMEs in Central Java. For this reason, this study chooses Surakarta City as the research area.

Another reason for choosing a research area in Surakarta City is that there are previous studies on MSMEs in Surakarta City, but no one has specifically examined the orientation of MSMEs in implementing CSR. These studies, among others, were conducted by [12] which examined the effectiveness of the performance of the RKB of Bank BRI Solo in developing its fostered SMEs in terms of factors, indicators, measures, and the process of evaluating the effectiveness of the performance of the RKB of Bank BRI Solo. The research shows that the performance of the RKB of Bank BRI Solo in developing its fostered SMEs has been carried out effectively but has not been researched and provided further descriptions of the SMEs studied in terms of CSR practices and the orientation chosen. Subsequent research conducted by [13], who examined MSMEs in Surakarta with the ability to respond to the market, turned out to have a significant effect on product innovation excellence and marketing performance. These studies have not revealed CSR practices by MSMEs in Surakarta, including the chosen orientation in CSR practices.

The following study by Harto [14] examined the effect of CSR orientation on Corporate Social Performance. The CSR orientation in this study includes a corporate ethics orientation and a legal orientation. The results obtained are that the company's ethical orientation is positively related to the company's social performance, while the legal orientation is negatively related to the company's social performance. The study has paid attention to environmental uncertainty as a moderating variable, which indicates a significant interaction between the company's orientation towards social responsibility in MSMEs in Central Java. This research examines economic orientation, legal orientation, ethical orientation, and discretionary orientation as independent variables and is carried out on large, medium, small, and micro-scale companies. Unlike Harto's study, this focuses on the CSR orientation in employee, community, market, environmental, and an overall orientation perspective on MSMEs.

Another case is the study conducted by [15] which shows waste treatment as one of the CSR activities at Batik SMEs in Sukoharjo Regency. The CSR practice of waste treatment was chosen as a strategy carried out sustainably, but from this study, there has not been a specific picture of the orientation chosen by MSMEs in the practice of CSR in waste treatment.

Based on the previous studies and the phenomena that occur in MSMEs above, the purpose of this study is to explore (pooling) empirical data on the CSR orientation of MSMEs in Surakarta in terms of CSR practices that are oriented towards employees, society, markets, the environment and as a whole.

2. Literature review

2.1 Corporate social responsibility (CSR)

CSR (Corporate Social Responsibility) by the general public is understood as a concept or action taken by a company in social activities as a form of corporate responsibility to the environment and the community around the company.

According to [16] the concept of CSR is a way of running a business that is not solely for doing business. Many MSMEs have done it even though they do not consider it a CSR or a Sustainability business. Nevertheless, they have implemented values in running the business by maintaining good relations with employees, customers, and stakeholders and contributing positively to society. Applying these values has demonstrated CSR as part of running a business to gain business profits while doing commendable things by providing important benefits for a better life for the community and providing protection to the environment.

In Surakarta Mayor Regulation Number 12-A of 2013 concerning the implementation of Corporate Social Responsibility (Corporate Social Responsibility) in Surakarta City [17], CSR is defined as the responsibility inherent in every company to continue to create a harmonious, balanced and appropriate relationship with the environment, values, norms, and culture of society.

In companies, especially the large ones, there are 5 (five) pillars in the CSR practices of the Prince of Wales International Business Forum, including:

1. HR capacity development (Building Human Capital).

   The company forms reliable human resources or empowers the community.

2. Strengthening the community economy (Strengthening Economies).

   The company contributes to economic improvement, including assisting in poverty alleviation.

3. Maintenance of community harmony (Assessing Social Cohesion).

The company maintains the stability and harmony of the community so that conflicts do not occur, especially those related to business activities.

4. Implementation of good governance (Encouraging Good Governance).

   The company implements good governance.

5. Environmental Conservation (Protecting The Environment).

   The company preserves the physical, social and cultural environment.

2.2 Micro, small, and medium enterprises (MSMEs)

The characteristics of MSMEs are different in each literature on several sources from agencies, institutions, government regulations, including laws. According to Hernita [18], the main characteristics of MSMEs that distinguish large and medium enterprises are that MSMEs are labor-intensive businesses, are found in all locations, especially in rural areas, are more dependent on local materials, and are the main providers of essential goods and services for low-income or poor people. According to him, the general description of MSMEs is described as follows:

1. As livelihood activities, small and medium-sized businesses are used as job opportunities to earn a living, more commonly known as the informal sector. Examples are "Meatball" traders, "Chicken Noodle dumplings", "Somai," "Pentol" and "Cilok," who are known as street vendors (PKL).

2. As Micro-enterprise, it is a small and medium business with the nature of artisans but does not yet have entrepreneurial characteristics.

3. As a small dynamic enterprise, it is a small and medium-sized enterprise with an entrepreneurial spirit and can accept subcontract and export work.

4. as a fast-moving enterprise, it is a small and medium business that already has an entrepreneurial spirit and will transform into a big business (industry/company).

Under Law No. 20/2008 [19] Concerning Micro, Small and Medium Enterprises, MSMEs are expected to increase the position, role, and potential of MSMEs in realizing economic growth, equity and increasing people's income, creating job opportunities, and alleviating poverty. The empowerment of MSMEs is carried out by the Government, Regional Governments, the Business World, and the community by developing a conducive climate, providing business opportunities, support, protection, and business development as widely as possible in a comprehensive, synergistic, and sustainable manner. The classification of MSMEs based on Law no. 20/2008 [19] is as in **Table 1** as follows.

2.3 CSR practice orientation

As stated by Cohen [16] that sometimes companies do not realize that they have implemented CSR. However, this does not mean that companies that provide contributions or other benefits by not solely seeking business profits can be considered not to have a CSR orientation. This is because CSR orientation and CSR practices are interrelated. In some companies, the orientation of CSR practices is carried out by preparing a Sustainability Report. Some literature explains that the Sustainability

Business size	Asset (in IDR)	Revenue (in IDR)
Micro Business	Minimum 50 Million	Maximum 300 Million
Small Business	>50 Million – 500 Million	Maximum 3 Billion
Medium Business	>500 Million – 10 Billion	>2.5–50 Billion

Source: law no. 20/2008.

Table 1.
The classification of MSMEs based on law no. 20/2008.

Report is known as the Triple Bottom Line Approach, which reflects the orientation of CSR practices in a simple way, namely Productivity issues, Environmental Responsibility, and Social Domains.

Another study on the orientation of CSR practices, as presented by Harto [14], is the orientation of CSR practices described based on four components of corporate social responsibility, including the first, namely the orientation based on economic responsibility, where businesses must be productive and profitable to face competition. The second CSR practice orientation is based on legal/legal responsibility, which views that the business must comply with applicable laws and regulations. The third orientation of CSR practices is based on ethical responsibility, which is the obligation to run a business ethically, obey the rules even though they are not written, follow the norms and values of the community. Furthermore, the fourth CSR practice orientation is based on discretionary responsibility in which companies must voluntarily provide benefits according to community expectations without any coercion. This CSR practice orientation has not been specific or focused on MSMEs but still applies to companies at all scales, large and medium, small and micro. In this study, further references are needed that only focus on the orientation of CSR practices for MSMEs.

Based on other literature [10], several orientations of CSR practices in Europe have been found that specifically focus on Small Medium Enterprises (SMEs), which can be equated with SMEs in Indonesia. The impact of CSR practices on MSME often cannot appear immediately, and sometimes it even takes a long time to obtain. The orientation of CSR practices is often carried out using a practical approach to not causing many costs. These approaches include employee-oriented, community-oriented, market-oriented, environment-oriented, and general or comprehensive oriented.

According to Das Maitreyee et al. [20], CSR practices in Asia are conceptualized based on recommendations derived from the Western context that have been developed. The reason is that they have started earlier than in Asia. However, in terms of economy and culture, CSR practices for MSMEs in Asia are unique in that they are still traditional and dominated by culture, which tends to be in the form of philanthropic activities that benefit their local communities. Asia is different from the West in terms of the economy, which tends to use market orientation to build a good brand image. In Asia, many family business owners feel that social value is more important to be remembered by their income and what they do for society. MSMEs practice CSR in donations to various charity hospitals, religious organizations, schools, or orphanages, or it can be said that CSR is more ethical and philanthropic oriented and dominated mainly by culture. The role of the Government is to determine the implementation of CSR practices in Asia; in this case, the Government understands the potential of MSMEs in economic development, thereby facilitating the control of loan interest rates, making environmental-related policies through regulations from local governments, and providing support through promotions to encourage SMEs to collaborate in running their business.

The implementation of CSR practices in Indonesia is not much different from that carried out in Asia, more specifically for the Surakarta City area; policies still influence the orientation of CSR practices through government regulations. These regulations include Central Java Provincial Regulation No. 2 of 2017 concerning Corporate Social and Environmental Responsibility (TJSLP) [21], in addition to Surakarta City Regulation No. 2 of 2015 concerning Corporate Social Responsibility [22], and Surakarta Mayor Regulation No. 3-A 2016 concerning Instructions for the Implementation of the Surakarta City Regulation No. 2 of 2015 [23].

In addition to the influence of government policies, the implementation of CSR is also influenced by organizations engaged in economic, social, and environmental changes. For example, as presented by Kharisma et al. [24], the Organization for Economic Co-operation and Development (OECD) sets international standards for various fields, from agriculture and taxes to chemical safety, which also influences flexible Good Corporate Governance related to conditions, culture, and traditions in their respective regions.

3. Method

This research is a descriptive study, which was carried out quantitatively cross-sectional. The population in this study were MSMEs located in Surakarta City as a center for developing national MSMEs, while the sample, which amounted to 90 MSMEs, was conducted using a simple random sampling method. Data collection with this number of samples was carried out by distributing questionnaires filled out by MSME business actors in Surakarta City.

Before being used, the questionnaire instrument was first tested in a pilot study of 10 business actors. The purpose of the pilot study was to test the validity and reliability of the instrument, which was carried out using the Pearson's product–moment and Cronbach's alpha methods, and a content test was conducted to improve the instrument's grammatical suitability so that the instrument is easy to understand and easy to fill out by respondents.

At the time of the survey, the research respondents were given an explanation of this research's aims and objectives, including the principle of confidentiality, and were given an overview of the number of questions, how to fill in, and urged them to fill in completely with the actual conditions.

Statistical analysis used is statistical data processing which is commonly used in descriptive research using the help of IBM SPSS Statistics Data Editor 25 software. The first step is a descriptive statistical analysis to determine the highest and lowest values, average values, and standard deviations. These values can be used for data analysis to reveal the orientation of CSR practices quantitatively. Then the normality test was conducted to determine whether the data from the questionnaires reflected a normal distribution or not. In the end, conclusions are drawn from the results of the analysis.

4. Result

After analyzing the data descriptively, the results are that the mean value of each question item is greater than the standard deviation number. Thus, it means that the study results have very low deviations so that the distribution of the data is normal and valid.

Based on each question item, it can be grouped according to the orientation of CSR practices and then analyzed the tendency of respondents' answers regarding whether or not the orientation is carried out in CSR practices. The grouping is as follows:

4.1 Employee orientation in CSR practices

Table 2 shows the frequency distribution results of the K1 questionnaire items, and the mean value is 3.69 with a median of 4 and mode 4, which means that, on average, MSMEs have provided compensation to employees. Then the results of the frequency distribution of the K2 questionnaire items with a mean value of 3.18 and a median of 3 and mode 4, on average, MSMEs have provided Social Security protection to employees. Furthermore, the frequency distribution results of the K3 questionnaire items with a mean value of 3.06 and a median of 3 and mode 4 indicate that, on average, MSMEs train employees to work according to the assigned tasks. In addition, results of the frequency distribution of K4 questionnaire items with a mean value of 3.46 and with a median of 4 and mode 4 indicate that, on average, MSMEs are involved in product or service improvement. The results of the frequency distribution of K5 questionnaire items with a mean value of 2.57 with a median of 3 and mode 3 indicate that there is a tendency for employers to provide opportunities for money loans for employees in need.

4.2 Community orientation in CSR practices

Table 3 shows the frequency distribution of M7, M8, M9, and M10 questionnaire items that describe community orientation in CSR practices. Item M7 with a mean value of 3.31, Median 3, and Mode 4 indicates that MSMEs have contributed special funds for the development of the surrounding community. Item 8 with a mean of 3.38, a median of 4, and a mode of 4 indicates that, on average, MSMEs contribute funds to village organizations around the district or village pillar. Item M9, with a mean of 3.29, a median, and mode of 3, indicates that MSMEs make donations to schools or religious institutions. Finally, **Table 3** indicated that MSMEs provide non-detailed donations to the surrounding community as shown in statistic value: Mean of 3.39 and a 4 for median and mode.

Statistics							
		K1	**K2**	**K3**	**K4**	**K5**	**K6**
N	Valid	90	90	90	90	90	90
	Missing	0	0	0	0	0	0
Mean		3,69	3,18	3,06	3,46	2,57	2,61
Median		4,00	3,00	3,00	4,00	3,00	3,00
Mode		4	4	4	4	3	2
Sum		332	286	275	311	231	235
Percentiles	25	3,00	2,00	2,00	3,00	1,00	2,00
	50	4,00	3,00	3,00	4,00	3,00	3,00
	75	4,00	4,00	4,00	4,00	3,00	3,00

Table 2.
Frequency distribution analysis and categorization on employee orientation in CSR practices.

Statistics					
		M7	M8	M9	M10
N	Valid	90	90	90	90
	Missing	0	0	0	0
Mean		3,31	3,38	3,29	3,39
Median		3,00	4,00	3,00	4,00
Mode		4	4	3	4
Sum		298	304	296	305
Percentiles	25	3,00	3,00	3,00	3,00
	50	3,00	4,00	3,00	4,00
	75	4,00	4,00	4,00	4,00

Table 3.
Frequency distribution analysis and categorization on community orientation in CSR practices.

4.3 Market orientation in CSR practices

Table 4 shows the results of the questionnaire items P11, P12, P13, P14, P15, P16, P17, P18, P19, and P20, all of which describe those related to market orientation in CSR practices of MSMEs. Questionnaire item P11 relates to business licenses, with a mean of 3.39 with a median of 3 and mode of 3 (three), indicating that, on average, MSMEs have a business license. Questionnaire item P12 relates to safe products, with a mean of 2.99 with a median of 3 and mode 4, indicating that, on average, MSMEs have safe products for consumers. Questionnaire item P13, related to advertising, has a mean of 3.36, with a median of 4 and mode 4, indicating that MSMEs place advertisements as promotional media on average. The questionnaire item related to product prices, P14, has a mean of 3.64, with a median of 4 and mode 4, indicating that, on average, MSMEs set their product prices following the market price in general. Questionnaire items P15 are those related to MSME suppliers, and those items have a mean of 3.54 with a median of 4 and a mode of 4, indicating that, on average, MSMEs have suppliers. The distribution aspect is contained in the P16 questionnaire item. This item has a Mean of 2.46 with a median

Statistics											
		P11	P12	P13	P14	P15	P16	P17	P18	P19	P20
N	Valid	90	90	90	90	90	90	90	90	90	90
	Missing	0	0	0	0	0	0	0	0	0	0
Mean		3,39	2,99	3,36	3,64	3,54	2,46	2,59	3,22	3,24	3,11
Median		3,00	3,00	4,00	4,00	4,00	2,00	3,00	3,50	4,00	3,00
Mode		3	4	4	4	4	2	3	4	4	3
Sum		305	269	302	328	319	221	233	290	292	280
Percentiles	25	3,00	2,75	3,00	3,00	3,00	1,75	2,00	3,00	3,00	3,00
	50	3,00	3,00	4,00	4,00	4,00	2,00	3,00	3,50	4,00	3,00
	75	4,00	4,00	4,00	4,00	4,00	3,00	3,00	4,00	4,00	4,00

Table 4.
Frequency distribution analysis and categorization on market orientation in CSR practices.

of 2 and mode 2, which indicates that MSMEs have no problems with suppliers. Questionnaire item P17, which relates to payments to suppliers, has a mean of 2.59 with a median of 3 and mode 3, which indicates that, on average, MSMEs have no problems in payment to suppliers. Aspects of the Government's role towards MSME service products are described in the questionnaire item P18, with a Mean of 3.22, a median of 3.5, and a Mode 4, which indicates that, on average, the Government utilizes MSME products or services. Questionnaire item P19, which relates to the cooperation of SMEs with large companies, has a Mean of 3.24 with a median of 4 and mode 4, indicating that MSMEs cooperate with large companies, whether they supply products or are supplied with products. Finally, the questionnaire item related to cooperation between MSMEs with similar companies, P20, with a mean of 3.11, a median of 3, and a mode of 3, shows that, on average, MSMEs cooperate with similar entrepreneurs, especially when they receive many orders.

4.4 Environmental orientation in CSR practices

Table 5 shows the questionnaire items L21, L22, L23, L23, L24, and L25 related to environmental orientation in CSR practices. Questionnaire item L21, relating to aspects of environmentally friendly products, has a mean of 3.39, with a median of 4 and mode 4, indicating that, on average, MSMEs produce environmentally friendly products. The questionnaire item related to plastic packaging, P22, which has a mean of 2.20 with a median of 2 and mode 2, shows that SMEs avoid plastic packaging, although not all of them. Questionnaire item L23 is closely related to production waste and has a mean of 2.36 with a median of 2 and mode 2 (two) and shows that, on average, MSMEs produce waste during their production. Furthermore, the disturbing waste aspect stated in the questionnaire item L24 has a mean of 2.57 with a median of 3 and mode 3, which indicates that there are still MSMEs whose production activities produce waste that disturbs the surrounding environment. Finally, the questionnaire item L25, which relates to waste treatment, has a mean of 3.04 with a median of 3 and mode 3, indicating that MSMEs have attempted to implement simple waste management.

The next step of data analysis is to find the most dominant practice orientation in implementing CSR practices by MSMEs, using confirmatory factor analysis (CFA). Before conducting confirmatory factor analysis, it is necessary to ascertain whether the existing correlation is sufficient to determine a correlation between

Statistics						
		L21	L22	L23	L24	L25
N	Valid	90	90	90	90	90
	Missing	0	0	0	0	0
Mean		3,39	2,20	2,36	2,57	3,04
Median		4,00	2,00	2,00	3,00	3,00
Mode		4	2	2	3	3
Sum		305	198	212	231	274
Percentiles	25	3,00	2,00	2,00	2,00	3,00
	50	4,00	2,00	2,00	3,00	3,00
	75	4,00	3,00	3,00	3,00	4,00

Table 5.
Frequency distribution analysis and categorization on environmental orientation in CSR practices.

Kaiser-Meyer-Olkin Measure of Sampling Adequacy		,705
Bartlett's Test of Sphericity	Approx. Chi-Square	1313,967
	df	300
	Sig.	,000

Table 6.
Kaiser Meyer Olkin test analysis measure of sampling adequacy and Bartlett's test of sphericity.

variables using the Bartlett of Sphericity Test. In addition, the level of intercorrelation between variables is also measured, including measuring whether factor analysis can be carried out/continued. These measurements used the Kaiser Meyer Olkin Measure of Sampling Adequacy (KMO MSA) test tool. The results of data analysis using SPSS obtained a KMO value of 0.705, which means that confirmatory

	Communalities	
	Initial	**Extraction**
K1	1,000	,700
K2	1,000	,488
K3	1,000	,717
K4	1,000	,700
K5	1,000	,682
K6	1,000	,733
M7	1,000	,759
M8	1,000	,876
M9	1,000	,699
M10	1,000	,864
P11	1,000	,818
P12	1,000	,691
P13	1,000	,689
P14	1,000	,689
P15	1,000	,772
P16	1,000	,808
P17	1,000	,840
P18	1,000	,824
P19	1,000	,719
P20	1,000	,705
L21	1,000	,554
L22	1,000	,762
L23	1,000	,824
L24	1,000	,813
L25	1,000	,558
Extraction method: principal component analysis.		

Table 7.
Analysis of how much the variable can explain the factor.

factor analysis can be carried out because the KMO value is more than 0.5. The Bartlett test of Sphericity with Chi-Square value is 1313,967, significant with a value of 0.000, so it can be concluded that the confirmatory factor analysis test can be continued (**Table 6**).

Confirmatory factor analysis begins with exploring how much a variable can explain the factor. The result is that almost all variables have extraction values of more than 0.5, which means that the existing variables can explain the factors. Achievements below 0.5 on K2 items are 0.488, as can be seen in **Table 7**. Of all the variables, based on the processed SPSS data, it was found that the maximum factors that can be obtained are 7 (seven) factors with Initial Eigenvalues of more than 1 (one) (**Table 8**).

The next step is to determine which factor each variable belongs to and how the variable correlates to the factor. Then the results are sorted from the largest to the smallest value for each factor so that it is easy to determine the grouping. The order of the most dominant factors and their variables based on the rotated component matrix table (**Table 9**) is as follows:

a. Factor I: variable M8(0.904), M10(0.868), M7(0.790), M9(0.677).

b. Factor II: variables P18(0.876), P19(0.777), P20(0.666), and P13(0.657).

c. Factor III: variable P17(0.858), P16(0.828) P12(−0.580), K2(0.563), K5 (0.495).

d. Factor IV: variable K1(0.773), K4(0.654), P14(0.601), P15(0.576), L21 (0.458).

e. Factor V: variable L23(0.890), L24(0.812),

f. Factor VI: variable P11(0.797), K6(0.698), K3(0.650),

g. Factor VII: variable L22(0.829), and L25(0.466).

5. Discussion

The data above shows that the most dominant orientation in MSME CSR practices in Surakarta is the Community Orientation and Market Orientation. Community-oriented CSR carried out by MSMEs in Surakarta provides donations or donors to the surrounding community, orphanages, or sponsors community activities such as youth organizations for youth and PKK activities for mothers and sponsoring road repair activities around business locations. It is also found in the study of SMEs in Europe [10] such as in Finland, some SMEs donate part of their profits to children getting a basic education. Furthermore, a medium-sized Romanian company carried out CSR by assisting in school laboratory equipment and humanitarian sponsorship for local children. Medium-sized companies in Spain also carry out CSR activities by providing sponsorship of sports activities to local youth. In line with MSMEs in Asia, mostly family businesses, CSR practices are also more community-oriented, with donations to hospitals, religious organizations, and schools [20].

The market-oriented CSR activities for SMEs in Surakarta also showed the most dominant results. The activities carried out included targeting market-oriented CSR activities, namely producing products that already have permits such as PIRT,

Total variance explained									
Component	Initial eigenvalues			Extraction sums of squared loadings			Rotation sums of squared loadings		
	Total	% of variance	Cumulative %	Total	% of variance	Cumulative %	Total	% of variance	Cumulative %
1	6,075	24,299	24,299	6,075	24,299	24,299	3,334	13,334	13,334
2	3,443	13,771	38,070	3,443	13,771	38,070	3,035	12,140	25,475
3	2,322	9,288	47,358	2,322	9,288	47,358	3,033	12,133	37,608
4	1,920	7,679	55,037	1,920	7,679	55,037	2,505	10,021	47,629
5	1,813	7,252	62,290	1,813	7,252	62,290	2,362	9,448	57,076
6	1,573	6,291	68,581	1,573	6,291	68,581	2,284	9,136	63,213
7	1,140	4,561	73,142	1,143	4,561	73,142	1,732	6,929	73,142
8	,887	3,548	76,690						
9	,785	3,142	79,832						
10	,700	2,800	82,632						
11	,654	2,617	85,249						
12	,519	2,075	87,324						
13	,504	2,016	89,340						
14	,465	1,859	91,199						
15	,399	1,594	92,793						
16	,286	1,142	93,935						
17	,270	1,080	95,016						
18	,238	,952	95,968						
19	,226	,904	96,872						
20	,191	,764	97,636						

DOI: http://dx.doi.org/10.5772/intechopen.99859

Total variance explained									
Component	Initial eigenvalues			Extraction sums of squared loadings			Rotation sums of squared loadings		
	Total	% of variance	Cumulative %	Total	% of variance	Cumulative %	Total	% of variance	Cumulative %
21	,155	,620	98,256						
22	,143	,574	98,830						
23	,126	,503	99,333						
24	,094	,375	99,708						
25	,073	,292	100,000						

Extraction method: principal component analysis.

Table 8.
Analysis of initial eigevalues.

Rotated component matrix[a]							
	Component						
	1	2	3	4	5	6	7
M8	,904	,141	,138	,089	,105	-,001	-,007
M10	,865	,271	,121	-,058	-,022	,111	,079
M7	,790	,029	,305	,120	,157	,051	-,007
M9	,677	,132	,027	,362	-,188	-,233	,034
P18	,213	,876	,077	,071	,004	,021	,018
P19	,335	,777	-,001	,012	,017	-,054	,015
P20	-,083	,666	-,179	,044	-,107	,189	,416
P13	,008	,657	,090	,290	-,128	-,024	-,386
P17	,188	-,127	,858	,142	,175	-,043	,007
P16	,114	-,104	,828	,153	222	-,157	,041
P12	-,283	-,251	-,580	-,084	,004	,447	-,070
K2	,135	,167	,563	,092	-,262	,169	,139
K5	-,180	-,311	-,495	-,121	-,293	,430	,149
K1	-,051	,066	,216	,773	-,114	,138	,129
K4	,340	-,135	,348	,654	-,052	-,099	-,073
P14	,294	,330	-,031	,601	,272	-,126	-,202
P15	,301	,449	-,050	,576	,252	-,031	-,286
L21	,017	,306	,177	,458	,428	-,150	-,119
L23	,002	-,012	-,067	,023	,890	,061	,153
L24	,101	-,107	,340	-,041	,812	-,119	,016
P11	-,061	,016	-,037	,160	,129	,797	-,368
K6	,061	,129	,066	-,245	-,301	,698	,265
K3	,135	,004	-,441	,054	,023	,650	,281
L22	,063	-,033	,134	-,093	203	-,035	,829
L25	,023	,091	-,015	,394	-,302	,322	,446

Extraction method: principal component analysis.
Rotation method: varimax with Kaiser normalization.
[a]Rotation converged in 19 iterations.

Table 9.
Rotated component matrix analysis.

ensuring the cleanliness and safety of the products produced, conducting promotions in the form of advertisements through various social media, establish good cooperation and relationships with suppliers as well as cooperation with other MSMEs and local governments. This finding is also in line with research that has been carried out in SMEs in Europe, such as several SMEs in Finland, Spain, and Poland who always pay attention to the quality of the products sold and maintain good relations with suppliers [10]. The more dominant community and market-oriented CSR practices are also in line with research [4] which sees CSR as a social response process, policy, program, and impact on the relationship between the company and society.

6. Conclusion

From the data analysis above, it can be concluded that the most dominant orientation in CSR practices by MSMEs is the Community Orientation and Market Orientation. The existence of MSMEs in society makes Community Orientation the most important thing, where MSMEs strive to provide benefits to the surrounding community. Furthermore, besides providing benefits to the community around MSMEs, they still consider CSR practices to improve their existence and maintain business continuity so that Market Orientation becomes the next choice for MSMEs in CSR practices. The two orientations are more dominant than the other orientations.

The limitation of this study is the limited sample obtained. Although it only obtained samples from the Surakarta area and its surroundings, the conclusions of this research can reflect the existing situation because Surakarta City and its surroundings are also a national barometer of business. Based on the above limitations, it is necessary to expand the research sample for future research by considering the diverse representation of the small and medium industry sector.

Acknowledgements

Susilaningtyas B. Kurniawati is a doctoral student in the Doctoral Program in Economics, Faculty of Economics and Business, Universitas Sebelas Maret.

Hasan Fauzi, PhD., is a Professor in Accounting & CSR, Accounting Department, Faculty of Economics and Business, Universitas Sebelas Maret, and Director of Indonesian Center for Social and Environmental Accounting Research and Development (ICSEARD). He is currently also as Deputy Chairman of the Board of Trustees, Universitas Sebelas Maret.

Author details

Susilaningtyas B. Kurniawati and Hasan Fauzi*
Universitas Sebelas Maret, Surakarta, Indonesia

*Address all correspondence to: hfauzi@icseard.uns.ac.id

References

[1] Fauzi, H. Svensson, G.; Rahman, A. A. 2010. *Triple Botton Line as Sustainable Corporate Performance: A Proposition for The Future*. Sustainability, 2, pp: 1345-1360.

[2] Carroll, A. B. 1979. *A Three-Dimension Conceptual Model Of Corporate Sosial Performace*. Academy of Management Review 4: 497-506.

[3] Wartick, S.L., and Cochran, P,L. 1985. *The Evolution of The Corporate Sosial Performance Model*. Academy of Management Review, 10:758-769.

[4] Wood, D. J. 1991. *Corporate Social Performance Revisited*. Academy of Management Review, 16:691-718.

[5] Clarkson, Max B E. 1995. *A Stakeholder Theory of The Corporation: Concept, Evidence, and Implications*. The Academy of Management.

[6] Republic of Indionesia, 2007. Law No. 40/ 2007 concerning Limited company

[7] https://www.kominfo.go.id/content/detail/11526/kemenkop-ukm-379-juta-umkm-sudah-go-online/0/sorotan_media, accessed on January 5, 2020

[8] https://www.wartaekonomi.co.id/read241156/big-data-ukm-dianggap-mampu-dongkrak-lini-usaha-lebih-besar.html accessed on January 10, 2020

[9] https://www.republika.co.id/berita/koran/industri/16/12/07/ohtak15-rikorsel-kembangkan-umkm-ramah-lingkungan accessed on January 8, 2020

[10] https://europa.eu.int/comm/enterprise/csr/ms_sme_projects.htm. CSR and Competitiveness European SMEs Good Parctice. www.csr-in-smes.eu, 2005.

[11] Central Bureau of Statistics (Badan Pusat Statistik)-bps.go.id

[12] Ayungga N.R, 2019. *Efektivitas Kinerja Rumah Kreatif Bumn Bank Bri Solo Terhadap Perkembangan Ukm Binaannya. Universitas Ailangga.*

[13] Fatonah Siti dan Nugroho A.P. 2017. Peran Mediasi Keunggulan Inovasi Produk Dalam Meningkatkan Kinerja Pemasaran Batik Skala Kecil dan Menengah Surakarta. Prosiding Seminar Nasional AIMI. ISBN 1234-5678-90-12-1.

[14] Harto, P. 2013. *Pengaruh Orientasi Corporate Social Responsibility (Csr) Terhadap Kinerja Sosial Perusahaan Dengan Ketidakpastian Lingkungan Sebagai Variabel Pemoderasi. Jurnal Akuntansi Indonesia, Vol. 2 No. 1 Januari 2013, Hal. 37-50*

[15] Kurniawan M. 2013 Wawan, Purwanto. P, Sudarno. S. *Strategi Pengelolaan Air Limbah Sentra UMKM Batik Yang Berkelanjutan Di Kabupaten Sukoharjo*. Jurnal Ilmu Lingkungan, Vol. 11 Issue 2:62-72. ISSN 1829-8907.

[16] Cohen. E. 2013. *Sustainability Reporting For SMEs*. Greenleaft Publishing Book.

[17] The Republic of Indonesia, 2013. Surakarta Mayor Regulation No. 12-A/ 2013 concerning Implementation of Corporate Social and Environmental Responsibility

[18] Sahban Hernita, 2015. Menembus Badai UMKM. Sahmediapre

[19] Republic of Indionesia, 2008. Law No. 20/2008 concerning Micro small and Medium Enterprises

[20] Das Maitreyee, Rangarajan. K, Dutta Gautam. 2019. *Corporate sustainability in SMEs: an Asian perspective.* Journal of Asia Business Studies. Vol. 14 No. 1.109-138.

[21] The Republic of Indonesia, 2017, Central Java Provincial Regulation No 2/ 2017 concerning Corporate Social and Environmental Responsibility (TJSLP)

[22] The Republic of Indonesia, 2015. Surakarta City Regulation No.2/ 2015 concerning Corporate Social Responsibility

[23] Surakarta Mayor Regulation Nomor 3-A/2016 concerning Instructions for the Implementation of the Surakarta City Regulation Number 2 of 2015

[24] Sudarwanto S, Triwahyudi P & Kharisma DB. 2018. *Implementation of Corporate Social Responsibility of Region-Owned Enterprises on The Conservation of Living Environmental Function of Bengawan Solo River*. Journal of Law, Policy and Globalization. ISSN 2224-3240. ISSN 2224-3259. Vol. 74.

Chapter 3

Role of Red Palm Oil in Male Obesity and Infertility Prevention

Folorunso A. Olabiyi, Yapo G. Aboua and Thomas K. Monsees

Abstract

The African continent has wide, varied, and rich plant diversity due to its climate. Some of these plants and their products have received tremendous attention due to their benefits in treating and managing ailments that plagues humanity. Red palm oil (RPO) is one of such natural products that have immense nutritional value with ability to ameliorate cardiac- and reproductive-related disorders. In this review article, the current knowledge on the potential of RPO as a phytomedicine to lessen or even prevent the negative impact of obesity on general health status and male fertility was evaluated. This study was done using electronic databases such as PubMed, Scopus, Science Direct, Google Scholar and Web of Science. The study revealed some controversies and inconsistent reports on the effect of RPO on obesity and male fertility which needs further research using appropriate experimental models of obesity. Obesity is known to disrupt male fertility by causing changes to the hypothalamic- pituitary-gonadal axis, thus impairing steroidogenesis and spermatogenesis. As spermatozoa are extremely sensitive towards oxidative stress, a carefully balanced daily supplementation of normal diet with antioxidant-rich RPO might be useful to protect spermatozoa and preserving male fertility. RPO was shown to be useful to protect against or ameliorate toxin- or medical condition-induced male infertility. Also, RPO is packed with powerful antioxidants like carotenoids and vitamin E which helps to prevent cell damage. However, its role in obesity prevention remains a debate.

Keywords: Obesity, male Infertility, Red palm oil (RPO), male Reproduction, medicinal plant

1. Introduction

In recent time, obesity is considered one of the most critical public health concerns due to its severe physiological and economic implications. The scourge was once considered a problem only in the developed nations [1]. Contrarily, it has been reported that people living in low income countries with less affluence are not spared. In this case, the obesity could be associated with stress, alcoholism and poor nutrition due to poverty in developing countries, with its prevalence rising with age in both childhood and adulthood, regardless of gender [2, 3]. Few decades ago, obesity prevalence in Western and Westernizing countries was more than doubled, the cause of which could be attributed to over-nutrition [4]. What then is obesity? Obesity is described as the accumulation of adipose tissue to excess and to an extent that impairs both physical and psychosocial health and well-being [3]. More so, obesity presents a specific concerns and challenges for the reproductive health of men and women, affecting fertility, reproduction, pregnancy and the resulting child's long-term health [4]. An epidemiological study by the world health organization (WHO) revealed that approximately 2 billion adults are overweight [5].

Palm trees (*Elaeis guineensis*) are native to Africa and other countries such as Malaysia and Indonesia [6]. They bear the palm fruits from which red palm oil, RPO is extracted. Despite its usefulness as an essential oil for cooking, there have been insufficient evidence and inconsistent reports on the effect of RPO in human health and wellness. The aim of this review article, therefore, is to provide an overview of the use of red palm oil in the amelioration of the negative health effects of obesity and male infertility.

2. Origin

Red palm oil (RPO), found in the tropical and sub-Saharan Africa is natural oil obtained from the fleshy red mesocarp of the fruits from the oil palm tree (*Elaeis guineensis*). RPO contains lipid-soluble antioxidants like carotenoids (α and β-carotenes, lycopenes), vitamin E (in the form of α, β, δ tocotrienols and tocopherol) and ubiquinone [7]. It has also been reported that the red-orange color of red palm fruits was due to its carotene and lycopene contents as present in tomatoes, carrots and other fruits and vegetables [7]. Red palm oil is different from other plant and animal oil because it contains 50% saturated fatty acids, 40% unsaturated fatty acids, and 10% polyunsaturated fatty acids [8–10].

3. Biological and medicinal usefulness of RPO

Studies indicate that RPO has beneficial effects on arterial thrombosis and hypertension associated with oxidative stress [11, 12] and is also protective against the consequences of ischemia/reperfusion injury [13, 14]. More so, Aboua et al. [15] showed that RPO could possibly inhibit apoptosis in rat sperm, while its role in reducing oxidative stress in HIV/AIDS and tuberculosis patients has also been reported [16]. Besides, RPO has been reported to be useful in wound healing [17]. Moreover, as palm oil is very rich in carotenes, it is also found useful for preventing and treating vitamin A deficiency [18]. Therefore, due to its antioxidant effect, RPO is recommended as a supplement to the diet of pregnant women. RPO also possesses potentials in preventing high blood pressure and breast cancer [19].

3.1 RPO in diabetes mellitus

The effects of palm oil (PO) and groundnut oil (GO) supplementation on the antioxidant status and diabetic indices in Alloxan (100 mg/kg) induced diabetic Wistar rats was investigated. In this study, blood glucose, plasma vitamin E, superoxide dismutase (SOD), total protein and albumin levels were measured. It was concluded that GO exhibited superior antioxidant activities and that the supplementation of red palm oil and ground nut oil as a source of antioxidant was beneficial in diabetic state as it reduced blood glucose and enhance antioxidant status [20]. A similar study investigated the antioxidant status in streptozotocin-induced diabetic male Wistar rats following intake of red palm oil and/or rooibos. The result revealed the anti-oxidative potentials of red palm oil, rooibos, and a synergistic effect of their combination in diabetic conditions, hence, they could be beneficial in the management of diabetes and its complications [21].

3.2 Linking metabolic syndrome, obesity and type 2 diabetes

Obesity, in excess of visceral adiposity, is closely related to insulin resistance, hyperglycemia, dyslipidemia and hypertension; taken together, these group of metabolic disorders are termed "metabolic syndrome [22]. Abdominal obesity is the key component of the metabolic syndrome, with a predominance of intra-abdominal visceral fat accumulation, indirectly measured by waist circumference in clinical practice. Research has shown that a chronic low-grade inflammation and an activation of the immune system are involved in the pathogenesis of obesity-related insulin resistance and type 2 diabetes [23]. The investigators noted that the systemic inflammatory markers are risk factors for the development of type 2 diabetes and its macrovascular complications. Besides, adipose tissue, liver, muscle and pancreas are themselves sites of inflammation when there is obesity. An infiltration of macrophages and other immune cells is observed in these tissues associated with a cell population shift from anti-inflammatory to a pro-inflammatory profile. It is noteworthy that these cells are crucial for the production of pro-inflammatory cytokines, which act in an autocrine and paracrine manner to interfere with insulin signaling in peripheral tissues or induce β-cell dysfunction and subsequent insulin deficiency. Most especially, the pro-inflammatory interleukin-1β is implicated in the pathogenesis of type 2 diabetes through the activation of the NLRP3 inflammasome.

4. Obesity

Obesity, by definition, is an excess of body fat that poses a risk to health. It is assessed clinically via the expression of body weight as a function of height – the body mass index (BMI). This is calculated by dividing an individual's weight (in kilograms) by the square of their height (in meters) (kg/m^2) [3, 24]. BMI is used in adults to delineate overall body fatness where a BMI of 18.5–24.9 kg/m^2 is considered to be normal. Abnormal BMI's are sub-categorized according to severity, where a BMI $\geq 25\ kg/m^2$ is considered to be within the overweight range, BMIs $\geq 30\ kg/m^2$ are considered as obesity and BMI ranges $\geq 40\ kg/m^2$ are classified as severe/morbid obesity. Obesity is, however, regarded as a disease of opulence, easily remedied by the reduction of food intake and considered to be irrelevant elsewhere. Furthermore, obesity was not a major concern for the WHO as their priority was to deal with malnutrition and the problems of the third world. By the mid-1990's, obesity had become a prominent problem for obesity specialists but was still not being taken seriously by most national governments. The prevalence is generally higher in women than in men, although the highest incidence was recorded in males aged 25–29 years residing in low-middle income countries.

4.1 Causes of obesity

Obesity is an unintended consequence of a 'Western lifestyle' [25] where economic, social and technological advances [26] have resulted in urbanization and reduced physical activity [27]. Diets have transitioned from natural, organic foods to refined, high fat and high sugar alternatives, leading to a nutrition transition parallel to the observed economic growth [28]. These changes have led to a rapid increase in the prevalence of obesity [25]. As a historically Western disease, obesity was largely localized to developed countries until recent decades where middle- and low income-countries began to experience rates comparable to those of high-income countries. The rapid economic expansion and modernization experienced by middle- and low-income countries, as well as changing diets, are the greatest contributors to this epidemic [28] as more and more of these countries adopt Western diets and lifestyles. Close to 30% of the populations within middle income countries are classified as overweight or obese, with South Africa (SA) having the highest prevalence in sub-Saharan Africa, where the prevalence was reported to be 27% in adults over the age of 18 years in 2016 [29]. While the predominant causes of the energy imbalance seen in obesity are inadequate physical activity and unhealthy diets, not all those who are exposed to these unfavorable behaviors will develop the disease. Obesity arises from the interactions between an at-risk genetic profile and environmental risk factors, such as physical inactivity, excessive caloric intake, the intrauterine environment, medications, socioeconomic status, and possibly novel factors such as insufficient sleep, endocrine disruptors, and the gastrointestinal micro biome [30]. Obesity is a multifactorial disease, and many underlying factors disproportionately predispose subsets of the population to its development, and several of these are non-modifiable factors including age, gender, race, and genetics. Specific structural and functional changes are observed in obese visceral adipose tissue, together with local inflammation and adipokine production, promoting metabolic disturbances [31, 32]. Obesity leads to the presence of ectopic fat surrounding organs and to the accumulation of lipids in tissues themselves. Peripancreatic adipose tissue is implicated in glucose homeostasis regulation and can be impaired in obesity.

4.2 Prevention of obesity

Obesity is not only a clinical problem, but also a public health concern. A deeper assessment of obesity requires a multidisciplinary and transdisciplinary approach with complementary knowledge of molecular, clinical, bioinformatic, and syndemic frameworks that affect the underlying mechanisms and factors that have resulted in the current epidemic of obesity [33].

Strong evidence exists that weight loss reduces blood pressure in both overweight hypertensive and non-hypertensive individuals; reduces serum triglycerides and increases high-density lipoprotein (HDL)-cholesterol and generally produces some reduction in total serum cholesterol and low-density lipoprotein (LDL)-cholesterol [34]. A variety of effective options exist for the management of overweight and obese patients, including dietary therapy approaches such as low-calorie diets and lower-fat diets; altering physical activity patterns; behavior therapy techniques; pharmacotherapy; surgery; and combinations of these techniques. In this chapter, red palm oil is being proposed as treatment regimen for obesity prevention.

4.3 Red palm oil in obesity prevention

The oil palm tree (*Elaeis guineensis*) from the family Arecaceae is a high oil-producing agricultural crop. Palm oil is available in different forms, which include crude or red palm oil and refined palm olein (refined PO) [35]. Despite the contentious issues surrounding dietary fats, they are considered essential nutrients because they are required to perform critical functions in the body including serving as a carrier of preformed fat-soluble vitamins, enhancing the bioavailability of fat-soluble micronutrients and providing essential substrate for the synthesis of metabolically active compounds (such as the steroid hormones, testosterone, estrogen and progesterone) among other useful functions. These benefits of fats notwithstanding, diets that are high in fat are strongly associated with an increased prevalence of obesity and an increased risk of developing coronary artery disease, high blood pressure, diabetes mellitus, and certain types of cancer [36]. RPO consumption's effects on health are still debated. Two of the most important edible oils in the sub- Saharan Africa, are coconut oil and palm oil. Along with palm kernel oil, they are often referred to collectively, as the tropical oils and are typically known to be rich in saturated fats [36]. According to Boateng and co-workers [37], RPO contains α, β and γ carotenes, phytosterols such as sitosterol, stigmasterol and campesterol. These lipophilic sterols are easily absorbed in the gastrointestinal tract, and then converted through a series of enzymatic reactions into cholesterol, which is a major precursor of steroid hormones. A moderate use of palm oil is likely to be beneficial for blood lipid profiles. The RPO rich vitamin E, composed mainly of tocopherols and tocotrienols act as potent antioxidants that make it relatively stable to oxidation. Both animal and human studies show that tocotrienols could reduce plasma cholesterol, apolipoprotein B, thromboxane B2, and platelet factor IV. They could also inhibit or delay the oxidative deterioration of cellular membranes. The benefits of RPO to health include a reduction in the risk of arterial thrombosis and/or atherosclerosis, inhibition of endogenous cholesterol biosynthesis, platelet aggregation, a reduction in oxidative stress and a reduction in blood pressure. It has also been shown that dietary red palm oil, taken in moderation in animals and humans, promotes the efficient utilization of nutrients, activates hepatic drug metabolizing enzymes, facilitates the hemoglobinization of red blood cells and improves immune function [7]. All the above-mentioned as well as hyperlipidemia and hypertension are underlining conditions that characterize obese male.

In a systematic review on animal intervention studies, Syarifah-Noratiqah et al. [38] concluded on the evidence that palm oil and palm olein possess high potential as lipid-lowering agents. In another study on pharmacological potential of Oil Palm Phenolics (OPP), Syarifah-Noratiqah et al. [39] also concluded that individual components of OPP (Caffeoylshikimic Acid (CFA), p-Hydroxybenzoic Acid, Protocatechuic Acid (PCA) and Hydroxytyrosol, have unique pharmacological potential including neuroprotection, anti-cancer, cardioprotection and hypolipidemic effects. Single or in combination of all three phenolic acids into one OPP liquor would produce high pharmacological potential OPP liquor for the nutraceutical and pharmaceutical market. OPP extracted from bio-wastes of oil palm industry would provide an opportunity to transform a biowaste burden into a range of potential applications for health and wellness.

5. Male infertility

The WHO defines infertility as 'the inability of a sexually active, non-contracepting couple to achieve pregnancy in one year [40]. According to the WHO, about 9% of couples worldwide have fertility problems and around 70 million people are regarded as infertile [41]. The male factor varies and many publications contribute a factor of 30–40% to the man, 40% to the woman and the remainder is said to be idiopathic. The production of spermatozoa (spermatogenesis) in the testis and their subsequent maturation to physiological functionality in the epididymis are complex processes that needed to be strictly controlled in a timely and spatial manner. A disturbance of any of the individual steps involved may lead to impaired male fertility.

5.1 Causes of male infertility

There are many internal and external causes that can lead to male infertility. They span from genetic mutations or variations, medical conditions to lifestyle choices. Genetic mutations may distort e.g., hormonal levels of Follicle Stimulating

Hormone (FSH), Luteinizing Hormone (LH) or testosterone or their correspondent receptor sensitivities whereas genetic variations in the number of sex chromosomes can lead to clinical syndromes such as Turner (X0) or Klinefelter (XXY). Medical conditions cover a wide range and include distorted hormonal levels of the hypothalamic–pituitary-testicular (HPT) axis, immunological infertility (production of sperm antibodies), obstruction of the ductus deference, retrograde ejaculation (i.e. ejaculation into the bladder), erectile disfunction, varicoceles (swollen veins in the scrotum that block blood drainage), sperm disorders (mostly very low numbers or not made at all, odd shape, no straight motility, not fully mature thus unable to fertilize the female egg). External factors may involve exposure to toxicants (such as pesticides, heavy metals, PCB etc.) [42], heat [43], radiation [44] or xenoestrogens etc. [45]. Lifestyle choices potentially leading to male infertility may embrace excessive alcohol, illicit drugs or medication abuse. Furthermore, medical comorbidities such as cystic fibrosis, chronic obstructive pulmonary disease, and obesity can lead to male infertility [41].

The group of medical conditions including dyslipidemia, hypertension, insulin resistance, and obesity is often referred to as 'metabolic syndrome' (MetS) [46]. Each of these disorders can affect male fertility in its own way but when combined as in metabolic syndrome, additive effects have been observed [46]. Lotti et al. [47] noticed a decline in age-adjusted testosterone levels in male MetS patients without changes in LH and FSH levels. Furthermore, after adjusting for age and total testosterone, a negative correlation between the number of MetS components and normal spermatozoa morphology as well as erectile dysfunction was found [47]. In infertile men, a positive correlation between MetS and prostatic abnormalities has been shown [48]. Sertoli cells of the testis use glucose to produce lactate that is needed by the developing germ cells [49]. The transport of glucose from the blood capillaries through the basal compartment of the Sertoli cells and then through the blood-testis barrier towards the adluminal compartment of the seminiferous tubules is tightly controlled. Furthermore, this glucose transport is also regulated by the Hypothalamic–Pituitary-Gonadal (HPG) axis [50]. Hyperglycemia conditions in testicular cells such as in diabetic men interfere with these transport mechanisms and may thus compromise spermatogenesis. For males, obesity has a severe impact on the development and function of the testicles, epididymis, prostate, and other male reproductive organs [51]. With the increase of body mass index and abdominal circumference, ejaculation volume gradually reduces, and the total sperm count in semen also decreases [52]. In addition, obesity damages sperm chromatin or inhibits chromatin condensation [43]. In men, obesity may cause a gradual decline in sperm quality and thus reduce fertility [53]. Obesity also led to DNA fragmentation, increases apoptosis and epigenetic changes that can be transferred to the offspring [54].

5.2 Prevention and treatment

A healthy diet and lifestyle combined with regular physical activity can help to avoid MetS and obesity and thus may prevent male infertility. Spermatozoa are very susceptible to oxidative stress. Because of their small amount of cytoplasm, spermatozoa have very limited capacity to defend reactive oxygen species. Moreover, the plasma membrane of spermatozoa contains a high amount of unsaturated fatty acids which makes them an easy target for ROS-induced damages such as lipid peroxidation and DNA damage. Both can lead to reduced semen quality and thus affect male fertility. Food that is rich in anti-oxidants, vitamins (A, E and C) and omega-3 fatty acids such as many plant products, fish and other seafood has been shown to be beneficial for male fertility [55, 56]. The high provitamin A content of RPO may be beneficial for fertility because vitamin A plays a role in spermatogenesis [57]. RPO also contains vitamin D which is a known regulator of enzymes involved in the production of steroids (steroidogenesis) [58].

5.3 Red palm oil in male infertility prevention

As shown in the previous section in this chapter, RPO is rich in antioxidants such as carotenoids, vitamin E derivatives, and ubiquinone and should thus be a good candidate to prevent male infertility. Though, scientific literature on the effect of RPO on male reproduction is scarce.

Overall, a direct effect of RPO on spermatogenesis is inconclusive. Some in vivo studies reported poor sperm functions and morphology after exposure to RPO, whereas other noticed no significant effects. However, many studies demonstrated that RPO can protect against or ameliorate toxin- or medical condition-induced male infertility. Aboua et al. [59] showed that long-term RPO supplementation to rats had no effect on enzymes activities or substrates involved in the antioxidant defense system (GSH, CAT, SOD), lipid peroxidation or intracellular ROS levels in spermatozoa. Similarly, RPO exposure did not significantly change concentration or motility of spermatozoa. RPO did, however, significantly lower or even prevent ROS-induced changes to peroxide-injected animals. Peroxide alone led to lipid membrane peroxidation and higher intracellular ROS levels and consequently significantly lowered sperm concentration and motility and reduced enzyme activities. But peroxide exposure in combination with RPO prevented these changes and thus protected the sperm from ROS-induced damage [59]. In a similar approach Jegede et al. [60] demonstrated that RPO can attenuate heavy metal induced testicular damage in rats. Here, the administration of lead acetate led to a significant rise in reactive oxygen species that in turn decreased GSH concentrations and led to reduced spermatozoa numbers and motility and an increase in abnormal sperm morphology. Coadministration of RPO was able to partially protect against these changes as shown by significantly increased GSH levels and improved spermatozoa parameters. Peroxide in higher concentration induces DNA breaks and programmed cell death (apoptosis) in spermatozoa which consequently leads to lower sperm counts and reduces male fertility [61]. Aboua et al. [15] demonstrated that a 60-day oral supplementation of RPO can prevent peroxide-induced apoptosis in rat spermatozoa. Here, RPO supplementation prevented the activation of caspases 3 and 7 and, therefore, apoptosis. In addition, RPO seems to reduce the activation of p53 which will also lessen apoptosis.

As stated before, obesity can lead to diabetes type 2 which is known to impair male fertility by inter alia reducing sperm quality [53]. In diabetic rats RPO supplementation was shown to elevate the percentage of progressive motile spermatozoa [9]. The authors further demonstrated that RPO in combination with an aqueous extract from fermented rooibos exerted no negative effect on sperm motility parameters as measured by Computer Assisted Sperm Analysis but improved these parameters in diabetic rats. There is a significant linear relationship between abdominal obesity and prostate cancer [62]. Treatment of prostate cancer often involves radiation therapy or surgery that will remove the prostate and the seminal vesicles. Both options greatly impair the production of semen and thus often led to male infertility. RPO can lower the unpleasant effects of chemotherapy and is thus used in cancer management [63]. However, RPO or more precisely, a tocotrienol-rich fraction (TRF) from RPO may have potential as a remedy for prostate cancer. Using three different prostate cancer cell lines [64] noticed that TRF selectively inhibited cell proliferation and induced apoptosis in these cells.

As shown above, the antioxidative capacity of RPO can help to reduce or even prevent ROS-induced damage to the male germ cell. On the other hand, a certain physiological amount of ROS activity is essential for spermatogenesis and normal sperm functions. Thus, excessive use of antioxidants can have a negative effect on spermatogenesis and male fertility [65, 66]. This might be the reason as to why Aboua et al. [67] noted a significant decline in the motility of spermatozoa in vitro after exposure to RPO. A more recent in vivo study investigated the effects of a specific high-fat diet on the male rat reproductive performances. Male Wistar rats were fed either with a 15% palm oil diet or a standard diet for 16 weeks. The authors concluded that the palm oil significantly impaired sexual behavior, ejaculatory activities and sperm motility, viability, and morphology of male rats [68]. Overall, a precisely tuned balance between ROS level needed for physiological sperm function and sufficient antioxidants to combat oxidative cellular damage is vital for male fertility.

6. Conclusion

Obesity is a worldwide occurring problem that causes many medical disorders and diseases. Obesity is known to disrupt male fertility and the capacity to impair reproduction through alteration in the hypothalamic- pituitary-gonadal axis. The obesity induced disruption of testicular steroidogenesis and metabolic dysregulation, cytokines and adipokines negatively impact on semen parameters such as sperm concentration, motility, viability, and normal morphology. In addition, obesity inhibits chromatin condensation, DNA fragmentation, increases apoptosis and epigenetic changes that can be transferred to the offspring. A balanced diet and healthy lifestyle combined with regular physical activity can help to avoid metabolic syndrome and obesity and thus, may prevent obesity-induced male infertility. Spermatozoa are extremely sensitive towards oxidative stress. Therefore, a sensible daily supplementation of normal diet with an antioxidant-rich natural product such as RPO might be useful to protect spermatozoa from oxidative stress and helps to preserve male fertility.

7. Future research

No doubt, RPO should be considered as one of the healthy and nutritional oils. However, there are inconsistent and controversial reports on its role in amelioration of obesity and by extension, male infertility prevention. Therefore, further studies to investigate the effect of RPO on the renal, hepatic, and cardiac functions as well as spermatogenesis in dietary-induced obese male Wistar rats are hereby proposed. Besides, human volunteers can be recruited for a longitudinal testing on the effect of the oil consumption over a period of 6–12 weeks.

Conflict of interest

The authors declared no conflict of interests.

Author details

Folorunso A. Olabiyi[1]*, Yapo G. Aboua[2] and Thomas K. Monsees[3]

1 Clinical Chemistry Unit, Department of Medical Laboratory Science, Afe Babalola University, Ado-Ekiti, Nigeria

2 Faculty of Applied Sciences, Department of Health Sciences, Namibia University of Science and Technology, Windhoek, Namibia

3 Department of Medical Biosciences, University of the Western Cape, South Africa

*Address all correspondence to: folabiyi@abuad.edu.ng; folabiyi@gmail.com

References

[1] Patel D. Pharmacotherapy for the management of obesity. Metabolism. 2015; 64(11):1376-1385. DOI 10.1016/j.metabol.2015.08.001

[2] Haththotuwa RN, Wijeyaratne CN, Senarath U, Mahmood T, Arulkumaran S. Worldwide epidemic of obesity. Obesity: A ticking time bomb for reproductive health. 2013; 3-11.

[3] James PT. Obesity: the worldwide epidemic. Clinics in dermatology. 2004; 22(4):276-80. DOI 10.1016/j.clindermatol.2004.01.010.

[4] Ng M, Fleming T, Robinson M, Thomson B, Graetz N, Margono C, Mullany EC, Biryukov S, Abbafati C, Abera SF, Abraham JP. Global, regional, and national prevalence of overweight and obesity in children and adults during 1980-2013: a systematic analysis for the Global Burden of Disease Study 2013. The lancet. 2014; 384(9945):766-781. DOI 10.1016/S0140-6736(14)60460-8

[5] "Obesity and overweight". *World Health Organization [Internet] 2003*. Available from: https://www.who.int/news-room/fact-sheets/detail/obesity-and-overweight# (Accessed: 2021 April 30)

[6] Mahmood TA, Arulkumaran S, editors. Obesity: A ticking time bomb for reproductive health. Newnes; 2012. ISBN-13 978-0124160453

[7] Oguntibeju OO, Esterhuyse AJ, Truter EJ. Red palm oil: nutritional, physiological and therapeutic roles in improving human wellbeing and quality of life. British journal of biomedical science. 2009; 66(4):216-222. DOI 10.1080/09674845.2009.11730279

[8] Atawodi SE, Yusufu L, Atawodi JC, Asuku, Yakubu OE. Phenolic compounds and antioxidant potential of Nigerian red palm oil (Elaeis guineensis). International Journal of Biology. 2011; 3(2):153-161. DOI 10.5539/ijb.v3n2p153

[9] Ayeleso AO, Oguntibeju OO, Aboua YG, Brooks NL. Effects of red palm oil and rooibos on sperm motility parameters in streptozotocininduced diabetic rats. African Journal of Traditional, Complementary and Alternative Medicines. 2014; 11(5):8-15. DOI 10.4314/ajtcam.v11i5.2

[10] Mancini A, Imperlini E, Nigro E, Montagnese C, Daniele A, Orru` S, et al. Biological and nutritional properties of palm oil and palmitic acid: effects on health. Molecules. 2015; 20. (9):17339-61. DOI 10.3390/molecules200917339

[11] Edem D. Palm oil: Biochemical, physiological, nutritional, hematological and toxicological aspects: A review. Plant Foods for Human Nutrition. 2002; 57(3):319-341. DOI 10.1023/a:1021828132707

[12] Narang D, Sood S, Thomas M, Dinda A, Maulik S. Effect of dietary palm olein oil on oxidative stress associated with ischemic-reperfusion injury in isolated rat heart. BMC Pharmacology. 2004; 4(1): 29. DOI 10.1186/1471-2210-4-29

[13] Esterhuyse A, Du Toit E, Benade A, Van Rooyen J. Dietary red palm oil improves reperfusion cardiac function in the isolated perfused rat heart of animals fed a high cholesterol diet. Prostaglandins, Leukotrienes and Essential Fatty Acids. 2005; 72(3):153-161. DOI 10.1016/j.plefa.2004.10.014

[14] Bester D, Van Rooyen J, Du Toit E, Esterhuyse A. Red palm oil protects against the consequences of oxidative stress when supplemented with dislipidaemic diets. Medical Technology SA. 2006; 20(1):3-10.

[15] Aboua YG, Brooks N, Awoniyi DO, Plessis SS. Red palm oil: A natural good Samaritan for sperm apoptosis. Medical Technology SA. 2009; 23(1):8-10. DOI 10.1038/aja.2009.55

[16] Oguntibeju O, Esterhuyse A, Truter E. Possible role of red palm oil supplementation in reducing oxidative stress in HIV/AIDS and TB patients: A Review. Journal of Medicinal Plants Research. 2010; 4(3):188-196. Available online at http://www.academicjournals.org/JMPR

[17] United States Department of Agriculture. Agricultural Statistics. 2004, 48-51.

[18] Souganidis E, Laillou A, Leyvraz M, Moench-Pfanner R. A comparison of retinyl palmitate and red palm oil β-carotene as strategies to address vitamin A deficiency. Nutrients. 2013;5(8):3257-3271. DOI 10.3390/nu5083257

[19] Sylvester PW, Shah S. Antioxidants in dietary oils: their potential role in breast cancer prevention. Malaysian journal of nutrition. 2002;8(1):1-1. PMID **22692436**

[20] Olabiyi FA, Olatunji OI, Makinwa TT, Oguntibeju OO. Palm Oil and Groundnut Oil Supplementation Effects on Blood Glucose and Antioxidant Status in Alloxan – induced Diabetic Rats. *Pak. J. Pharm. Sci.* 2016: *29*(1): 83-87. PMID 26826842.

[21] Ayeleso A, Brooks N, Oguntibeju O. Modulation of antioxidant status in streptozotocin-induced diabetic male Wistar rats following intake of red palm oil and/or rooibos. Asian Pac J Trop Med. 2014 Jul;7(7):536-544. DOI 10.1016/S1995-7645(14)60090-0

[22] Alberti KG, Eckel RH, Grundy SM, Zimmet PZ, Cleeman JI, Donato KA, Fruchart JC, James WP, Loria CM, Smith Jr SC. Harmonizing the metabolic syndrome: a joint interim statement of the international diabetes federation task force on epidemiology and prevention; national heart, lung, and blood institute; American heart association; world heart federation; international atherosclerosis society; and international association for the study of obesity. *Circulation*. 2009; 120(16):1640-5. DOI 10.1161/CIRCULATIONAHA.109.192644

[23] Nathalie Esser, Sylvie Legrand-Poels, Jacques Piette, André J. Scheen, Nicolas Paquot, Inflammation as a link between obesity, metabolic syndrome and type 2 diabetes, Diabetes Research and Clinical Practice, 2014, 105 (2):141-150. DOI 10.1016/j.diabres.2014.04.006

[24] World Health Organization [Internet]. 2018a. Non-communicable diseases country profiles 2018. Available from: https://www.who.int/nmh/publications/ncd-profiles-2018/en/ ISBN: 978 92 4 151462 0

[25] Martinez JA. Body-weight regulation: Causes of obesity. In Vol. 59. CAB International *Proceedings of the Nutrition Society*. 2000; 337-345. DOI 10.1017/s0029665100000380

[26] Pi-Sunyer FX. The Obesity Epidemic: Pathophysiology and Consequences of Obesity. Obesity Research. 2002; 10(S12):97S-104S. DOI 10.1038/oby.2002.202

[27] Walker ARP, Adam F, Walker BF. World pandemic of obesity: The situation in Southern African populations. Public Health. 2001; 115(6):368-372. DOI 10.1038/sj/ph/1900790

[28] Alaba O, Chola L. Socioeconomic inequalities in adult obesity prevalence in South Africa: a decomposition analysis. International journal of environmental research and public health. 2014; *11*(3), pp.3387-3406. DOI 10.3390/ijerph110303387

[29] World Health Organization [Internet]. 2018b. Obesity and overweight. Available from: https://www.who.int/en/news-room/fact-sheets/detail/obesity-and-overweight [Accessed: 2019, September 22].

[30] Goodarzi MO. Genetics of obesity: what genetic association studies have taught us about the biology of obesity and its complications. The Lancet Diabetes & Endocrinology. 2018;6(3):223-236. DOI 10.1016/S2213-8587(17)30200-0

[31] Lysaght J, van der Stok EP, Allott EH, Casey R, Donohoe CL, Howard JM, McGarrigle SA, Ravi N, Reynolds JV, Pidgeon GP. Pro-Inflammatory and Tumour Proliferative Properties of Excess Visceral Adipose Tissue. Cancer Letters. 2011; 312, 62-72. DOI 10.1016/j.canlet.2011.07.034

[32] Tchkonia T, Thomou T, Zhu Y, Karagiannides I, Pothoulakis C, Jensen MD, Kirkland JL. Mechanisms and Metabolic Implications of Regional Differences among Fat Depots. Cell Metabolism. 2013; 17, 644-656. DOI 10.1016/j.cmet.2013.03.008

[33] Frühbeck G, Kiortsis DN, Catalán V. Precision medicine: diagnosis and management of obesity. The lancet Diabetes & endocrinology. 2018; 6(3):164-166. DOI 10.1016/S2213-8587(17)30312-1

[34] Zelissen PM, Mathus-Vliegen EM. Treatment of overweight and obesity in adults: proposal for a guideline. Nederlands tijdschrift voor geneeskunde. 2004; 148(42):2060-6. PMID **15532327**

[35] Mba OI, Dumont MJ, Ngadi M. Palm oil: Processing, characterization and utilization in the food industry–A review. Food bioscience. 2015; 10:26-41. DOI 10.1016/j.fbio.2015.01.003

[36] Sowers JR. Obesity as a cardiovascular risk factor. The American journal of medicine. 2003;115(8):37-41. DOI 10.1016/j.amjmed.2003.08.012

[37] Boateng L, Ansong R, Owusu W, Steiner-Asiedu M. Coconut oil and palm oil's role in nutrition, health and national development: A review. Ghana medical journal. 2016; 50(3):189-196. PMID: 27752194

[38] Syarifah-Noratiqah SB, Fairus S, Zulfarina MS, Nasrullah Z, Qodriyah HM, Naina-Mohamed I. The Effects of Palm Oil on Plasma and Serum Lipid Parameters: A Systematic Review on Animal Intervention Studies. Frontiers in veterinary science. 2020;7:303. DOI 10.3389/fvets.2020.00303

[39] Syarifah-Noratiqah SB, Zulfarina MS, Ahmad SU, Fairus S, Naina-Mohamed I. The pharmacological potential of Oil Palm Phenolics (OPP) individual components. International journal of medical sciences. 2019;16(5):711. DOI 10.7150/ijms.29934

[40] WHO. https://www.who.int/teams/sexual-and-reproductive-health-and-research/areas-of-work/fertility-care/infertility-definitions-and-terminology [Accessed 2021-04-21].

[41] Fainberg J, Kashanian JA. Recent advances in understanding and managing male infertility F1000Research. 2019,8:670. DOI 10.12688/f1000research.17076.1

[42] Wong EWP, Cheng CY. Impacts of environmental toxicants on male reproductive dysfunction. Trends Pharmacol Sci. 2011;32: 290-299. DOI 10.1016/j.tips.2011.01.001

[43] Lin C, Shin DG, Park SG, et al.: Curcumin dose-dependently improves spermatogenic disorders induced by scrotal heat stress in mice. Food and Function. 2015;6:3770-3777. DOI: 10.1039/c5fo00726g

[44] Kesari KK, Agarwal A, Henkel R. Radiations and male fertility. Reproductive Biology and Endocrinology. 2018;16:118. DOI 10.1186/s12958-018-0431-1

[45] Cooke PS, Manjunatha X, Nanjappa K, Ko CM, Prins GS, Hess RA. Estrogens in male physiology. Physiol Rev 2017;97:995-1043. DOI 10.1152/physrev.00018.2016

[46] Martins AD, Majzoub A, Agarwal A. Metabolic syndrome and male fertility. The World J of Men's Health 2019;37:113-127. DOI 10.5534/wjmh.180055

[47] Lotti F, Corona G, Degli Innocenti S, Filimberti E, Scognamiglio V, Vignozzi L, Forti G, Maggi M. Seminal, ultrasound and psychobiological parameters correlate with metabolic syndrome in male members of infertile couples. Andrology 2013;1:229-239. DOI 10.1111/j.2047-2927.2012.00031.x

[48] Lotti F, Corona G, Vignozzi L, Rossi M, Maseroli E, Cipriani S, Gacci M, Forti G, Maggi M. Metabolic syndrome and prostate abnormalities in male subjects of infertile couples. Asian J Androl 2014;16:295-304. DOI 10.4103/1008-682X.122341

[49] Monsees TK, Franz M, Gebhardt S, Winterstein U, Schill WB, Hayatpour J. Sertoli cells as a target for reproductive hazards. Andrologia 2000;32:239-246. DOI 10.1046/j.1439-0272.2000.00391.x

[50] Maresch CC, Stute DC, Alves MG, Oliveira PF, de Kretser DM, Linn T. Diabetes-induced hyperglycemia impairs male reproductive function: a systematic review. Human Reproduction Update 2018;24:86-105. DOI 10.1093/humupd/dmx033

[51] Matsuzawanagata N, Takamura T, Ando H, et al.: Increased oxidative stress precedes the onset of high-fat diet–induced insulin resistance and obesity. Metabolism. 2008;57:1071-1077. DOI 10.1016/j.metabol.2008.03.010.

[52] Eisenberg ML, Kim S, Chen Z, Sundaram R, Schisterman EF, Louis GM: The relationship between male BMI and waist circumference on semen quality: Data from the LIFE study. Human Reproduction. 2014;29:193-200. DOI 10.1093/humrep/det428

[53] Barkabi-Zanjani S, Ghorbanzadeh V, Aslani M, Ghalibafabbaghi A, Chodri L. Diabetes mellitus and the impairment of male reproductive function: possible signaling pathways. Diabetes & Metabolic Syndrome: Clinical Research & Reviews 2020;14:1307-1314. DOI 10.3390/biology10030241

[54] Leisegang K, Sengupta P, Agarwal A, Henkel R. Obesity and male infertility: Mechanisms and management. Andrologia 2021;53:e13617. DOI 10.1111/and.13617

[55] Hosseini B, Nourmohamadi M, Hajipour S, Taghizadeh M, Asemi Z, Keshavarz SA, Jafarnejad S. The Effect of Omega-3 Fatty Acids, EPA, and/or DHA on Male Infertility: A Systematic Review and Meta-analysis. Journal of Dietary Supplements, 2019;16:2,245-2,256, DOI 10.1080/19390211.2018.1431753

[56] Torres-Arce E, Vizmanos B, Babio N, Márquez-Sandoval F, Salas-Huetos A. Dietary Antioxidants in the Treatment of Male Infertility: Counteracting Oxidative Stress. Biology. 2021 Mar;10(3):241 DOI 10.3390/biology10030241

[57] McArdle HJ, Ashworth CJ Micronutrients in fetal growth and development. Br Med Bull 1999;55:499-510. DOI 10.1258/0007142991902574.

[58] Lundqvit 2014 Lundqvist J. Vitamin D as a regulator of steroidogenic enzymes [version 1; peer review: 1 approved, 1 approved with

reservations]. *F1000Research* 2014,3:155 DOI 10.12688/f1000research.4714.1

[59] Aboua YG, Brooks N, Mahfouz RZ, Agarwal A, du Plessis SS. A red palm oil diet can reduce the effects of oxidative stress on rat spermatozoa. Andrologia. 2012;44:32-40. DOI 10.1111/j. 1439-0272.2010.01133.x

[60] Jegede AI, Offor U, Azu OO, Akinloye O. Red Palm Oil attenuates lead acetate induced testicular damage in adult male Sprague-Dawley rats. Evidence-Based Complementary and Alternative Medicine. 2015. DOI 10.1155/2015/130261

[61] Agarwal A, Saleh RA, Bedaiwy MA. Role of reactive oxygen species in the pathophysiology of human reproduction. Fertil Steril 2003;79: 829-843. DOI 10.1016/ s0015-0282(02)04948-8

[62] Choi JB, Myong JP, Lee Y, Kim I, Kim JH, Hong SH, Ha US. Does increased body mass index lead to elevated prostate cancer risk? It depends on waist circumference. MBC Cancer 2020;20;589. DOI 10.1186/ s12885-020-07089-5

[63] Loganathan R, Subramaniam KM, Radhakrishnan AK, Choo YM, Teng KT. Health-promoting effects of red palm oil: evidence from animal and human studies. Nutrition Reviews 2017;75:98-113. DOI 10.1093/ nutrit/nuw054

[64] Srivastava JK, Gupta S. Tocotrienol-rich fraction of palm oil induces cell cycle arrest and apoptosis selectively in human prostate cancer cells. BBRC 2006;346:447-453. DOI 10.1016/j. bbrc.2006.05.147

[65] Wagner H, Cheng JW, Ko EY. Role of reactive oxygen species in male infertility: An updated review of literature. Arab J Urol 2018;16:35-43. DOI 10.1016/j.aju.2017.11.001

[66] Opuwari C, Monsees T. Green tea consumption increases sperm concentration and viability in male rats and is safe for reproductive, liver and kidney health. Scientific Reports 2020:10:15269. DOI 10.1038/ s41598-020-72319-6

[67] Aboua YG, du Plessis SS, Reichgelt P, Brooks N. The in vitro effects of superoxide, some commercially available antioxidants and red palm oil on sperm motility. Asian J Androl. 2009;11:695-702. DOI 10.1038/ aja.2009.55

[68] Defo PB, Wankeu-Nya M, Ngadjui E, Fozin G, Kemka F, Kamanyi A, Watcho P. Palm oil diet-induced obesity impairs male rat reproductive performance. Annals of Reproductive Medicine and Treatment. 2017; 2(2):1012.

Chapter 4

Palm Oil Clinker as a Waste by-Product: Utilization and Circular Economy Potential

Ahmad Hussaini Jagaba,
Shamsul Rahman Mohamed Kutty,
Gasim Hayder Ahmed Salih, Azmatullah Noor,
Mohammad Fakhuma Ubaidillah bin Md Hafiz,
Nura Shehu Aliyu Yaro, Anwar Ameen Hezam Saeed,
Ibrahim Mohammed Lawal, Abdullahi Haruna Birniwa
and Abdullahi Usman Kilaco

Abstract

Conservation of natural resources to create ecological balance could be significantly improved by substituting them with waste by-products. Palm oil industry operations increases annually, thereby generating huge quantity of waste to be dumped into the landfill. Palm oil clinker (POC) is a solid waste by-product produced in one of the oil palm processing phases. This chapter is designed to highlight the generation, disposal problems, properties and composition of POC. The waste to resource potentials of POC would be greatly discussed in the chapter starting with the application of POC in conventional and geopolymer structural elements such as beams, slabs, columns made of either concrete, mortar or paste for coarse aggregates, sand and cement replacement. Aspects such as performance of POC in wastewater treatment processes, fine aggregate and cement replacement in asphaltic and bituminous mixtures during highway construction, a bio-filler in coatings for steel manufacturing processes and a catalyst during energy generation would also be discussed. Circular economy potentials, risk assessment and leaching behavior during POC utilization would be evaluated. The chapter also discusses the effectiveness of POC in soil stabilization and the effect of POC pretreatment for performance enhancement. Towards an efficient utilization, it is important to carry out technical and economic studies, as well as life cycle assessments, in order to compare all the POC areas of application described in the present review article. POC powder has proven to be pozzolanic with maximum values of 17, 53.7, 0.92, 3.87, 1.46, for CaO, SiO_2, SO_3, Fe_2O_3 and Al_2O_3. Therefore, the present chapter would inspire researchers to find research gaps that will aid the sustainable use of agro-industry wastes. The fundamental knowledge contained in the chapter could also serve as a wake-up call for researchers that will motivate them to explore the high potential of utilizing POC for greater environmental benefits associated with less cost when compared with conventional materials.

IntechOpen

Keywords: Palm oil clinker, Waste utilization, Circular economy, Palm oil mill, Sustainable environment

1. Introduction

Oil palm, commercially named as Elaeis quineensis, is one of the main agricultural crops that thrives in a hot tropical climate. It produces vegetable oil fit for human consumption. As can be seen in **Figure 1a**, the tree is naturally brown and seed reddish in color because of a high betacarotene content. The oil palm industry has been conveniently quoted as the main sector producing abundant biomass as renewable sources in different forms; these include empty fruit bunches (EFB), mesocarp fiber (MF), palm shell (PS), oil palm fronds (OPF) and oil palm trunks (OPT). Palm oil industries face significant challenges in meeting the increasingly stringent environmental regulations on waste disposal.

Palm oil clinker (POC) is a waste by-product generated in one of the oil palm processing phases [1]. It is a residue from the heating zone of a steam boiler during electricity generation produced in huge amount from oil palm fibers and shells calcination in a suitable proportion of (30:70) at 100–850°C [2, 3]. It is subsequently cooled at air atmosphere [4].

Physically, POC shown in **Figure 1b** are naturally porous, flaky, gray in color, irregular in shape, lighter in weight, with rough and sharp broken surfaces [5, 6]. They are mostly presented as solid lightweight materials between the sizes of 2 and 15 cm [7]. POC is made up of inorganic oxides, 3.35% organic carbon and minerals like halite, lysite, eglestonite, elatossite, quartz and cristobalite [8]. Chemical oxides composition of POC highlighted in **Table 1** varies according to several factors. Some of these are: fiber to shell ratio, applied incineration process temperature, palm tree location soil condition, POC form (nano, powder, fine or coarse particle size) etc. [7, 12]. POC is a pozzolanic aggregate capable of producing appropriate attachment in a geopolymer matrix because of alumina-siliceous compounds presence [22]. It is no more news that, pore structure has a close association with the resistance of cement-based substances to fluid infiltration. These include, pore size distribution, interconnectivity and porosity [23].

Oil palm extraction rate has rapidly increased due to the increasing oil palm global demand [24]. As a consequence, fresh porous lumped POC is continuously

(a) (b)

Figure 1.
(a) An oil palm tree and (b) palm oil clinker chunks.

State of POC	Chemical compositions											Ref.
	CaO	SiO_2	SO_3	Fe_2O_3	Al_2O_3	MgO	P_2O_5	K_2O	TiO_2	Mn_2O_3	Na_2O	Others	
Powder	6.37	59.90	0.39	6.93	3.89	3.30	3.47	15.10	0.29	—	—	0.36	[5]
Powder	3.20	61.29	0.10	4.31	5.89	3.16	3.12	10.79	0.12	—	—	—	[9]
Powder	3.26	60.29	0.11	4.71	5.83	3.76	3.10	7.79	0.13	—	—	—	[7]
Powder	3.28	60.29	0.31	4.71	5.83	4.20	—	—	—	—	—	—	[10]
Powder	6.93	63.90	0.21	3.30	3.89	3.37	2.12	10.20	—	—	—	—	[6]
Powder	3.89	65.30	0.09	5.65	4.23	3.72	0.78	13.65	0.13	—	—	—	[11]
Powder	8.00	60.00	—	4.00	4.00	5.00	5.00	12.00	—	—	—	2.00	[12]
Powder	3.28	60.29	0.31	4.71	5.83	4.20	3.78	7.24	0.10	0.12	0.20	—	[13]
Powder	6.37	59.90	2.60	6.93	5.37	3.13	0.07	15.10	0.12	0.12	0.24	—	[14]
Powder	4.82	63.10	0.15	9.00	3.20	3.50	3.09	12.50	0.21	0.12	0.16	0.13	[15]
Powder	6.09	42.30		3.288	3.09	2.41	2.95	15.20	0.17	—	—	—	[16]
Powder	7.05	55.39	0.19	10.81	2.18	2.00	3.97	17.7		—	—	0.28	[17]
Sand	6.37	59.90	0.39	6.93	3.89	3.30	3.47	15.10	0.29	—	—	0.36	[18]
Sand	5.74	65.40	0.64	2.71	1.95	6.40	6.56	9.52	0.11	0.17	0.32	—	[19]
Sand	3.28	60.29	0.31	4.71	5.83	4.20	3.78	7.24	0.10	0.12	0.20	—	[20]
Sand	17.00	53.7	0.92	3.87	1.46	2.37	5.29	13.9	—	—	—	—	[21]
Coarse	8.16	59.63	0.73	4.62	3.70	5.01	5.37	11.66	0.22	—	0.32	0.58	[2]

Table 1.
Chemical composition of POC.

generated [25]. The produced POC in the boiler is mixed in suspension, moved from the combustion boiler, and deposited in the factory yard [13]. It is a solid waste product of little to no economic benefit that causes pollution of the atmosphere, soil degradation and ground water contamination [8]. In recent times, it is mostly dumped in landfills that not only causes soil pollution but go as far as contaminating ground water [26]. It is important to note that the continuous dumping activity would cause waste accumulation at the dump site and creates the need to allocate new disposal area. Continuous disposal would result in waste rack up at the dumpsite, necessitating the allocation of new space for landfills. This would have negative consequences for the environment as fertile land is converted into a refuse collection area [27].

Besides overcoming waste disposal problems, integrating low cost and environmentally-friendly waste materials in new and sustainable product development would help in environmental pollution control, appropriate land use and promoting sustainability [14, 28]. Therefore, reusing POC for different applications would assist in natural resources preservation, reduce greenhouse gas emission, pave way for proper consumption and producing cleaner environment [23, 29].

With the technological advancement, there is a need for a change in using traditional old materials for industrial applications. Raw materials used by industries are affecting the environment to a larger extent. There is a dire need to change the current scenario especially for developing countries [21]. To achieve the concept of green technology, many attempts have been carried out to develop low-carbon footprint products or techniques. Due to their high mechanical properties and environmental benefit, POC appear as a future prospective industrial material and have applications in different areas. This article reviews the physical, chemical and microstructural properties of palm oil clinker (POC) by-products of palm oil. It aims to give a comprehensive survey of already-well-established or future potential energy applications of POC. A critical comparison of their use in different area is reported and their modification by various physical and chemical routes is detailed. The new direction beyond the state of the art is the application of POC in Nano form. This is why only one [1] article is found in this chapter.

2. Waste to resource potentials of POC

2.1 POC in geopolymer made structural elements

Geopolymer is an inorganic material that can be formed through the use of a binder. According to [30], any material that contain silica and alumina can be utilized as a binder. Alkali activators are also important for the production of geopolymers. Numerous high silica and alumina containing waste materials could be utilized for geopolymer production due to their pleasing size, shape and chemical composition. POC, considered a pozzolanic aggregate, has the capacity to create good bond in geopolymer matrix as it possesses the aforementioned characteristics. In contrast to POC with OPC concrete, the use of a geopolymer binder increases the workability and strength of POC concrete thus lowering its water absorbability. A green and long-lasting structural lightweight concrete can be produced by combining POC with Fly ash-based geopolymer binder [31]. Utilizing POC particles in the geo-polymeric specimen results in structural elements with good resistance to water absorption.

Sustainability in high strength concrete production can be achieved by combining POC with fly ash as a geopolymer based binder. Designing and mixing concrete with 100% POC aggregate can give rise to a concrete with compressive strength >30 MPa and a density of 1821 kg/m^3. However, 32% strength reduction was experienced as natural aggregate was substituted by OPC. 75% POC aggregate replacement in geopolymer concrete mixtures has been proven by [31] to be the most effective one. As POC concentration increases in geopolymer concrete mixtures, water absorption increases as density decreases.

POC sand was used for full sand replacement in a geopolymer mortar and achieved comparable mechanical properties showing high resistance to MgSO4 and HCl solutions. 53 MPa was recorded as the 28-day compressive strength with 17% density reduction [32].

A geopolymer concrete that contain 100% POC as coarse aggregate was designed and evaluated. According to the results, 41.5 MPa was the highest compressive strength achieved at 28 days curing with a density range of 1910–2172 kg/m3. Splitting tensile strength increased and UPV values were also good. POC also improved the compressive toughness of the geopolymer mortar. The study concluded that, structural grade lightweight geopolymer concrete could be produced by using POC [22].

2.2 POC in conventional structural elements

The physical characteristics of POC used by several researchers are shown in **Table 2**.

2.2.1 POC as a coarse aggregate

The application of commercial aggregates was minimized due to high production costs emanating as a result of excessive raw materials and energy consumption. They also increase the dead weight of structures. Therefore, introducing POC, being a porous and lightweight material that contain high volume of solid waste materials are used to produce structural lightweight aggregate with potentials for high strength and good workability concrete. POC density is said to be less than that of normal aggregate [33]. Even though substituting normal weight coarse aggregate with POC wrecks the splitting tensile strength and modulus of elasticity, it however improves the concretes compressive strength [43].

The physical properties of POC aggregate have a notable influence on produced concrete properties. An equivalent of 1 m^3 of soil is saved when 1 m^3 of POC aggregates is utilized for concrete production instead of being discarded in a landfill. This would substantially lead to a safer and more productive climate [29]. CO2 emissions were said to have decreased by 20% when natural aggregate was totally replaced with POC coarse aggregate [44]. POC aggregates are lightweight and porous by nature, and they contribute to the reduction in concrete structural density [1].

POC increases concrete mixtures porosity and permeability. Compressive strength reduction of about 65% was recorded at full POC replacement. Nonetheless, concretes with lower strength could be used for pedestrian trials and walkways construction [38].

POC aggregate crushing value is three times less than that of gravel aggregate, thereby indicating higher energy consumption [14]. Having a density of 1990.33 kg/m3, makes POC aggregates ideal for use in lightweight concrete mix proportions [40].

Experimental investigation was carried out on concrete substituted by POC as a filler and an aggregates material for high strength concrete (HSC) creation. The permeable nature and uneven form of POC coarse had a negative impact on the fresh concrete mix's workability. Nonetheless, adding POC powder as a filler improved the workability. Adding POC powder in POC concrete mixes improved compressive, splitting tensile and flexural strengths by 0–13%, 2–10% and 1–9%, respectively compared to POC mix without POC powder. According to Rapid Chloride Permeability Test (RCPT) carried out, both POC concrete mixes, with and without POC powder, have a strong resistance to chloride penetration with very low permeability category <100°C [7].

Size of aggregate (mm)	Bulk dry density (kg/m^3)	Saturated density (kg/m^3)	Specific gravity	Water absorption (%)	Aggregate impact value	Aggregate crushing value (%)	Fineness modulus	Moisture content (%)	Los Angeles abrasion (%)	Compressive strength (MPa) @ 28 days	Splitting tensile strength (MPa)	Flexural strength (MPa)	Ref.
—	2050	2076	—	—	—	—	—	—	—	57	—	—	[33]
5–14	823	—	1.62	4.43	—	18.04	—	—	26.21	>30	—	—	[31]
—	1471	65	—	3.91	—	—	5.88	—	—	57.8	3.79	—	[34]
4.75–10	732	—	1.81	4.35	—	56.44	—	0.28	—	55.2–83.5	—	4–7.8	[7]
5–14	781.08	1769.2	1.82	4.35	25.36	18.08	27.09	0.07	27.09	30.9	2.29	—	[35]
—	860	—	1.69	7	36.3	21.2	—	—	23.9	—	3.05–3.31	4.48–5.38	[36]
—	1419	1875–1995	1.9	4.23	26.01	18.04	4.99	—	27.08	50–60	3.2–4.6	—	[37]
—	782	—	1.8	3.56	—	—	6.32	—	—	61.67	—	—	[2]
4.75–14	732	—	1.73	3 ± 2	—	56.44	—	0.5	—	33.01–39.32	2.61–3.28	3.75–4.42	[5]
4.75–9.5	732	—	1.88	3 ± 2	—	56.44	—	—	—	3.43–9.52	—	—	[38]
5–14	732	—	1.73	3 ± 2	—	56.44	—	1 ± 0.5	—	33–49	—	—	[27]
5–12.5	781.08	—	1.82	4.35	25.36	—	—	—	27.09	46	—	—	[39]
—	817	—	1.92	—	—	—	2.6	1.3	38.7	—	—	—	[40]
—	568	—	1.75	5.67	27.31	—	—	0.08	25.05	—	—	—	[41]
—	793	—	1.76	4.67	48.6	47.9	—	—	—	—	—	—	[25]
—	—	—	—	—	—	—	—	—	—	27.51	—	2.54	[4]
< 4.75	811	—	2.15	5.75	—	—	—	0.11	—	—	—	—	[7]
< 5	1118.86	—	2.01	26.45	—	—	3.31	0.11	—	—	—	—	[35]
<5	811	—	2.15	10 ± 5	—	—	—	0.5 ± 0.25	—	—	—	—	[28]
—	918	—	1.98	—	—	—	2.3	1.27	—	—	—	—	[40]

DOI: http://dx.doi.org/10.5772/intechopen.97312

Size of aggregate (mm)	Bulk dry density (kg/m^3)	Saturated density (kg/m^3)	Specific gravity	Water absorption (%)	Aggregate impact value	Aggregate crushing value (%)	Fineness modulus	Moisture content (%)	Los Angeles abrasion (%)	Compressive strength (MPa) @ 28 days	Splitting tensile strength (MPa)	Flexural strength (MPa)	Ref.
≤ 4.775	—	—	1.92	3.3 ± 1	—	—	3.52	1.5 ± 0.5	—	53	—	—	[32]
—	113	—	2.08	3.6	—	—	3.12	—	—	—	—	—	[42]
≤ 4.75	835.2	—	1.92	3.3 ± 1	—	—	3.52	1.5 ± 0.5	—	—	—	—	[20]
—	1085	—	1.94	9.77	—	—	2.6	0.27	—	—	—	—	[16]

Table 2.
Physical properties of POC.

POC concrete beams have been known to provide sufficient notice of impending failure by exhibiting traditional structural ductile behavior. At service loads, the crack width (0.24–0.3 mm) of POC concrete beam was found to be within the BS8110 overall permissible value for durability requirements [45].

In an oil palm shell (OPS) lightweight concrete, OPS aggregates were partly replaced with POC coarse aggregates from 0 to 50%. The slum value, density (2–4%), compressive strength and modulus of elasticity (18–24%) of the OPS concrete increases as POC coarse aggregate increases in the mix. More so, at 20–50% POC coarse aggregate addition, grade 35 OPS concrete was upgraded to grade 40. As a result, it's classified as a high-strength lightweight concrete [36]. In a related study, authors reported a positive impact on both workability, UPV and compressive strength. Highest compressive strength of ~63 MPa which is about 43% higher than the control mix was obtained for the OPS:POC mixture. This may be due to the efficient POC and mortar interlocking. With maximum obtainable stress between 0.00173–0.00401 > the normal weight concrete (NWC), the OPS:POC mixture could have better shrinkage crack resisting capacity. Furthermore, a 2.5 fold rise in elasticity modulus could remarkably control deflection [45].

POC aggregate could be used to develop high-strength lightweight concrete with a 28-day compressive strength of 50–60 MPa and an oven-dry density of 1875–1995 kg/m3. At full water curing and air-drying curing conditions, equivalent compressive strengths were recorded. This proves that, POC lightweight concrete is not too delicate to curing method. The study suggests the use of regular sand with a nominal grain size not more than 2 mm. This is to improve elastic modulus of the concrete [36]. Ultrasonic pulse value (UPV) tests value for POC concrete was good with a compressive strength and hardened density of 33–49 MPa and 2074–2358 kg/m3 at 28 days respectively. At 10% POC replacement for coarse aggregate, grade 40 concrete was obtained. However, increasing POC replacement ratio with coarse aggregate reduces the concrete workability. The advantage of applying POC as a lightweight aggregate is to decrease concrete structures dead load by up to 35% without much loss in structural strength. The decrease in dead load can save construction cost without compromising structural integrity. Therefore, applying lighter waste materials such as POC can greatly reduce concrete costs, due to its low cost of RM 0.020 per kg. This will go a long way in reducing the need for non-sustainable natural resources. For structural application, shear failure mode of POC concrete beams were found to be close to that of regular weight concrete beams, and as well in line with ASTM: C330 [28].

Despite the concrete's higher porosity, self-compacting lightweight concrete (SCLWC) had strong UPV values. Tensile splitting strength, compressive strength and flexural to compressive strength ratio also met the strength requirement for SCLWC [25]. Therefore, SCLWC is classified as a form of lightweight concrete with a high strength because 28-day compressive strength >40 N/mm2. As an actively mobilized material, POC was also able to amplify the filling and passing ability of self-compacting concrete. The concrete showed less segregation resistance due to low POC coarse density. Although obtained density values were in an acceptable range, coarse aggregates replacement with POC in SCLWC reduced density in oven-dry and saturated surface-dry conditions by 16% and 18%, respectively.

Lightweight aggregate concretes made of POC with 12% less dead load compared to the conventional concrete mix showed an acceptable splitting tensile strengths and workability without any segregation or floating at an average water to cement ratio. Interestingly, even after 28 days of curing, POC concretes did not achieve their maximum strength [34]. Testing the efficiency of POC in concrete slabs, the mechanical interlock (m) and friction (k) between the steel and concrete were found to be 117.67 N/mm2 and 0.0973 N/mm2, respectively. It was also discovered that horizontal shear-bond strength and structural behavior are satisfactory, nearly comparable to the conventional concrete slabs and could be used for composite slabs construction. Compared to conventional concrete slabs, POC concrete slab possess a reduction in weight of 18.3% [35]. Under absolute air, water, and 3 days water curing, the abrasion resistance and strength properties of concrete comprising POC coarse aggregate were investigated. The compressive strength of POC concrete cured in air and in water for 3 days displayed comparable conduct, with a maximum loss in strength of about 5% and an acceptable abrasion resistance. Interestingly, abrasion resistance was improved when cured in full water [44]. Air curing application in a tropical environment permit POC concrete to achieve the desired strength due to the surroundings high humidity. However, water curing is the most appropriate curing method for POC light aggregate concrete, because it contains enough water to ensure proper hydration and pozzolanic reactivity [41].

2.2.2 POC as a fine aggregate

Palm oil clinker (POC) has in recent times being used for partial replacement of fine aggregates in structural elements. This was possible due to the grading features and particle size distribution similitude between sand and POC fine aggregate [5]. The particle size distribution of POC ranging from 100 to 400 mm, indicates that they are suitable for use as fine aggregates. A study by [4] found the compressive and flexural strength of concrete to be increasing with the sand replacement with POC. The study further confirms that fine aggregate replacement with POC had no remarkable impact on compressive strength. However, it decreasing concrete workability [28].

Sand was totally replaced by POC in a mortar designed using the volume-based approach. At 28-day curing, 41 MPa was recorded as the compressive strength of the mortar. POCS aids the gain of early-stage strength development for up to 77%. With 4.09 km/s as the POC mortar velocity, well-compacted specimens were obtained. The poriferous structure and rough nature of POC aids in the formation of a stronger bond with cement paste. The price of the mortar could be reduced by 16% when POC is utilized [20]. When POC was partially replaced with sand from 0 to 40% by weight of sand to investigate its effect on fly ash cement sand brick engineering properties, it was found that up to 30% POC usage enhanced the brick strength due to the pozzolanic effect of the fine clinker. Calcium hydroxide and silicon dioxide were responsible for the pore refinement and higher brick strength development [27]. Replacing OPC with fine aggregate increased the mortar sorptivity, initial and final water absorption because of its high porosity. OPC replacement changes the cement mortar thermophysical properties. At 100% sand replacement, compressive strength development ((7th day)/ (28th day)) was higher than samples containing lesser amounts of OPC. Under the same conditions, the specific heat capacity of mortar boosted by ~41%. Thermal conductivity and diffusivity lessened by 72% and 76% respectively. This shows that, OPC replaced mortar has the potential to lower heat transfer and energy consumption in buildings [46].

In a related study at 100% sand replacement, it was reported that POC fine has the potential to produce 86% and 78% compressive strength at 28 and 56 days curing respectively and providing almost 97% durability when compared to the conventional mix. POC fine durability showed a satisfactory outcome with good resistance against corrosion risk. POC fine is capable of lowering the carbon emissions of mortar by 50%. More so, POC fine can improve the engineering economic index and engineering environmental index by 11% and 95%, respectively. Life cycle impact assessment (LCIA) shows POC's potential to encourage a healthier and safer community with substantial reduction in ecotoxicity [18].

Incorporating POC sand in OPS concrete is beneficial to reduce its sensitivity to lack of curing. OPC was used as a replacement for sand at 0–50% in an oil palm shell (OPS) lightweight aggregate concrete. It was discovered that the replacement does not affect the drying shrinkage strain. High percentage of POC replacement increased the water absorption of the concrete. The concrete was proven to possess high splitting tensile strength [42]. In comparison with normal mining sand, POC fine aggregates have lower density and higher water absorption. Surprisingly, the slump value of concrete containing 25% POC fine showed good workability. The POC fine replaced concrete was classified as high strength because 69–76 MPa was obtained as compressive strength for 28-day curing. 12.5% POC fine replacement in concrete is said to be practical and cost effective [47].

2.2.3 POC powder

POC powder is obtainable by grinding dry POC for ~8 h in a controlled ball mill at 150 RPM. It has been confirmed by several authors through microstructure analysis that POCP particles are blackish in color, irregular in shape and contain small pores with fibrous materials present [6]. SiO_2, Al_2O_3, Fe_2O_3, MgO and CaO are the major components found in POC powder with oxide composition >71.09%. This proves that the powder satisfies the chemical requirement of Class F fly ash [6]. POCP and cement generally have similar fineness. However, the suitability of using them in concrete relies on their pozzolanic activity [7]. Strength activity index result proved that POCP is a pozzolanic material [48]. To ensure that the required workability can be attained when used for partial replacement of cement in mortars, POCP being a pozzolanic material would require more water [49]. The crystallinity index of quartz in POC powder utilized by [6] was 0.97 indicating partial disorderliness of quartz and pozzolanic reactivity of the powder. The major component in POCP present in quartz and cristobalite phases at 2θ angle of 26.87° and 20.45°, respectively is SiO_2 [6]. A significant hump in XRD pattern from 10–35°C demonstrates the presence of an amorphous fraction that is reactive due to pozzolanic activity [48, 49].

The addition of POC powder to replace cement and quarry dust has greatly increased the fresh and hardened density and compressive strength of produced blocks. Classified as thermally efficient and light weight blocks, the properties of the produced blocks meet the required thresholds and were higher than those of the common stabilized compressed earth blocks [16]. The use of POCP for cement replacement of about 40% in a cement-lime masonry mortar is recommended based on fresh density, consistency and air content requirements. Split tensile strength at 90 days of curing was greatly improved due to pozzolanic reactivity of POCP at longer duration. Flexural bond strength of the POCP mortar attained about 70% of control mortar. It also reduced 32% carbon footprint, 20% cost saving and save reasonable amount of energy [13].

A study attempts to investigate the durability performance and microstructure behavior of masonry mortars where POCP was used for cement replacement. With a compressive strength of 12.5 MPa, 40% cement replacement appeared to be a reliable mortar in terms of durability front with similar 28-day drying shrinkage to control mortar mix. The mixture possess extremely good electrical resistivity [49].

POC powder significantly enhances concrete compactness. At 15% increment, it improves the modulus of elasticity for up to 60% as compared to normal concrete. This could be attributed to concrete stiffness enhancement. At same increment, highest splitting tensile and flexural strengths in the range of normal weight concrete were recorded. Also, 15% and 30% strength enhancements were obtained for flexural and compressive strengths (65 MPa). The study also found that utilizing

POC powder of ~15–20% as a filler or cementitious materials in producing 45 grade lightweight concrete, CO_2 was reduced [12]. In a similar study trying to improve concrete strength, authors used varying proportions of nano-palm oil clinker powder (NPOCP) for cement replacement. It was discovered that, as NPOCP content is increased in the concrete mix, density decreases. This is because, cement has higher specific gravity than NPOCP. However, increase of NPOCP content increases concrete workability. The highest and lowest compressive values were obtained at 10% and 40% NPOCP replacement levels [16].

POCP replacement level of up to 30% enhanced the resistance of recycled aggregate-based concrete against water absorption and risk of corrosion decreased to a "moderate" level after 90 days curing period. In terms of compressive strength, POCP optimal replacement level to attain satisfactory result is 20% in comparison with the normal mix [23].

In a study by [5], the surface voids of POC coarse were filled and coated with POCP as a filler material. This mixture could decrease the quantity of aggregates derived from primary sources that are continuously exploited. It also increases the paste content necessary to make the mixes more cohesive. A notable increment in flexural strength was attained between 5 and 25% higher as compared to the POC concrete with 20–30% attained for compressive strength. However, supplementing POCP led to a decrease in water absorption value by decreasing the pore size, thereby producing highly densified paste. Specimens that contain POCP were reported to exhibit greater chloride-ion resistance.

POC powder can reduce the cost of mortar by 41%, save 3.3% of cement production, 52% carbon emission reduction. 50% POC powder replaced mortar could achieve 70% strength and 60% structural efficiency as compared to normal mortar [14]. The pozzolanic reactivity, microstructure properties investigation and strength activity index result confirmed that POC powder has pozzolanic property and good for utilization in cement-based applications.

2.3 POC in wastewater treatment

Domestic and industrial activities discharge wastewater containing high concentrations of various contaminants into water bodies [50]. Wastewater usually full of contaminants is considered as any water that is not safe for the intended use. Wastewater as a hazardous substance/material is a by-product resulting from human activities [51]. However, it is a source of chemical and thermal energy. Industrial operations in different mining fields, battery manufacturing, tannery, smelting, electroplating, textile, leather, petroleum processes, etc. are described as the major sources of wastewater. Surface runoff, sewer infiltration and poor management of urban solid waste also generate wastewater. Discharging all these without treatment into watercourses exhausts the good quality of freshwater water bodies. Wastewater is known to contain toxic pollutants like heavy metals, organic substances (dyes, PAHs etc.) posing a great environmental threat for all living organisms [52, 53]. Therefore, reduction in effluent quantity and improving the quality would have major positive effects on land use and human health [51]. To achieve that, compliance with acceptable limits is required prior to discharging effluents into the environment. Researchers have engaged in developing safe, functional, cost-effective, and appropriate wastewater treatment technologies to improve the ecosystem, lessen pollutants' detrimental effects, and minimize the risk of global warming and climate change. Unfortunately, some technologies and materials have shortcomings. As a result, it is imperative to develop safe, cost-effective, and long-lasting wastewater treatment materials.

POC as a waste material have recently been utilized by several researchers for wastewater treatment using techniques such adsorption, biological system etc. Arsenic adsorption with palm oil clinker sand (POCS) was studied by [54]. They found out that pH, arsenic concentration, POCS (mg), and temperature are the four significant variables that control arsenic adsorption. Similar to several other adsorbents, solution initial pH portrays the most prominent influence on adsorption. Water absorption, fitness modulus and specific gravity were said to be the POC properties responsible for arsenic adsorption and process stability. Furthermore, the rich microporous structure and surface functional groups POCs play vital role in the marvelous arsenic adsorption.

In a conventional activated sludge system, POC acted as a submerged attached growth media for the treatment of domestic wastewater. Performance efficiency of the POC in the extended aeration system (EAS) was evaluated by COD, TSS, MLSS, and MLVSS. Comparing the performance of POC submerged system to a biological activated sludge system, it could be concluded that using POC as an attach growth system can reduce the organic contaminant in effluent discharge [55].

POC as a filter media in a sequence batch reactor system is capable of extending its useful life, and reduce the demand for manufacturing new and sustainable media. In a comparative analysis between two SBR reactors with and without POC as a submerged fixed media, the former has higher ammonia removal efficiency of about 90% while the latter has 85% [56]. In a related study treating domestic wastewater in an SBR system, the average removal rate for ammonia and COD were 0.001 mg ammonia/mg MLVSS and 0.0069 mg COD/mg MLVSS respectively. This amount to ammonia and 90% and 70% removal efficiencies for ammonia and COD respectively [57].

2.4 Soil stabilization

Deep foundations specifically for soft soil has been a problem for long time. This pushed geotechnical engineers to opt for Lightweight Concrete Pile (LCP) due to their peculiar properties such low density, surface roughness, low strength and high porosity. Different materials have been utilized to improve the concrete pile properties for performance enhancement. POC incorporated concrete pile (p-LCP), foamed concrete pile (f-LCP) and normal concrete pile (NCP) for floating foundation were investigated by conducting static and dynamic load tests. Findings revealed that, higher compressive stress and driving resistance values were obtained for p-LCP and f-LCP when compared with NCP. Correlating the compressive stress and driving resistance values of p-LCP and f-LCP with the pile ultimate load carrying capacity, the applied load for p-LCP and f-LCP can be increased by 4.5% and 27.3% respectively. The driving resistance could also be increased by 27.6% and 16.5% for p-LCP and f-LCP, respectively. Therefore, the study concluded that, p-LCP or f-LCP are better than NCP for deep foundation of particular structure in soft soil [58].

2.5 Highway construction

In the underdeveloped, developing and developed nations, highway construction is vital for the well-being of citizens. This result to the over utilization of natural resources for the construction. However, most of these resources have different environment and financial implications to the immediate community. Therefore, few researchers come up with the idea of using POC waste material for highway pavement construction application and help solve POC disposal issues.

A study assessed the effect of using palm oil clinker (POC) as a substitute to fine aggregate on the mechanical properties of stone mastic asphalt (SMA) mixtures. The results proved the suitability of 100% POC replacement for fine aggregate in SMA mixture, as it enhanced the drain down, resistance to moisture damage, resistance to rutting, and resilient modulus when compared to that of control mixture. However, 40% and 60% replacement are considered as the optimum because of their outstanding mechanical properties. It also possesses higher indirect tensile strength for wet and dry conditions. Cantabro loss (durability performance) for POC- 80 and POC-100 exceeded that of the control sample as all mixtures fulfilled the standard requirement of the maximum value (20%) for weight loss. Authors concluded that, the use of POC as fine aggregates can greatly improve asphalt mix performance in flexible pavement construction [19]. In a related study by same authors, using the Marshall mix design, to select the optimum binder content, asphalt mixture samples with different percentages of asphalt binder content (5.0%, 5.5%, 6.0%, 6.5%, and 7.0%) were prepared. The results showed that POC could satisfy the mix design requirements in terms of Marshall stability, flow, quotient, and volumetric properties. However, POC has less effect on optimum binder content. The length of the elastic stage POC replaced mixture is higher than that of the control mixture, thereby, enhancing the elastic properties and making them more inclined towards plastic fracture. The fracture life of asphalt mixtures increases by increasing the POC content in the mix. As a result, the asphalt mixtures are strong and stiff enough to withstand permanent deformation following traffic loads [59].

A study undertaken by [21] employed a high shear mixer to determine the appropriateness of utilizing palm oil clinker fine (POCF) as bitumen modifiers by material characterization tests. The impact of modification mixing parameters was also evaluated. The result from characterization confirmed its pozzolanic property. Thus, suggesting the feasibility of utilizing it as a bitumen modifier. The optimum mixing parameters obtained were 900 rpm at 160°C for 30 min with 6.3% of POCF as the optimum dosage. The study gathered that the incorporation of POCF enhances conventional bitumen properties.

2.6 POC as a catalyst

In the downstream petrochemical industries, ethylene (C_2H_4) is one of the most highly sought raw materials. C_2H_4 is a primary precursor for surfactant fabrication, plastic manufacturing and polyethylene production. Rather than landfilling POC, the current work attempted the valorization of silica-rich POC into POC derived SBA- 15 (POC-SBA-15) catalysts and modulation of its surface acidity for C_2H_4 production via ethanol dehydration. 400oC temperature, 50 wt% ethanol concentration, 16 mL/g.h LHSV were found to be the optimal conditions for ethanol dehydration over POC-SBA-15 [5] with the lowest strong and highest weak acidity. The POC mix catalyzes the process for up to 105 h [26].

2.7 POC as a bio-filler

To improve the mechanical strength, water resistance and fire protection performance of steel structures, it is essential to use appropriate and cost-effective materials as bio-fillers in solvent-borne intumescent coatings. To that effect, waste by-products like chicken eggshells (CES), rice husk and POC are being used to lessen the use of synthetic fillers. To produce intumescent coatings, POC and hybrid fillers are homogeneously mixed with an acrylic binder and subsequently blended with flame-retardant additives. POC have the advantages of large volume availability and direct usage without further processing.

Study by [3] revealed that, the optimum composition of POC and hybrid fillers results in intumescent coating with the greatest fire retardancy with the lowest equilibrium temperature (171.3°C) because of its high thermal stability, high water resistance and excellent adhesion strength/mechanical properties. POC as a fire-retardant filler let the binder to mix appropriately, resulting in a more homogeneous coating with better interfacial bonding. It was discovered that combining POC with $Mg(OH)_2$ fillers also enhances the adhesion strength of intumescent coating.

In a related study by same authors, hybrid fillers with POC were mixed in appropriate quantity of additives and acrylic binder to produce intumescent coatings. Findings revealed that, specimen with POC as a sole filler greatly enhanced the fire protection efficiency of the intumescent coating, with <10% temperature difference when compared to specimen with hybrid fillers. For hybrid fillers composition, specimen consisting of POC/$Al(OH)_3$/TiO_2 greatly improved the coatings water resistance due to $Al(OH)_3$ low solubility in water, while specimen containing $Mg(OH)_2$ had higher mechanical strength because of the strong bond that exist between the acrylic binder/$Mg(OH)_2$ filler and metal surface [60].

2.8 Risk assessment and leaching in POC utilization

Heavy metal leaching from waste depends on the matrix's bonding energy and perhaps even the leaching state. In terms of POC, the concentration of heavy metals present depends on the palm oil mill boiler burning condition and geological condition of the location where palm oil tree grew [8]. POC solubility relies on different bonding energy in solid matrix. If the hydration energy exceeds existing bonding strength of POC matrix, POC dissolves into the solution; otherwise, the metals of POC are deposited as residue at the bottom of the vessel. Under normal environmental condition, heavy metals do not leach from solid matrix of POC, because the leaching values of heavy metals are well below the standard limit of risk. With POC acid soluble fraction in the range of 0.0–9.27%, risk assessment code (RAC) analysis by [8] confirms the safe incorporation of POC in cement-based applications because RAC values are <1%. Therefore, there is no potential threat to environment and health safety [8].

It's crucial to understand human exposure to ionizing radiation because radiation from natural sources can cause cancer and genetic mutations that influence future generations. Knowing the radiological health hazards caused by the incorporation of POC in building elements is very important. The radioactivity levels were measured by [9] and the activity concentration in POC was found to be less than 50 Bq kg-1 world average values for building materials. To evaluate the potential radiological hazards, radiological parameters and hazard indices, such as absorbed dose rate, radium equivalent activity, and annual effective dose were determined. Obtained results were within the recommended standard limit, precisely less than unity. This implies that POC is safe to be used in concrete construction.

3. Pretreatment of POC for performance enhancement

3.1 Effect of hydrochloric (HCl) acid and magnesium sulfate ($MgSO_4$) attack

The effect of hydrochloric acid and magnesium sulfate attack on POC supplemented concrete was evaluated by [10]. The outcome proved that 30% POC addition minimizes concrete deterioration and loss in compressive strength when dipped in a HCl solution and a 30 MPa strength 90 days curing. The concrete containing POC showed higher performance against deterioration, mass and strength loss. This could be due to low quantity of calcium hydroxide, well known as weak in acid attack resistance. However, when the concrete got exposed to $MgSO_4$ attack, less micro-cracks were seen.

3.2 Effect of thermal and chemical treatment on POC structural elements

Fire resistance of any structural element greatly depends on the stability of concrete ingredients at elevated temperatures. Therefore, researchers usually conduct thermal activation to evaluate its effect on physical and mechanical properties, crystalline structure, minerals, organic carbon content, morphology and chemical composition. Investigating these properties in necessary as they directly or indirectly affect the structural elements compressive strength. This was why Karim et al. [11] studied the effect of temperature on microstructure change and compressive strength of cement paste incorporated with POC. It was reported that, thermal activation at 600°C, and 800°C for a duration of 3 h yield higher residual compressive strength for POC specimen than that obtained for OPC specimen. This could mainly be due to the pozzolanic reaction of POC specimen when heated at elevated temperature. Also, C-S-H gels were more stable in POC containing cement paste after an elevated temperature exposure. This signifies that POC incorporated specimen has higher fire resistance. Crack formation was also higher in OPC paste surface, which is an indication of higher superiority of POCPC in making fire resistant concrete.

In a related study by same authors [61], it was gathered that 580°C for 3 h is proven to be the appropriate condition for thermal activation effect on POCP as the compressive strength of mortar was significantly increased, organic carbon content in POC reduced as inorganic oxides content increases, with an increment rate of 3.4%, 3.5% and 3.4% for SiO_2, Al_2O_3 and Fe_2O_3 respectively at °C. Porosity reduced as fibers were eliminated and POC color transformed from black to gray. It was also discovered that thermal activation has no significant influence on POC crystalline structure. Therefore, it has been proven that thermal treatment can enhance POC pozzolanic reactivity by elevating the maturing process of hardened specimens and unburned carbon removal.

In concrete specimen prepared with 25%, 50%, 75% and 100% replacement of Oil Palm Shell (OPS) with POC as coarse aggregate at an elevated temperature up to 500°C for 30–60 min, POC aggregate experienced negligible weight loss of <1% with excellent resistance. As the POC content is being increased, number of cracks and crack width decreases. At 100% OPS replacement with POC, the loss of residual compressive strength of only 9% indicates the vast improvement of OPS concrete using POC [2].

In other to investigate the pozzolanic reactivity of POC powder by chemical pretreatment, the powder was replaced at 2.5–15.0% by weight of cement for pre-treated and untreated POC powder in mortar mixtures. POC impregnation with low HCl acid concentration was able to enhance its pozzolanic reactivity through the hike of active silica proportion and reduced impurities and traces of metallic elements. The combination of 0.1 M of HCl acid and 1 h of impregnation time was selected as the optimum pre-treatment parameters. The strength activity index of up to 7.5% of cement content replacement with pre-treated POC increased in the hardened mortar. Authors also concluded that, the pre-treatment process would enhance the pozzolanic reactivity of POC powder up to 170% higher, increase the proportion of amorphous silica up to 9.6%, and contribute more to the strength development of mortar compared to the untreated POC powder [17].

4. Circular economy

The circular economy seeks to sustainably merge economic activities with environmental protection. It stresses the utilization of solar, wind, biomass, and waste-derived energy in the product lifecycle. It also encourages material, product, and components re-use, repair, remanufacturing, upgrading, refurbishing, and cascading [62]. It is termed as a remedy for increasing positive environmental effects while increasing economic growth by incorporating alternative manufacturing, utilization, and disposal systems. It strives to step away from the 'make, use, dispose' approach and supports the cyclical application and utilization of processes. However complicates life for people because it requires consumers behavioral changes in terms of perception to values, patterns, and relationships [63]. The main foundation of circular economy is built on the foundations of structures that encourage the responsible and cyclical utilization of materials and energy to preserve the economic value of resources for as long as humanly possible. To accomplish sustainable development goals (SDGs), circular economy has been portrayed as an accelerator towards enhancing in areas of sustainability, resource management, social equality, social responsibility and productivity [64]. It not only stimulates economic development but also shifts demand from a linear "extract-produce-use-dump" model to a cyclical flow model. It is also said to reduce carbon footprint [65]. Interestingly, companies are now integrating the idea into their everyday operations.

4.1 Principles of a circular economy

Three principles of a circular economy have been described by the Ellen MacArthur Foundation, namely;

- By dematerializing and virtualizing service delivery, as well as promoting green technology and processes, a limited stock of natural resources will be conserved and optimized.

- Recycling, refurbishment, and remanufacturing of goods and services to regenerate and recirculate capital without lowering their value.

- Priorities for industrial prosperity are; damage reduction, waste elimination, and the use of sustainable and resilient resources [64].

4.2 Approaches to foster circular economy

The following have been identified as approaches for nurturing circular economy:

- Regeneration: ecosystem renewal and repair

- Sharing: asset reusing, updating, and sharing.

- Optimization: improved efficiency, remote sensing and control, waste elimination, and big data utilization.

- Loop: organic waste is processed, remanufactured, and biochemically extracted, resulting in outputs used as inputs in the economy.

- Virtualization: dematerialization by using digital and virtual services
- Exchange: implementing innovative service and business models, as well as new technical innovations [65].

According to literature, there are a variety of business models for adopting the circular economy. They are circular economy and:

- Manufacturing
- Supply Chain Management
- Energy (energy transition, renewable energy, and biogas for electrification)
- Consumer
- Waste Management e.g. (agricultural and industrial waste) [64].

Relating circular economy to the oil palm industry, it has been reported by researchers that quantities of various dry palm oil biomass wastes can be obtained for 1 ton of crude palm oil produced from fresh fruit bunches (FFB). They could be: six tons of palm fronds, five tons of empty fruit bunches, one ton each of mesocarp fiber and palm trunks, 250 kg of Palm kernel cake, and 500 kg of palm kernel shell Palm oil mill effluent (POME) (100 tons) [24, 66, 67].

As a commodity, palm oil acting as a feed, energy, and chemical source has proven to be successful in creating a sustainable and healthy circular economy. The development of the new circular economy paradigm pave way for proper utilization of palm oil clinker. In light of the circular economy strategy, the use of POC for many industrial applications earlier discussed in this chapter is gratifying for the environment and community's well-being.

5. Conclusion

This chapter was designed to highlight the generation, disposal problems, properties and composition of POC. The waste to resource potentials of POC were greatly discussed starting with the application of POC in conventional and geopolymer structural elements such as beams, slabs, columns made of either concrete, mortar or paste for coarse aggregates, sand and cement replacement. Aspects such as performance of POC in wastewater treatment processes, fine aggregate and cement replacement in asphaltic and bituminous mixtures during highway construction, a bio-filler in coatings for steel manufacturing processes and a catalyst during energy generation were also discussed. Circular economy potentials, risk assessment and leaching behavior during POC utilization would be evaluated. The chapter also discussed the effectiveness of POC in soil stabilization and the effect of POC pretreatment for performance enhancement. During the study, it was discovered that POC utilization for intumescent coating can contribute to environmental conservation and reduce production cost. 37% of waste materials from palm industry are used in the development of green concrete and with the global significant rise in vegetable oil production, it is projected to grow even further. This is anticipated to rise further with the global increase in vegetable oil demand. Thus, the incorporation of POC as an alternative raw material for concrete work, with or without pre-treatment, will help to maintain the construction industry's long-term viability. POC has been shown to function in a variety of concretes, including self-compacting, natural, lightweight, pervious, and supplementary cementitious materials. The present chapter could be used for researchers' foundational awareness that will motivate them to explore the high potential of utilizing POC for greater environmental benefits associated with less cost when compared with conventional materials. Finally, this chapter suggest future researchers to investigate the feasibility of utilizing micro, ultrafine and nano POC powder for various applications that will promote environmental sustainability.

Acknowledgements

The authors would like to express their gratitude to Universiti Teknogi PETRONAS (UTP) for supporting the research under the Fundamental Research Grant Scheme (FRGS) grant (015MA0-092). The authors would like to acknowledge also the support from Universiti Tenaga Nasional (UNITEN) under BOLD Research Grant 2020. The authors are also thankful to technologists; Mr. Zaaba bin Mohammad, Mrs Norhayama binti Ramli and Mr. Meor Asniwan Bin Mew Ghazali from the Department of Civil and Environmental Engineering, Universiti Teknologi PETRONAS, Malaysia for their assistance.

Conflict of interest

The authors declare no conflict of interest.

Author details

Ahmad Hussaini Jagaba[1,2*], Shamsul Rahman Mohamed Kutty[1,3], Gasim Hayder Ahmed Salih[4,5*], Azmatullah Noor[1], Mohammad Fakhuma Ubaidillah bin Md Hafiz[1], Nura Shehu Aliyu Yaro[1,6], Anwar Ameen Hezam Saeed[7], Ibrahim Mohammed Lawal[2,8], Abdullahi Haruna Birniwa[9] and Abdullahi Usman Kilaco[10]

1 Department of Civil and Environmental Engineering, Universiti Teknologi PETRONAS, Bandar Seri Iskandar, Perak Darul Ridzuan, Malaysia

2 Department of Civil Engineering, Abubakar Tafawa Balewa University, Bauchi, Nigeria

3 Centre of Urban Resource Sustainability, Institute of Self-Sustainable Building, Universiti Teknologi PETRONAS, Seri Iskandar, Perak Darul Ridzuan, Malaysia

4 Institute of Energy Infrastructure (IEI), Universiti Tenaga Nasional (UNITEN), Kajang, Selangor Darul Ehsan, Malaysia

5 Department of Civil Engineering, Universiti Tenaga Nasional (UNITEN), Kajang, Selangor Darul Ehsan, Malaysia

6 Department of Civil Engineering, Ahmadu Bello University, Zaria, Nigeria

7 Department of Chemical Engineering, Universiti Teknologi PETRONAS, Bandar Seri Iskandar, Perak Darul Ridzuan, Malaysia

8 Department of Civil and Environmental Engineering, University of Strathclyde, Glasgow, UK

9 Department of Chemistry, Sule Lamido University, Kafin-Hausa, Nigeria

10 Civil Engineering Department, University of Hafr Al-Batin, Hafr Al-Batin, Saudi Arabia

*Address all correspondence to: ahmad_19001511@utp.edu.my and gasim@uniten.edu.my

References

[1] Kanadasan J, Razak HA. Mix design for self-compacting palm oil clinker concrete based on particle packing. Materials Design. 2014;56:9-19.

[2] Jumaat MZ, Alengaram UJ, Ahmmad R, Bahri S, Islam AS. Characteristics of palm oil clinker as replacement for oil palm shell in lightweight concrete subjected to elevated temperature. Construction Building Materials. 2015;101:942-951.

[3] Mustapa S, SULONG NR. Performance of palm oil clinker as a bio-filler with hybrid fillers in intumescent fire protective coatings for steel. Sains Malaysiana. 2017;46(12):2489-2496.

[4] Ibrahim MHW, Mangi SA, Ridzuan MB, Burhanudin M, Jamaluddin N, Li KH, et al. Compressive and flexural strength of concrete containing palm oil biomass clinker with hooked-end steel fibers. International Journal of Integrated Engineering. 2018;10(9).

[5] Abutaha F, Abdul Razak H, Ibrahim HA. Effect of coating palm oil clinker aggregate on the engineering properties of normal grade concrete. Coatings. 2017;7(10):175.

[6] Karim MR, Hashim H, Razak HA, Yusoff S. Characterization of palm oil clinker powder for utilization in cement-based applications. Construction Building Materials. 2017;135:21-29.

[7] Abutaha F, Razak HA, Ibrahim HA, Ghayeb HHJR, conservation, Recycling. Adopting particle-packing method to develop high strength palm oil clinker concrete. Resources, conservation Recycling. 2018;131:247-58.

[8] Karim MR, Yusoff S, Razak HA, Chowdhury FI, Zabed H. Heavy metals leaching behaviour assessment of palm oil clinker. Sains Malaysiana. 2018;47(3):523-530.

[9] Karim MR, Khandaker MU, Asaduzzaman K, Razak HA, Yusoff S. Radiological risks assessment of building materials ingredients: palm oil clinker and fuel ash. Indoor Built Environment. 2019;28(4):479-491.

[10] Alnahhal MF, Alengaram UJ, Jumaat MZ, Alsubari B, Alqedra MA, Mo KH. Effect of aggressive chemicals on durability and microstructure properties of concrete containing crushed new concrete aggregate and non-traditional supplementary cementitious materials. Construction Building Materials. 2018;163:482-495.

[11] Karim MR, Chowdhury FI, Zabed H, Saidur M. Effect of elevated temperatures on compressive strength and microstructure of cement paste containing palm oil clinker powder. Construction Building Materials. 2018;183:376-383.

[12] Ahmmad R, Alengaram UJ, Jumaat MZ, Sulong NR, Yusuf MO, Rehman MA. Feasibility study on the use of high volume palm oil clinker waste in environmental friendly lightweight concrete. Construction Building Materials. 2017;135:94-103.

[13] Nayaka RR, Alengaram UJ, Jumaat MZ, Yusoff SB, Alnahhal MF. High volume cement replacement by environmental friendly industrial by-product palm oil clinker powder in cement–lime masonry mortar. Journal of Cleaner Production. 2018;190:272-284.

[14] Kanadasan J, Abdul Razak H. Utilization of palm oil clinker as cement replacement material. Materials. 2015;8(12):8817-8838.

[15] Hamada H, Alattar A, Yahaya F, Muthusamy K, Tayeh BA. Mechanical

properties of semi-lightweight concrete containing nano-palm oil clinker powder. Physics Chemistry of the Earth, Parts A/B/C.102977.

[16] Shakir AA, Ibrahim MW, Othman N, Mohammed AA, Burhanudin M. Production of eco-friendly hybrid blocks. Construction Building Materials. 2020;257:119536.

[17] Ismail AH, Kusbiantoro A, Chin SC, Muthusamy K, Islam M, Tee KF. Pozzolanic reactivity and strength activity index of mortar containing palm oil clinker pretreated with hydrochloric acid. Journal of Cleaner Production. 2020;242:118565.

[18] Kanadasan J, Razak HA, Subramaniam V. Properties of high flowable mortar containing high volume palm oil clinker (POC) fine for eco-friendly construction. Journal of Cleaner Production. 2018;170: 1244-1259.

[19] Babalghaith AM, Koting S, Sulong NHR, Karim MR, AlMashjary BM. Performance evaluation of stone mastic asphalt (SMA) mixtures with palm oil clinker (POC) as fine aggregate replacement. Construction Building Materials. 2020;262:120546.

[20] Darvish P, Alengaram UJ, Poh YS, Ibrahim S, Yusoff S. Volume based design approach for sustainable palm oil clinker as whole replacement for conventional sand in mortar. Journal of Building Engineering. 2020;32:101660.

[21] Yaro N, Napiah M, Sutanto M, Usman A, Saeed S, Kaura J. Influence of modification mixing parameters on conventional properties of palm oil clinker fine (POCF)-modified bitumen. Materials Today: Proceedings. 2021.

[22] Kabir SA, Alengaram UJ, Jumaat MZ, Yusoff S, Sharmin A, Bashar II. Performance evaluation and some durability characteristics of environmental friendly palm oil clinker based geopolymer concrete. Journal of cleaner production. 2017;161:477-492.

[23] Alnahhal MF, Alengaram UJ, Jumaat MZ, Abutaha F, Alqedra MA, Nayaka RR. Assessment on engineering properties and CO2 emissions of recycled aggregate concrete incorporating waste products as supplements to Portland cement. Journal of cleaner production. 2018;203:822-835.

[24] Jagaba A, Kutty S, Hayder G, Baloo L, Ghaleb A, Lawal I, et al. Degradation of Cd, Cu, Fe, Mn, Pb and Zn by Moringa-oleifera, zeolite, ferric-chloride, chitosan and alum in an industrial effluent. Ain Shams Engineering Journal. 2020.

[25] Feen OS, Mohamed RN, Mohamed A, Khalid NHA. Effects of coarse palm oil clinker on properties of self-compacting lightweight concrete. Jurnal Teknologi. 2017;79(6).

[26] Cheng YW, Chong CC, Cheng CK, Ng KH, Witoon T, Juan JC. Ethylene production from ethanol dehydration over mesoporous SBA-15 catalyst derived from palm oil clinker waste. Journal of Cleaner Production. 2020;249:119323.

[27] Muthusamy K, Budiea AMA, Mohsin SMS, Zam NSM, Nadzri NEA. Properties of fly ash cement brick containing palm oil clinker as fine aggregate replacement. Materials Today: Proceedings. 2020.

[28] Abutaha F, Razak HA, Kanadasan J. Effect of palm oil clinker (POC) aggregates on fresh and hardened properties of concrete. Construction Building Materials. 2016;112:416-423.

[29] Kanadasan J, Razak HA. Engineering and sustainability performance of self-compacting palm

oil mill incinerated waste concrete. Journal of Cleaner Production. 2015;89:78-86.

[30] Nazari A, Riahi S, Bagheri A. Designing water resistant lightweight geopolymers produced from waste materials. Materials Design. 2012;35:296-302.

[31] Malkawi AB, Habib M, Alzubi Y, Aladwan J. Engineering properties of lightweight geopolymer concrete using palm oil clinker aggregate. International Journal. 2020;18(65):132-139.

[32] Darvish P, Alengaram UJ, Poh YS, Ibrahim S, Yusoff S. Performance evaluation of palm oil clinker sand as replacement for conventional sand in geopolymer mortar. Construction Building Materials. 2020;258:120352.

[33] Aslam M, Shafigh P, Jumaat MZ, editors. Effect of replacement of oil-palm-boiler clinker with oil palm shell on the properties of concrete. AIP Conference Proceedings; 2016: AIP Publishing LLC.

[34] Shafigh P, Chai LJ, Mahmud HB, Nomeli MA. A comparison study of the fresh and hardened properties of normal weight and lightweight aggregate concretes. Journal of building Engineering. 2018;15:252-260.

[35] Mohammed BS, Al-Ganad MA, Abdullahi M. Analytical and experimental studies on composite slabs utilising palm oil clinker concrete. Construction Building Materials. 2011;25(8):3550-3560.

[36] Aslam M, Shafigh P, Jumaat MZ, Lachemi M. Benefits of using blended waste coarse lightweight aggregates in structural lightweight aggregate concrete. Journal of Cleaner Production. 2016;119:108-117.

[37] Chai LJ, Shafigh P, Bin Mahmud H. Production of high-strength lightweight concrete using waste lightweight oil-palm-boiler-clinker and limestone powder. European Journal of Environmental Civil Engineering. 2019;23(3):325-344.

[38] Ibrahim HA, Razak HA. Effect of palm oil clinker incorporation on properties of pervious concrete. Construction Building Materials. 2016;115:70-77.

[39] Huda MN, Jumat MZB, Islam AS. Flexural performance of reinforced oil palm shell & palm oil clinker concrete (PSCC) beam. Construction Building Materials. 2016;127:18-25.

[40] Nazreen MS, Mohamed RN, Ab Kadir MA, Azillah N, Shukri NA, Mansor S, et al., editors. Characterization of lightweight concrete made of palm oil clinker aggregates. MATEC Web of Conferences; 2018: EDP Sciences.

[41] Muthusamy K, Mirza J, Zamri NA, Hussin MW, Majeed APA, Kusbiantoro A, et al. Properties of high strength palm oil clinker lightweight concrete containing palm oil fuel ash in tropical climate. Construction Building Materials. 2019;199:163-177.

[42] Shafigh P, Mahmud HB, Jumaat MZB, Ahmmad R, Bahri S. Structural lightweight aggregate concrete using two types of waste from the palm oil industry as aggregate. Journal of Cleaner Production. 2014;80:187-196.

[43] Chai LJ, Shafigh P, Mahmud H, Aslam MJSM. Effect of substitution of normal weight coarse aggregate with oil-palm-boiler clinker on properties of concrete. Sains Malaysiana. 2017;46(4):645-653.

[44] Ibrahim HA, Razak HA, Abutaha F. Strength and abrasion resistance of palm oil clinker pervious concrete under different curing method. Construction Building Materials. 2017;147:576-587.

[45] Ahmmad R, Jumaat MZ, Alengaram UJ, Bahri S, Rehman MA, bin Hashim HJJoCP. Performance evaluation of palm oil clinker as coarse aggregate in high strength lightweight concrete. Journal of Cleaner Production. 2016;112:566-574.

[46] Asadi I, Shafigh P, Hashemi M, Akhiani AR, Maghfouri M, Sajadi B, et al. Thermophysical properties of sustainable cement mortar containing oil palm boiler clinker (OPBC) as a fine aggregate. Construction Building Materials. 2020:121091.

[47] Soofinajafi M, Shafigh P, Akashah FW, Mahmud HB, editors. Mechanical properties of high strength concrete containing coal bottom ash and oil-palm boiler clinker as fine aggregates. MATEC Web of Conferences; 2016: EDP Sciences.

[48] Karim MR, Hashim H, Razak HA. Assessment of pozzolanic activity of palm oil clinker powder. Construction building materials. 2016;127:335-343.

[49] Nayaka RR, Alengaram UJ, Jumaat MZ, Yusoff SB. Microstructural investigation and durability performance of high volume industrial by-products-based masonry mortars. Construction Building Materials. 2018;189:906-923.

[50] Jagaba A, Kutty S, Hayder G, Baloo L, Abubakar S, Ghaleb A, et al. Water quality hazard assessment for hand dug wells in Rafin Zurfi, Bauchi State, Nigeria. Ain Shams Engineering Journal. 2020;11(4):983-999.

[51] Jagaba A, Kutty S, Lawal I, Abubakar S, Hassan I, Zubairu I, et al. Sequencing batch reactor technology for landfill leachate treatment: A state-of-the-art review. Journal of Environmental Management. 2021;282:111946.

[52] Abubakar S, Lawal I, Hassan I, Jagaba A. Quality water analysis of public and private boreholes (a case study of Azare Town, Bauchi, Nigeria). Am J Eng Res. 2016;5(2):204-208.

[53] Jagaba AH, Abubakar S, Nasara MA, Jagaba SM, Chamah HM, Lawal IM. Defluoridation of drinking water by activated carbon prepared from tridax procumbens plant (A Case Study of Gashaka Village, Hong LGA, Adamawa State, Nigeria). International Journal of Computational Theoretical Chemistry. 2019;7(1):1.

[54] Rehman MA, Yusoff I, Ahmmad R, Alias YJW, Air,, Pollution S. Arsenic adsorption using palm oil waste clinker sand biotechnology: an experimental and optimization approach. Water, Air, Soil Pollution. 2015;226(5):1-13.

[55] Al-dhawi BN, Kutty SR, Almahbashi NM, Noor A, Jagaba AH. ORGANICS REMOVAL FROM DOMESTIC WASTEWATER UTILIZING PALM OIL CLINKER (POC) MEDIA IN A SUBMERGED ATTACHED GROWTH SYSTEMS.

[56] Chew Y, Yap Z, Kutty S, Ghaleb A, Almahbashi N, editors. Removal of ammonia by palm oil clinker (POC) as submerged fixed media in sequence batch reactor (SBR) mode. IOP Conference Series: Materials Science and Engineering; 2020: IOP Publishing.

[57] Kuan YZ, Kutty SR, Ghaleb AA. Kinetics Coefficient of Palm Oil Clinker Media for an Attached Growth Media in Sequencing Batch Reactor Mode. Journal of Ecological Engineering. 2019;20(9).

[58] Sulaeman A, Fulazzaky MA, Haroen M, Bakar I. Field test results of palm oil clinker concrete pile and foamed concrete pile for floating foundation in soft soil. KSCE Journal of Civil Engineering. 2018;22(7): 2232-2240.

[59] Babalghaith AM, Koting S, Sulong NHR, Karim MR,

Mohammed SA, Ibrahim MR. Effect of Palm Oil Clinker (POC) Aggregate on the Mechanical Properties of Stone Mastic Asphalt (SMA) Mixtures. Sustainability. 2020;12(7):2716.

[60] Mustapa S, Sulong NR, editors. Performance of solvent-borne intumescent fire protective coating with Palm oil clinker as novel bio-filler on steel. IOP Conference Series: Materials Science and Engineering; 2017: IOP Publishing.

[61] Karim MR, Hashim H, Razak HA. Thermal activation effect on palm oil clinker properties and their influence on strength development in cement mortar. Construction Building Materials. 2016;125:670-678.

[62] Muñoz E, Navia R. Circular economy in urban systems: How to measure the impact? : SAGE Publications Sage UK: London, England; 2021.

[63] de Morais LHL, Pinto DC, Cruz-Jesus F. Circular economy engagement: Altruism, status, and cultural orientation as drivers for sustainable consumption. Sustainable Production Consumption. 2021;27:523-533.

[64] Mutezo G, Mulopo J. A review of Africa's transition from fossil fuels to renewable energy using circular economy principles. Renewable Sustainable Energy Reviews.137:110609.

[65] Upadhyay A, Mukhuty S, Kumar V, Kazancoglu Y. Blockchain technology and the circular economy: Implications for sustainability and social responsibility. Journal of Cleaner Production. 2021:126130.

[66] Sadeghi O, Fazeli A, Bakhtiarinejad M, Che Sidik NA, editors. An overview of waste-to-energy in Malaysia. Applied Mechanics and Materials; 2015: Trans Tech Publ.

[67] Jagaba A, Kutty S, Hayder G, Latiff A, Aziz N, Umaru I, et al. Sustainable use of natural and chemical coagulants for contaminants removal from palm oil mill effluent: A comparative analysis. Ain Shams Engineering Journal. 2020;11(4):951-960.

Chapter 5

Sustainability of the Oil Palm Industry

Dickson Osei Darkwah and Meilina Ong-Abdullah

Abstract

The oil palm (*Elaies guineensis* Jacq) is the largest produced and highly traded vegetable oil globally yet has the lowest cost of production and significantly higher productivity compared to other oil crops. The crop has the potential of alleviating poverty for smallholders and lifting the economies of countries with large scale production notably, Malaysia and Indonesia and currently on high demand for use as biofuel feedstock. Irrespective of these advantages of the oil palm, there is a global concern on the devastating impact of the crop on the environment and ecosystem during plantation developments and expansions. Deforestation, biodiversity loss, water and air pollution and toxic compounds from palm oil mill effluents (POME) are some of the negative impacts of the oil palm. For the industry to be more beneficial and impactful globally, sustainability strategies becomes urgent need. Sustainability strategies such as increasing the yield of oil palm, precision agriculture, sustainability certification, support for smallholders and circular economy have been put across to curtail the negative impacts of oil palm expansion.

Keywords: environmental issues, sustainability certification, circular economy, precision agriculture, increasing the yield of oil palm

1. Introduction

Oil palm (*Elaeis guineensis* Jacq), a palm of African origin produces palm oil accounting for 30.8% of today's global production of oils and fats [1]. The versatility of the oil palm in oleo-chemical applications, food and biodiesel production makes it a sought after commodity in food and non-food industry. Globally, it is by far the most productive oil-bearing crop per land use and is capable of fulfilling the large and growing demand of vegetable oil that is estimated to reach 240 million by 2050 [2, 3]. While 56.2 million tons of palm oil were extracted from 17.24 million hectare of land under oil palm cultivation, only 23.6 million tons were produced from rapeseed grown on 36.4 million ha and cost of production for these two commodities stood at US$ 700 and US$ 850 per metric tons respectively, indicating lower cost of production for oil palm [4].

The oil palm is significantly more productive, versatile and most economically viable among the leading vegetable and fat crops due to its cost-effective production, wide use, extreme yields and profitability. It is the ideal crop to address several of the United Nation's Sustainable Development Goals (SDGs). Economically, it has the capacity to eradicate extreme hunger and poverty, lifting millions of people out of poverty in Indonesia and other parts of the world where it is produced on a large scale. In Indonesia, the livelihood of 25 million people depend on oil palm cultivation while in Malaysia, the number of people employed increased from 92, 352 in 1980 to 570,000 in 2009.

The oil palm is a profitable crop and when coupled with good government policies have the capacity to transform the livelihood of millions of people and as well improve health care and education in rural areas [5–7]. In Brazil specifically, Paha, the average yearly investment return on oil palm cultivation is US$ 2,000 per hectares [8].

The production of palm oil is dominated by Indonesia and Malaysia which together accounts for 85–90% of total palm oil production. Indonesia, the top producer and exporter of palm oil has over 14 million hectares of oil palm plantations in 2018, out of which 55.09% are from the large private companies, 40.62% from small holder plantations and 4.39% from state large estates. Indonesia produced 48.3 million metric tons of palm oil in 2020 [9, 10]. Malaysia, the second producer of oil palm has over 5 million hectares of oil palm plantations and also produced 19.14 million tons of palm oil in 2020 [11].

Countries that have devoted large tracts of land for oil palm production have increased their economies greatly. Indonesia and Malaysia made over USD 18.7 billion and USD 9.8 billion from the export of palm oil in 2017 respectively.

Globally, the area planted with oil palm is 27 million ha with Africa accounting for 2 million ha and a recorded production of 0.27 gigatons of fruits and 71.4 megatons of palm oil produced globally [12–14]. The economic impact of the oil palm industry in general is not contributed solely by large plantation companies, it is estimated that smallholders cultivate about 50% of the oil palm area globally [15]. However, in Africa, the smallholders account for approximately 70% of the oil palm cultivated [16]. The cultivation of oil palm by small holders have resulted in increased farm and employment incomes, reduced poverty rates at local, regional and national levels [17].

Irrespective of the socioeconomic impact of the oil palm, great criticism has emanated from international groups such as Green peace, Rainforest Action Network and World Wildlife Fund (WWF) for unsustainable practices that have led to deforestation, increased greenhouse gas (GHG) emissions and the loss of biodiversity [18, 19]. Lim et al. [20] defined sustainable palm oil production as the production that protects the natural environment, promotes intra and inter-generational equity, while enhancing commercial operations and sharing economic growth with the local community through employment and fair trade. The negative impacts raised by these pressure groups may include forced labour, food safety and environmental issues etc. However, this chapter aims at bringing some of the negative impacts on the environment on oil palm cultivation and expansion to light and as well suggest some strategies that can be adopted to ensure the sustainability of the oil palm industry.

1.1 Environmental concerns

Globally, the area under oil palm has increased from 5 million ha in 1980 to more than 20 million ha in 2018 [21] with this expansion coming from Malaysia and Indonesia although 240 million ha of available land are suitable for oil palm cultivation, excluding intact forest landscapes, high carbon stock forest conservation hot spots and protected areas. Although the oil palm is the driver of the growth of most economies of countries producing on a large scale, the expansion has also led to tropical deforestation and associated biodiversity loss, greenhouse gas emissions, land degradation, forest and peat land fires as well as air and water pollution [18, 19]. Overall expansion of annual crops e.g. soybean, maize, paddy rice, and sorghum has been more rapid and more wide spread than expansion of perennial crops especially in South America, Africa and tropical Asia from 1999 to 2008 creating more biodiversity loss with oil palm noted to be the fifth on the list of biodiversity loss [22].

1.1.1 Biodiversity loss

Loss of biodiversity are directly linked to natural forest loss. Reduced habitat structures as a result of oil palm expansion provides fewer niches for flora and fauna. Endangered species such as tiger, elephants and orangutan are on the verge of extinction. As new lands are cleared from the forest, there is an increased access to lands which leads to increased hunting pressure as well as habitation by humans resulting in increased conflicts between human and these animals, example is the flood plains of the Kinabatangan River in Sabah, Malaysia and in Riau and Bengkulu provinces in Indonesia. The population and survival of these species are seriously endangered due to forest degradation and deforestation, illegal logging, illegal hunting and trade, forest fires, subsistence agriculture and development of agricultural plantations especially oil palm, rubber and acacia plantations. Species diversity, density and biomass of invertebrate communities is estimated to suffer at least 45% decrease from land use transformation of tropical forests to oil palm plantations [23].

1.1.2 Deforestation and green house gas emissions

An assessment of deforestation and forest degradation from 1982 to 2007 [24] showed a 65% loss of forest cover over the last 25 years period or a loss of about 4.2 m ha of forest. While the development of timber plantations contributed to 24%, oil palm cultivation contributed to about 29% forest loss following initial exploitation of the timber source. Deforestation has significant negative impact on loss of biodiversity, changes in climate and rainfall pattern and distribution due to alteration in precipitation retention and rainfall rates. Estimation of the proportion of deforestation to expansion of oil palm cultivation in Indonesia ranged from 11–16%. In Ghana, Forest Carbon Partnership Facility reported that agricultural expansions contribute to 50% overall deforestation however, it was later discovered that only 7% of deforestation is associated with citrus, oil palm and rubber expansion combined [25].

Aside from global warming which initially focused on combustion of fossil fuels for heat and transportation and the subsequent release of CO_2, other anthropogenic activities have also contributed significantly to the release of CO_2 and that conversion of Carbon dense tropical forests is likely to be an important part of these. Stern [26] reported that deforestation contributes to about 18% of the global greenhouse gas (GHG) emissions. Development of plantation on tropical peat lands which are drained leads to oxidation which results in significant CO_2 release over an extended period. Use of fires for land clearing and the emission of methane gas from the effluent treatment ponds of palm oil all contribute to GHG emissions.

1.1.3 Environmental and aquatic pollutions caused by palm oil mill effluent (POME)

POME is the wastewater produced by processing oil palm which has a higher biochemical oxygen demanded (BOD) and chemical oxygen demand (COD) and contains higher concentration of organic nitrogen, phosphorus and different supplement substance [27]. Oil palm trunks, oil palm fronds, empty fruit bunches, palm pressed fibers and palm kernel shells, less fibrous materials such as palm kernel cake and liquid discharge are waste products generated during oil processing [28]. The boom in the oil palm industry has resulted in the setting up of many oil mills for processing of the fleshy mesocarp and kernel. In Malaysia, mills increased from 10 in 1960 to 410 in 2008. For every ton of FFB produced, 600–700 kg of

POME is generated. POME has contributed to environmental pollution due to the production of large quantities of by-products during the process of oil extraction.

POME is generated mainly from oil extraction, washing and cleaning processes in the mill and these contain cellulosic material, oil and grease [29]. The oily waste which forms part of the POME are hazardous pollutants to aquatic environments because they are highly toxic to aquatic organisms when discharged in large quantities into watercourses e.g. river, streams. These goes a long way to contribute to human health hazards and environmental pollution. In Africa, many people depend on rivers as source of water. Sequestering POME into these watercourses creates a great danger for those that depend solely on these water bodies as a source of drinking water and other domestic and irrigation purposes. POME contains high amount of oil and grease (4000 mg/L), COD (5000 mg/L), BOD (25000 mg/L) and total solids (40, 500 mg/L) [29]. In Malaysia, about 44 million tons of POME are produced and are increasing every year [30].

Efficient POME treatment is very necessary to avoid continuous contribution to human health hazards and environmental pollution and also ensure the sustainability of the oil palm industry. Although treatment methods such as waste stabilization ponds, activated sludge systems, closed anaerobic digester and hand application harvester have been employed in the treatment of POME, the use of microalgae has received much attention in current times due to its ability to remove pollutants and also breakdown the organic compounds present in it [31, 32]. Micro algae has the potential of treating waste water such as removal of CO_2 and NOx and high capacity of nutrients absorption [33]. Undoubtedly, microalgae can use the nitrogen and phosphorus compounds in water waste to produce micro algae biomass for various kinds of lipid generation, which can serves as substrate for biofuel production [31].

2. Strategies for sustainability of the oil palm industry

2.1 Increasing the yield of oil palm

Considering the numerous negative impacts associated with oil palm expansion, priority should be given to increasing the yield of oil palm in a unit area rather than expansion of the unit area or if expansion should be perceived then should be done on marginal, degraded or abandoned land to safeguard the forest. There is a significant gap between actual yield and potential yield. Globally, potential oil yield with simulation models are 18.5 t/ha/yr. with commercial plantations actualizing 12 t/ha/yr. and average stagnated productivity of 3 t/ha/yr. by smallholders [34]. In Ghana, potential yield stands at 21 t/ha/yr. of FFB yet 11 t/ha/yr. and 6.0 t/ha/yr. are actualized in commercial plantations and smallholder farms respectively, a decline in yield of about 50% and 71% for commercial and smallholder farms respectively [35].

Increasing the yield of oil palm by 20% in Malaysia and Indonesia will add 7.7 million tonnes of palm oil and this is equivalent to the production from 1.9 million hectare of new plantings [36]. In Ghana, about 327,600 ha of oil palm is under cultivation with 16,600 ha by commercial plantations and 311, 000 ha by smallholder with actual yields of 11 t/ha/yr. and 6.0 t/ha/yr. respectively. Acquisition of quality planting materials coupled with best management practices may increase yields to 17.9 t/ha and 17.6 t/ha for commercial plantations and smallholders respectively and this will avoid the expansion area of 452,533 ha. Should the gap be completely closed i.e. obtaining 21 t/ha/yr. will spare an expansion area of 597, 636 ha [35].

Closing the gap between actual yield and potential yield will have positive impact on the environment since the increase in oil palm productivity will reduce pressure on new lands. There are numerous constraints leading to the suboptimal yields obtained in Ghana and other parts of the world. These includes planting of uncertified seeds (high *dura* and *pisifera* contaminations), unsuitable climatic conditions, poor soil fertility, non-performance of good agricultural practices or best management practices. IOI Corporation Berhad [37] were able to close the gap in excess of 6.0 tonnes oil per ha in 2008.

In other to bridge the gap, the following can be applied

1. Access to high quality (high yielding, early bearing, disease and drought resistant etc) oil palm planting materials

2. Strict adherence to best management practices

3. A switch from conventional agriculture to precision agriculture

4. Support to smallholder and commercial plantations.

2.1.1 Access to high quality planting materials

High quality oil palm planting materials coupled with best management application could significantly and immensely increase the yield and incomes of smallholder farmers and commercial plantations. The switch to *tenera* planting materials increased oil yield about 30% over the *dura* materials which were then planted due to the thicker mesocarp [38]. The advent of biotechnology tools such as the tissue culture have also led to the production of clonal palm. Planting clonal palm as against commercial D x P at the same time and under the same area has also increased yields by approximately 25% over the *tenera* materials [39]. However, clonal palms are limited to few industries and countries but commercial D x P are available in most oil palm growing countries. Unfortunately, many smallholder farmers especially do not plant *tenera* materials and thus have hampered productivity significantly. At plantations and institutional levels, admixture and other human errors such as unintentional use of pollen from a non *pisifera* palms, self-pollination of *dura* parental palms, open pollination of *dura* parental palms by surrounding *dura* palms and imprecise selection of seed or seedlings [40] have contributed to contaminations. Ooi et al. [41] reported an average non *tenera* contamination of 10.7% in independent planting sites surrounding the MPOB's 6 research stations located in Peninsular, Malaysia. While 9.2% were contaminations from *dura*, 1.5% was from *pisifera* contamination. The unintentional planting of *dura* or *pisifera* oil palm seedlings reduces the overall yield and impacts land utilization that would otherwise devoted to more productive *tenera* palms.

In Ghana, [42] reported about 70% *dura* and *pisifera* contamination, comprising of 69.54% *dura* and 0.69% *pisifera* after surveying 97 smallholder farms which were planted on mined lands in three Districts of Central region with the seedlings supplied to the smallholder farmers by a Contracted nursery operator. It is believed that the seedlings supplied to the farmers were uncertified.

The advent of molecular markers have enabled marker assisted selection (MAS). Deployment of these tools could be used as a certification procedure with proper enforcement in place such that germinated seed nuts (at the seed production unit) as well as seedlings (at the nursery stage) to be planted are checked and those that are *tenera* planted. This will assist in bridging the gap between potential and actual yield.

2.1.2 Strict adherence to best management practices

Best management practices (BMP) are cost effective and practical agronomic techniques that focus on reducing yield gaps in oil palm by using production inputs and resources efficiently [43]. The aim of BMP's in oil palm is to increase the productivity of oil palm through improvements in agronomic practices and increased crop recovery. The application of BMP's are site specific because they are structured to address a specific production constraints and biophysiological conditions of individual locations [44]. BMP's can be 'yield taking' and 'yield making' practices [45]. The 'yield taking' increases yield in the short term by improving crop recovery operations with activities including regular harvesting within an interval of 7–10 days, bunches harvested should have a maximum of 5 'loose fruits' on the ground, access roads should be created within the plantations (harvesting paths, foot bridges to help cross drains, ring weeding (about 5 feet) around the palms to allow unhindered access to harvesting and collection of loose fruits and fruit bunches, pruning of dead, diseased and unproductive fronds for air circulation and quick identification of ripe fruits, quick transportation of harvested fruits to the mill within 24 hours after harvesting to help reduce fatty acids in the crude palm oil produced.

The 'yield making' practices also include but not limited to replanting of dead palms as well as removal of non-*tenera* palms to ensure there is optimum planting density (148 palms/ha) with the hope of minimizing excessive inter-palm competition for sunlight and nutrients, construction of drains especially in lowland areas to aid drainage during the wet season and ensure the availability of water during the dry season.

In addition, intercropped plants that are closer to the palm should also be exterminated to avoid intense competition for sunlight, nutrients and moisture. Ideally, plants used as intercrops should be planted about 3 m from the oil palm tree. Regular integrated weed control measures (3 times in a year), planting of cover crops such as *Peuraria phaseoloides* to reduce soil erosion, improve soil tilth, increase soil biological activity and fix nitrogen into the soils. Finally, BMP's should include fertilizer application using crop residues such as empty fruit bunch and pruned frond as well as inorganic fertilizers. Although fertilizer application in oil palm contributes about 30% of the cost of production in oil palm [46] its dividend is great. However, fertilizer application should be preceded by soil nutrient analysis to help decipher the sufficiencies and deficiencies to know the sources of nutrients to be used, the right amount to be applied, the right time and right place to be applied. Globally, the recommended fertilizer application is 260 kg N, 50 kg Phosphorus and 220 kg Potassium ha/yr. although reduced fertilizers at a rate of 136 kg N, 17 kg Phosphorus and 187 kg Potassium ha/yr. may be used [47] for adult palms. In Africa specifically Ghana, a general recommendation of fertilizer for application in matured palm (beyond 5 years) is 6 kg/palm of NKP 10:10:30 in addition to magnesium and boron. Notwithstanding, there is the need to establish multifactorial, multi locational nutrient response trials across different agro-ecological zones in areas where fertilizer will be applied to guide future recommendations for current materials and new materials been bred for optimum yield and productivity.

2.1.3 A switch from conventional agriculture to precision agriculture

Fertilizer application, harvesting, transportation of gathered fresh fruits bunches and loose fruits, weed control, and sanitary control are all potential sources of high production costs. Within the framework of ecologically sustainable development, these processes can be optimized and competitiveness increased.

Precision agriculture is a modern production management system in which new technologies are used to collect, analyze, and manage data in a sustainable manner [48]. Agriculture field machinery such as automated steering systems, data-driven targeted application of fertilizers and pesticides, field robots and drones, soil analysis sensors, and autonomous driving are all part of precision agriculture.

Precision agriculture is based on climatic, edaphic, and agronomic factors that influence yield, and its implementation can improve product quality, yield per unit area, production cost, and environmental impact. Precision agriculture works under three perspectives: agronomic (fertilization and irrigation, ensuring that optimum levels of rainfall and plant nutrients are applied to the palms), environmental (precise fertilization applied such that lower quantities are emitted into the atmosphere, fertilization above the economically optimum levels may lead to detrimental environmental effects such as GHG emissions, nutrient leaching losses, soil acidification, ground water pollutions [49, 50] and finally under economic perspective (increase in the production per unit area, the reduction of inputs or increase in efficiency).

On the basis of spatial and temporal variability, information about different types of soil properties, rainfall patterns, and availability of water courses such as dams, rivers, and streams, as well as the productivity of oil palm within a particular plot, can be electronically retrieved from field record files in real time. With this background knowledge, satellite-controlled precise agricultural machinery and intelligent sensors can be used for targeted seed planting, fertilizer and pesticide application, irrigation, and other agricultural tasks.

2.1.4 Support for smallholder farmers

The RSPO defines smallholders as farmers who grow oil palm with other subsistence crops, where the family contributes the majority of labor and the farm is the primary source of income, and where the planted area of oil palm is less than 50 hectares. However, the Ghanaian interpretation of the RSPO's P & C uses a 40-hectare barrier to designate smallholders, whereas most smallholders produce on plots of less than 10 ha, sometimes in conjunction with other crops.

Smallholders are divided into two groups: those who are assisted and those who are self-sufficient. Supported smallholders have a contractual commitment to sell their FFB to a mill or corporation in exchange for help, but independent smallholders have no contractual duty to sell their FFB or CPO to a mill or buyer [51]. As a result, independent smallholders are responsible for the growth, management, harvesting, and transportation of their FFB to milling centers. Smallholder farmers face major challenges around the world, including access to funds, land tenure issues, agricultural input access, a lack of extension services and supervision support, market access, low bargaining power, opaque pricing mechanisms, and insufficient or lack of capacity to implement certification requirements [51].

Smallholder farmers' yields have been reduced as a result of the aforementioned restraints. Furthermore, many smallholders have been denied certification by the RSPO because they do not meet the RSPO Principles and Criteria. The empowerment of smallholders is required to remedy these deficiencies. These goals could be reached by bolstering current small-holder mechanisms to improve access to financing, productivity, milling efficiency, sustainability, and market access. It would also be ideal if independent smallholders were encouraged to enter into contractual agreements with companies to gain access to improved seedlings, agricultural inputs, and technical extension services who would oversee most farm activities such as fertilizer application, record keeping, lining and pegging, pruning, and other activities to help increase oil palm productivity and increase the income of the smallholder. This is recommended since, in some oil-producing areas, neither the government nor civil society organizations have a mechanism in place to assist small-scale farmers.

2.2 Sustainability certification

Sustainability certification is a market-based process in which consumers pay a higher price for products that meet particular environmental and social requirements during manufacturing and throughout the value chain. The Roundtable Sustainable Palm Oil (RSOP) is the most well-known international certification for palm oil, but national schemes such as the Malaysia Sustainable Palm oil (MSPO) and Indonesia Sustainable Palm Oil (ISPO) exist in Malaysia and Indonesia respectively [52]. Oil palm sustainability certification includes RSPO, MSPO, ISPO, International Sustainability and Carbon Certification (ISCC) and the Roundtable on Sustainable Biomaterials and Sustainable Agriculture Networks (RSBAN) (SPOTT, 2015). MSPO and ISPO are mandatory government-led certification schemes, whilst the RSPO and ISCC are implemented on a voluntary basis [53].

The RSPO, non-profit organization, which was formed in 2004 develops and implements global standards for sustainable palm oil by bringing together stakeholders from the seven sectors of the palm oil industry: oil palm producers, processors or traders, consumer goods manufacturers, retailers, banks/investors, and environmental and social non-governmental organizations (NGOs).

In order to produce Certified Sustainable Palm Oil (CSPO), enterprises must meet a set of environmental and social criteria specified by the RSPO. These criteria, when appropriately followed, can assist to reduce the detrimental effects of oil palm cultivation on the environment and communities in palm oil-producing areas. The RSPO is a global organization with over 4,000 members from 92 countries who represent all aspects of the palm oil supply chain. They have pledged to produce, source, and/or utilize sustainable palm oil certified by the RSPO.

MSPO as at 31st December, 2019 had certified 3.6 million hectares or about 62.26% of the total area under oil palm cultivation using the certification standards. The standards contains principles, criteria and indicators generally aimed at sustainable production with the intention of mitigating the negative social and environmental impacts and encompasses traceability system, good agricultural practices, improved natural resource management and environmental responsibility and compliance with existing regulations [54]. Before the certification of oil palm producers, these standards must have been complied and audited by the RSPO, MSPO and ISPO. The principles of RSPO, MSPO and ISPO are shown in **Table 1**. The principles for certification involves transparency, legal operation, efficient production systems, respect for community and human right, small holder inclusion, rights of workers should be respected and operation that conserve the environment. Within this framework, the MSPO and ISPO have these'd out their own principles to suit them as indicated in **Table 1**.

Carlson et al. [55] reported that certification reduced deforestation by 33%. Saswattecha et al. [56] also reported that RSPO certified producers at the Tapi Basin in Thailand caused lowering of environmental impact (23–34%) compared to non-certified producers (58–75%). Studies in Indonesia indicated that smallholders who adopted the RSPO certification improved their livelihoods and increased their economic returns; they were also better trained in fertilizer, herbicide and pesticide application and handling than non-certified independent smallholders [54].

Although the certification of smallholders are increasingly challenging due to increased oil palm plantation management cost and fees resulting in less adoption of the RSPO certification, the need to put efficient and viable system such

DOI: http://dx.doi.org/10.5772/intechopen.100156

Principles	RSPO	MSPO	ISPO
Principle 1	Behave ethically and transparently	Management commitment and responsibility	Legality of plantation businesses
Principle 2	Operate legally and respect rights	Transparency	Plantation management
Principle 3	Optimize productivity, efficiency, positive impacts and resilience	Compliance with legal requirements	Protection of the utilization of primary forest and peatlands
Principle 4	Respect community and human rights and deliver benefits	Social responsibility, safety and employment conditions	Environmental management
Principle 5	Support smallholder inclusion	Environment, natural resources, biodiversity and ecosystems services	Responsibility for workers
Principle 6	Respect workers' rights and conditions	Best practice	Responsibility for social and economic empowerment
Principle 7	Protect, conserve and enhance ecosystems and the environment	Development of new plantings	Continuous business improvement

Source: https://rspo.org/about (Accessed: 8 May 2021).

Table 1.
Principles of RSPO, MSPO and ISPO.

as support system for small holders may help the oil palm production to be more sustainable and also go a long way to achieving the UN's Sustainable Development Goals (SDG).

2.3 Circular economy

Oil palm fronds (OPF) and trunks (OPT) from oil palm plantations, as well as EFB, MF, PKS, and POME from palm oil mills, are all biomass waste produced by the oil palm. For every ton of crude palm oil produced, 9 tons of biomass are created [57]. Conversion technologies, on the other hand, have been created and commercialized in order to better convert waste into a more usable product. While biorefineries may be able to convert these wastes into biofuel and other biochemicals, the oil palm industry may require fertilizer, steam, fuel, and electricity for their plantation and manufacturing facilities [58].

The linear economy, in which solid wastes such as OPF and OPT are disposed of to plantations for natural decomposition and liquid wastes such as POME are treated in open pond systems without biogas capture, fails to capitalize on the oil palm's potential economic profits [59]. Through the recycling of biomass waste, the oil palm industries and biorefineries, as consumers of oil palm products, can be targeted to achieve a sustainable circular economy. **Figure 1** depicts the difference between linear and circular economies. In this figure, the linear economy (from i to iv) encompasses harvesting and processing of FFB into CPO after which all the by-products are disposed of at dumping sites and landfills with the consequence of environmental pollution, health hazards and no economic return. However, in the circular economy, there is the treatment, further processing, recycling and re-use of

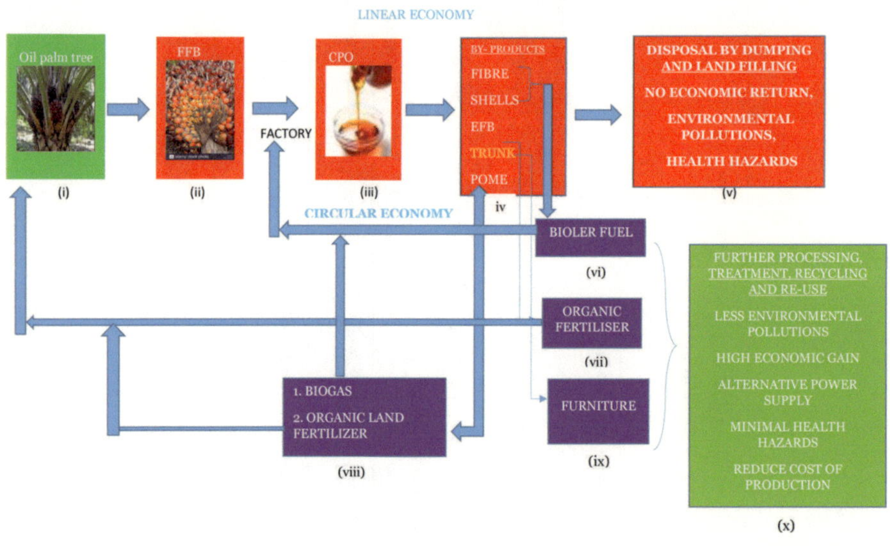

Figure 1.
Distinction between linear and circular economy. (i) Oil palm tree produces (ii) Fresh fruit bunch which is processed at the factory into (iii) CPO. By products (iv) such as fibre, shells, empty fruit bunch, shells, trunk, POME etc. are obtained. All these waste are disposed of by dumping and landfills with consequences of environmental pollution and health hazards. This whole process constitutes the linear economy. In the circular economy, fibre and shells as parts of the by-products (iv) are recycled as boiler fuel (iv) to produce power at the factory for continual processing. Similarly, the EFB from (iv) is processed into organic fertilizer (vii) which is used to fertilize the oil palm tree (i) to increase yield. In addition, POME from (iv) is treated to produce biogas and organic land fertilizer (viii) which goes back to the factory as energy and to the oil palm tree as fertilizer respectively. The trunk in (iv) is further processed into furniture (ix) for office and residential uses. This circular economy leads to less environmental pollution, minimal health hazards, reduce cost of production in (x).

the waste into boiler fuel, biogas, land organic fertilizers, furniture's etc. to produce power for the continuous running of the factory thereby reducing electricity bill, fertilization of oil palm plants to improve the yield and reduce cost of production and furniture for offices.

POME, EFB, decanter cake, and palm pressed fiber can be turned into new value-added goods (fertilizers) or used as an alternative energy source to power the manufacturing plant, thereby converting waste into wealth. Mattresses and cushions made from dried long fiber, renewable energy pellets and briquettes, and animal feed are examples of value-added products. Integrating the circular economy into palm oil production is an effective way to maximize the use of resources (raw materials) while reducing waste, pollution, and energy inefficiencies. This is accomplished by the reduction, reuse, and recycling of materials that are currently underutilized in the linear economy.

The circular economy is without a doubt one of the greatest sustainable development frameworks for the oil palm industry to effectively tackle biomass wastage issues while also avoiding high production costs, lowering negative environmental impact, and conserving resources for future generations. To maintain a competitive edge over the linear economy, better facilities, the incorporation of more advanced technology, and significant investment in the industry are required.

3. Conclusion

Globally, the economic impact of the oil palm cannot be underestimated. It is more productive and versatile among the leading vegetable and fat crops due to its cost effective production, wide use and extreme yield. Notwithstanding, there are environmental challenges that threatens the sustainability of the oil palm notably deforestation, biodiversity loss, increased greenhouse gas emissions and environmental and aquatic pollutions. In this chapter, strategies such as increasing the oil palm yield per unit area (intensification rather than extensification), precision agriculture adoption, support for smallholder farmers, sustainability certification, and circular economy are discussed. Adoption of these measures would greatly increase the oil palm's worldwide sustainability.

Author details

Dickson Osei Darkwah[1*] and Meilina Ong-Abdullah[2]

1 CSIR – Oil Palm Research Institute, Kade, Ghana

2 Malaysia Palm Oil Board, Selangor, Malaysia

*Address all correspondence to: oseidarkwah@yahoo.com

References

[1] Meilke T. Oil World. International Statistical Agricultural Information (ISTA). Hamburg, Germany: Meilke GmbH; 2018. p. 10.

[2] Corley R. How much palm oil do we need? Environmental Science and Policy 2009;12:134-139.

[3] Eycott AE, Advento AD, Waters HS, Luke SH, Aryawan AAK, Hood AS, and Turner EC. Resilience of ecological functions to drought in an oil palm agroecosystem. Environmental Research Communications 2019;10:001-015.

[4] Barcelos E, Rios SA, Cunha RNV, Lopes R, Motoike SY, Babiychuk E, Skirycz A, and Kushnir S. Oil palm natural diversity and the potential for yield improvement. Frontier Plant Science 2015;6:190-206.

[5] Norwana AAB, Kunjappan R, Chin M, Schoneveld G, Potter L, and Andriani R. The local impacts of oil palm expansion in Malaysia. Working Paper 2011;78:1-26.

[6] Sayer J, Ghazoul, J, Nelson P, and Klintuni Boedhihartono A. Oil palm expansion transforms tropical landscapes and livelihoods. Global Food Security 2012;1:114-119.

[7] Murphy DJ. The future of oil palm as a major global crop: Opportunities and challenges. Journal of Oil Palm Research 2014;26:1-24.

[8] Villela AA, Jaccoud D, Rosa LP, and Freitas MV. Status and prospects of oil palm in the Brazilian Amazon. Biomass Bioenergy 2014;67:270-278.

[9] Directorate General Plantation. Tree Crop Estate Statistic of Indonesia 2018-2020. Jakarta: Directorate General Plantation; 2019.

[10] Herdiansyah H, Negoro H, Adi RN, and Siti S. Palm oil plantation and cultivation: Prosperity and productivity of smallholders. Open Agriculture 2020;5(1):617-630.

[11] MPOC. Malaysia Palm Oil Statistic 2021. Available from: http://www.mpob.gov.my [Accessed: 20 July 2021]

[12] FAO (Food Agric. Organ.). Crops. FAOSTAT Statistical Database, Rome. 2020. Available from: http://www.fao.org/faostat/en/#data/QC [Accessed: 4 May 2021]

[13] Murphy DJ. Growing palm oil on former farmland cuts deforestation, CO2, and biodiversity loss. 2019. Available from: www.conversation.com [Accessed: 27 July 2021]

[14] Angelucci F. Analysis of incentives and disincentives for palm oil in Ghana. Technical Notes Series, MAFAP, FAO, Rome. 2013. Available from: http://www.fao.org/3/a-at550e.pdf

[15] Byerlee D, Falcon WP, and Naylor R. The Tropical Oil Crop Revolution: Food, Feed, Fuel, and Forests. New York: Oxford Univ. Press; 2017.

[16] Edwards RB. Export Agriculture and Rural Poverty: Evidence from Indonesian Palm Oil. Work Paper. Hanover, NH: Dartmouth Coll.; 2019.

[17] Ordway EM, Naylor RL, Nkongho RN, and Lambin EF. Oil palm expansion and deforestation in Southwest Cameroon associated with proliferation of informal mills. Nat. Commun. 2019;10:114.

[18] Dislich C, Keyel AC, Salecker J, Kisel Y, Meyer KM, et al. A review of the ecosystem functions in oil palm plantations, using forests as a reference system. Biol. Rev. Camb. Philos. Soc, 2017;92:1539-1569.

[19] Clough Y, Krishna VV, Corre MD, Darras K, Denmead LH, et al. Land-use choices follow profitability at the expense of ecological functions in Indonesian smallholder landscapes. Nat. Commun. 2016;7:13137.

[20] Lim CI, Biswas W, and Samyudia Y. Review of existing sustainability assessment methods for Malaysian palm oil production. Procedia CIRP 2015;26:13-18.

[21] FAO (Food Agric. Organ.). Crops. FAOSTAT Statistical Database, Rome. 2019. Available from: http://www.fao.org/faostat/en/#data/QC [Accessed: 4 March 2020]

[22] Phalan B, Bertzky M, Butchart SH, Donald PF, Scharlemann JP, Stattersfield AJ, et al. Crop expansion and conservation priorities in tropical countries. PLoS ONE 2013;8:51759.

[23] Barnes AD, Jochum M, Mumme S, Haneda NF, Farajallah A, et al. Consequences of tropical land use for multitrophic biodiversity and ecosystem functioning. Nat. Commun. 2014;5:5351.

[24] Uryu Y, Mott C, Foead N, et al. Deforestation, degradation, biodiversity loss and CO2 emission in Riau, Sumatra, Indonesia. Washington, DC: World Wildlife Fund (WWF); 2008.

[25] Forest Carbon Partnership Facility. Emission Reductions Program Document (ER-PD). ER Program Name and Country: Ghana Cocoa Forest REDD+ Programme (GCFRP). 2017. Available from: https://www.forestcarbonpartnership.org/system/files/documents/Ghana%20advanced%20draft%20ER-PD.pdf

[26] Stern N. The Economics of Climate Change: The Stern Review. Cambridge, UK: Cambridge University Press; 2006.

[27] Hadiyanto MC, and Soetrisnanto D. Phytoremediations of palm oil mill effluent (POME) by using aquatic plants and microalgae for biomass production. Journal of Environmental Science and Technology 2013;6:79-90.

[28] Singh RP, Ibrahim MH, Esa N, and Iliyana MS. Composting of waste from palm oil mill: A sustainable waste management practice. Reviews in Environmental Science and Bio Technology 2010;9(4):331-344.

[29] Bala JD, Lalung J, and Ismail N. Studies on the reduction of organic load from palm oil mill effluent (POME) by bacterial strains. Int. J. Recycle Org Waste Agric. 2014;4:1-10.

[30] Kamyab H, Chelliapan S, Mohd FMD, Rezania S, Tayebeh K, and Kumar A. Palm oil mill effluent as an environmental pollutant. DOI:10.5772/intechopen.75811

[31] Kamyab H, Md Din MF, Ponraj M, Keyvanfar A, Rezania S, Taib SM, and Abd Majid MZ. Isolation and screening of microalgae from agro-industrial wastewater (POME) for biomass and biodiesel sources. Desalination and Water Treatment 2016;57(60):29118-29125.

[32] Munoz R, and Guieysse B. Algal–bacterial processes for the treatment of hazardous contaminants: A review. Water Research 2006;40(15):2799-2815.

[33] Park KC, Whitney CGE, Kozera C, O'Leary SJB, and McGinn PJ. Seasonal isolation of microalgae from municipal wastewater for remediation and biofuel applications. Journal of Applied Microbiology 2015;119(1):76-87.

[34] Woittiez LS, van Wijk MT, Slingerland M, van Noordwijk M, and Giller KE. Yield gaps in oil palm: A quantitative review of contributing factors. European Journal of Agronomy 2017;83:57-77.

[35] Rhebergen T, Fairhurst T, Whitbread A, Giller KE, and Zingore S.

Yield gap analysis and entry points for improving productivity on large oil palm plantations and smallholder farms in Ghana. Agric. Syst. 2018;165:14-25.

[36] Cheng HT. Key sustainability issues in the palm oil sector. A discussion paper for multi-stakeholders consultations. 2010. Available from: www.ifc.org/palmoilstrategy [Accessed 10 July 2020]

[37] IOI Corporation Berhad. "Giving Back: Annual Report 2008". IOI Corporation; 2008.

[38] Hardon JJ, Corley RHV, and Lee CH. Breeding and selecting the oil palm. In: Abbot AJ, Atkin RK, editors. Improving Vegetatively Propagated Crops. Florida: Academic Press Inc.; 1987. pp. 64-81.

[39] Kushairi A, Tarmizi AH, Zamzuri I, Ong-Abdullah M, Samsul Kamal R, Ooi SE, Rajanaidu N. Production, performance and advances in oil palm tissue culture. International Seminar on Advances in Oil Palm Tissue Culture, 29 May 2010, Yogyakarta; 2010.

[40] Corley RHV. Illegitimacy in oil palm breeding – A review. J. Oil Palm Res. 2005;17:64-69.

[41] Ooi LC-L, Low E-TL, Abdullah MO, Nookiah R, Ting NC, Nagappan J, Manaf MAA, Chan K-L, Halim MA, Azizi N, Omar W, Murad AJ, Lakey N, Ordway JM, Favello A, Budiman MA, Van Brunt A, Beil M, Leininger MT, Jiang N, Smith SW, Brown CR, Kuek ACS, Bahrain S, Hoynes-O'Connor A, Nguyen AY, Chaudhari HG, Shah SA, Choo Y-M, Sambanthamurthi R, and Singh R. Non-tenera contamination and the economic impact of SHELL genetic testing in the Malaysian Independent Oil Palm Industry. Front. Plant Sci. 2016;7:771-781.

[42] Danso I, and Darkwah DO. Report on oil palm dura contamination and agronomic practices on Minerals Commission Sustainable Livelihood oil palm project in the Central region, Ghana. 2020. CSIR-OPRI/DI/ERR/178/2020.

[43] Donough C, Witt C, and Fairhurst T. Yield intensification in oil palm plantations through best management practice. Better Crops Plant Food 2009;93:12-14.

[44] Pauli N, Donough C, Oberthür T, Cock J, Verdooren R, Rahmadsyah AG, Indrasuara K, Lubis A., Dolong T, and Pasuquin JM. Changes in soil quality indicators under oil palm plantations following application of 'best management practices' in a four-year field trial. Agric. Ecosyst. Environ, 2014;195:98-111.

[45] Fairhurst T, and Griffiths W. Oil Palm: Best Management Practices for Yield Intensification. Penang, Malaysia: International Plant Nutrition Institute and Tropical Crop Consultants Limited; 2014. p. 180.

[46] Rajanaidu N, Kashairi A, and Mohd Din A. Monograph Oil Palm Genetic Resource. Perpustakaan Negara Malaysia; 2017. pp. 1-268.

[47] Darras KFA, Corre MD, Formaglio G, Tjoa A, Potapov A, Brambach F. Reducing fertilizer and avoiding herbicides in oil palm plantations-ecological and economic valuations. Front. For. Glob. Change 2019;10:3389.

[48] Romero HM. Guidelines for the use of precision agriculture for palm oil sustainability and competitiveness. Palmas 2008;29:1-4.

[49] Hassler E, Corre MD, Tjoa A, Utami SNH, and Veldkamp E. Soil fertility controls soil atmosphere carbon dioxide and methane fluxes in a tropical landscape converted from lowland forest to rubber and oil palm

plantations. Biogeosciences 2015;12:5831-5852.

[50] Kurniawan S, Corre, MD, Matson, AL, Schulte-Bisping H, Rahayu Utami SNH, and van Straaten O. Conversion of tropical forests to smallholder rubber and oil palm plantations impacts nutrient leaching losses and nutrient retention efficiency in highly weathered soils. Biogeosciences 2018;15:5131-5154.

[51] Proforest. Characterizing the oil palm small holder in Africa. 2014. Accessed from: https://www.proforest.net/fileadmin/uploads/proforest/Documents/Publications/baseline_study_smallholders_ghana.pdf

[52] Morgans CL, Meijaard E, Santika T, Law E, Budiharta S, et al. Evaluating the effectiveness of palm oil certification in delivering multiple sustainability objectives. Environ. Res. Lett. 2018;13:64032.

[53] Norhana AM, Ramli Z, Sum SM, and Awang AH. Sustainable palm oil certification scheme frameworks and impacts: A systematic literature review. Sustainability 2021;13(6):3263.

[54] Hutabarat S, Singerland M, Rietberg PI, and Dries L. Costs and benefits of certification of independent oil palm smallholders in Indonesia. Int. Food and Agribusiness Mgt. Ass. 2018;21(6):1-20.

[55] Carlson KM, Heilmayr R, Gibbs HK, Noojipady P, Burns DN, et al. Effect of oil palm sustainability certification on deforestation and fire in Indonesia. PNAS 2018;115:121-126.

[56] Saswattecha K, Kroeze C, Hein WJL. Assessing the environmental impact of palm oil produced in Thailand. Journal of Cleaner Production 2015;100:30-37.

[57] Loh SK, and Choo YM. Prospect, challenges and opportunities on biofuels in Malaysia. In: Pogaku R, and Sarbatly RH, editors. Advances in Biofuels. Boston, MA, USA: Springer; 2013; pp. 3-14.

[58] Yeo JYJ, How BS, Teng SY, Leong WD, Ng WPQ, Lim CH, Ngan SL, Sunarso J, and Lam HL. Synthesis of sustainable circular economy in palm oil industry using graph-theoretic method. Sustainability 2020;12:8081-9010.

[59] Ellen MacArthur Foundation. Towards the Circular Economy: Economic and Business Rationale for an Accelerated Transition. Cowes, UK: Ellen MacArthur Foundation; 2018.

Chapter 6

Palm Oil Clinker as Noise Control Materials

Zaiton Haron, Suhaida Ghalip, Khairulzan Yahya, Nadirah Darus, Herni Halim and Roslli Noor Mohamed

Abstract

Palm oil clinker (POC) is a waste from the production process of palm oil, a hard and porous materials. Many studies have focused on the effect of POC use on strength while this study discusses the ability of POC in concrete to absorb sound and its relationship with concrete properties. The study was done by replacing natural river sand in stages of 25, 50, 75 and 100 percent in a mixture of 1: 4 (cement: sand). Sound absorption coefficient (SAC), strength and physical properties affect the SAC were measured. Although POC significantly reduced the compressive strength but all specimens poses good strength more than 5 N/mm^2. An interesting result is that POC reduces interconnected porosity and total porosity when replacement is 100% but increases interconnected and total porosity when replacement is between 50 and 75%. SAC at 315 Hz was found has good relationship with percentage of POC and density. It is obtained that POC 50% yield good strength and sufficient SAC that can address the middle frequency range problem, thus can be further suggested to be used for masonry block application for noise control materials.

Keywords: Palm oil clinker, mortar, sound absorption, sustainable concrete, sound absorber

1. Introduction

Oil palm (*Elaeis guineensis* Jacq.) is one of the world's most efficient and versatile crop in the world. It is cultivated in continents of Asia, Africa and South America. In Asia, Malaysia, Indonesia and Thailand produced 91% of total palm oil worldwide [1]. In Malaysia, oil palm is planted on 5.45 million hectares, Indonesia 11.95 million hectares, while Thailand 820,000 hectare in 2020 [2]. 1 hectare oil palm plantation annually produces about 55 ton of dry matter in the form of fibrous biomass while yielding 5.5 ton of oil [3]. These include shells and fibres (**Figure 1**) [4]. Shells and fibres of palm oil are burned together in mills as fuel for firing the furnace of the mill to heat up the boiler, thus producing more waste materials, such as palm oil clinker (POC) and palm oil fuel ash (POFA) (see **Figure 2**) [1, 5]. About 1.1 ton of POC per every ton of oil produced were generated [6]. Palm oil Clinker is a large grey chunk that resembles a porous stone with irregular and flaky shape [1, 5].

Figure 3 shows a literature review search in Scopus data base and mapping results using VOSviewer software [7] show most research in relation to palm oil clinker is focused on their usage as aggregate replacement and investigation on their

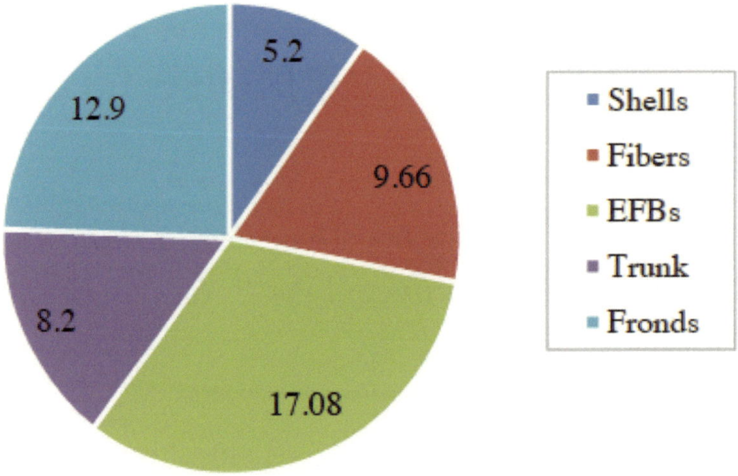

Figure 1.
Waste from oil production [4].

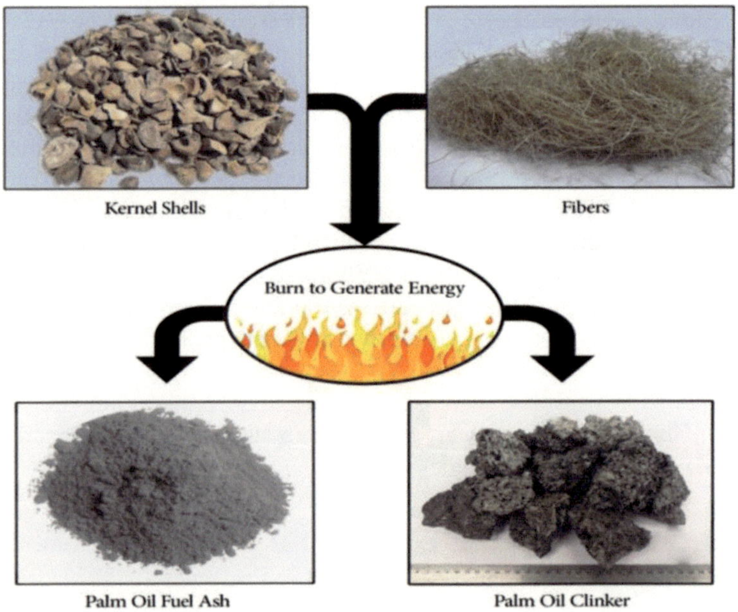

Figure 2.
POC production [1].

properties and strength for light weight concrete, mortar and sustainable concrete. Some of these research use combined dust POC and fly ash to replace cement and also used for self-compacting mortar [8]. The properties of acoustic concrete containing POC have just been initiated by [9].

POC is widely used as a lightweight aggregate due to its lightweight nature. POC is estimated to be 25% lighter than river sand and 48% lighter than crushed granite stone [10]. Thus, the density of mortar containing 100% POC sand is reduced by 7% compared to that of river sand [11]. The light nature of this POC aggregate is due to the physical properties of POC that contains micro-pores [12]. Due to the porosity of POC, concrete containing POC has lower compressive strength and tensile strength. POC also has an aggregate crushing value (ACV) of between 15 to 30 kN which is considerably lower than the values for the river sand. Therefore, there

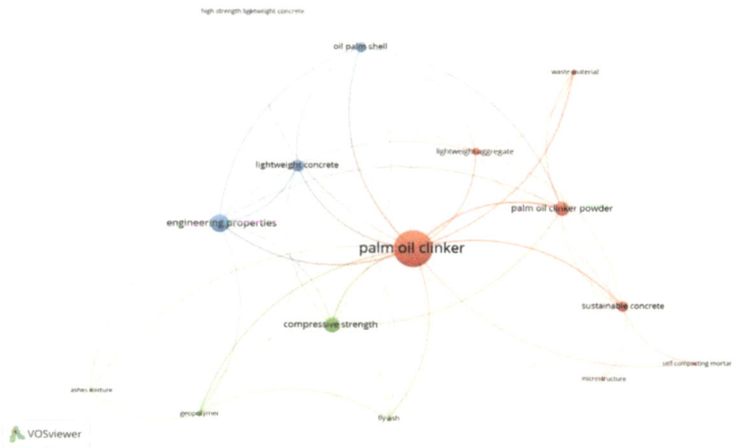

Figure 3.
VOSviewer mapping.

were researchers who coated POC to cover the macro pores of POC to slightly increase its compressive strength [12].

In fact, aggregate porosity can be utilised for the development of sound control materials. This has been stated by previous researchers where pores in aggregate is an important feature that influences the sound absorption [13–16]. For example porous-expended shale aggregate size of 12–19 mm increased the sound absorption value by 6% [14] compared with porous concrete using natural aggregate (lime stone) size 13–19 mm. This is because the extended shale aggregate has a porosity of 14.1% compared to the regular limestone aggregate of only 5.6%. Bottom ash also yield in a 13% increase in sound absorption [15] when replaced limestone aggregate with an aggregate-cement ratio of 20%. While, porous basalt stone with porosity 42% was found increased the porosity of concrete from 18 to 22% and caused an increase of sound absorption [16]. Preliminary studies of the sound absorption properties of concrete containing POC showed an increase in SAC at 1000 Hz [9].

Noise control materials are an important element component in reducing the environmental noise in urban areas such as noise barrier systems to reduce reflection from traffic noise. The reflective sound barrier system produces continuous reflections to create a "canyon" environment where users and the housing community near the road will be disturbed. According to the study, street canyon produces reverberance condition with RT30 between 1.2 to 1.4 s [17] which is a measure of annoyance. Road noise is also dominantly at 900 to 1100 Hz [18] which is in the range of human hearing sensitive between 20 Hz to 4000 Hz. Recently, it was found that middle frequency range between 200 and 630 Hz especially the 315 Hz produced high annoyance to resident, in particular on the elderly people [19].

The best noise control material is one that has porous properties because it can absorb sound and produce less reflection and at the same time avoid the 'street canyon' situation. Sound absorption is measured through a sound absorption coefficient (SAC) which indicates that the capability of material absorption between 0 to 1 in which the previous represented perfect reflection while the latter indicates perfect absorption. The nature of good sound absorption is when the value of SAC exceeds 0.35 [20].

The porosity of the aggregate causes an increase in the porosity of the concrete material and according to [16] interconnected porosity has a significant relationship with the sound absorption properties of the concrete. Further, Tie et al. [21] and Gonzalez et al. [22] stated the characteristic sound absorption properties related to the density of the material. Based on the sound absorption properties by concrete containing POC from a preliminary study by [9], it may be preferable for noise control materials. Therefore, this study aims to further investigate the potential of concrete POC as a noise control materials in alleviating the problem of noise pollution from roads and railways. In this study, further research on two main parameters related to SAC namely porosity and density and their relationship with sound absorption in POC concrete will be discussed further. By using regression analysis of the relationship between SAC, porosity and density can be established. Further, concrete POC mixtures suitable as sound absorbers can be identified.

2. Methodology

2.1 Material

POC sand as well as natural sand were utilised in this study. POC sand was used as replacement of natural sand. Palm oil clinker (POC) sand was obtained by crushing POC chuck obtained from the palm oil processing plants in Johor. The POC sand that passed 2.36 mm sieve according to ASTM C33 [23] was selected. **Figure 4** shows the grading of the POC compared with that of natural river sand. POC sand has a smaller size than the natural sand but both still well graded and can be used for the mixture. This is implying that surface area of PO is higher than that natural sand. In the SEM micrographs experiment, POC sand show craters between 14 μm to 61 μm and micro pores with diameters between 12 μm and 15 μm (**Figure 5**). POC sand has more porosity of 6% compared to natural river sand of only 3%.

2.2 Sample preparation and testing

POC was used as replacement of sand in mixture of 1:4 (one parts of cement to four parts of river sand by weight). The replacements were 25%, 50%, 75% and 100% in four mixtures using volume method. **Table 1** summarises the five mixes used including the reference sample (without replacement). During mixing, cement and fine sand aggregate were first mixed for about two minutes, followed by

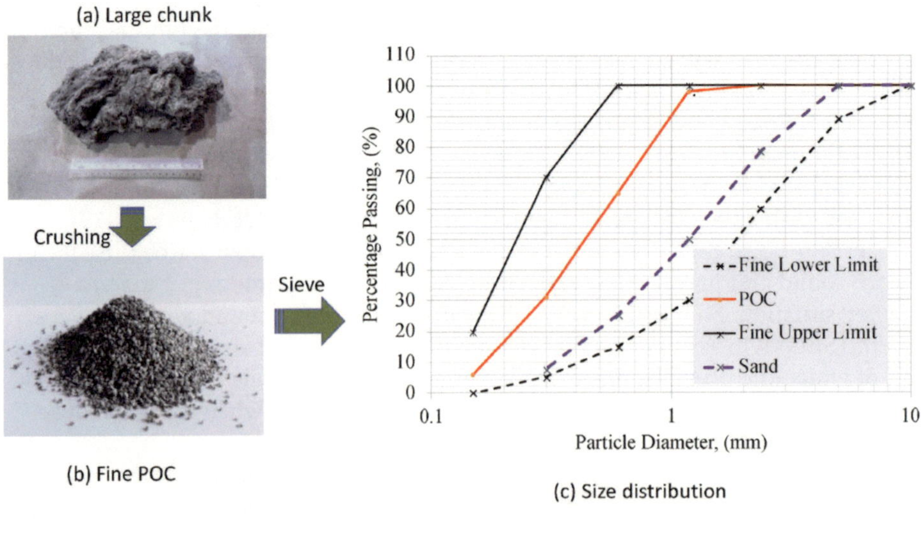

Figure 4.
Palm oil clinker (a) large chunk of POC (b) fine POC (c) size distribution.

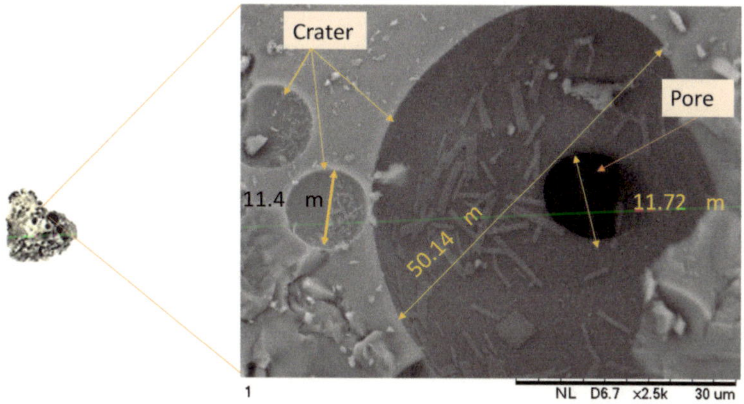

Figure 5.
SEM micrograph.

Mixture	Reference specimen	(25% POC)	(50% rep.)	(75% rep.)	(100% rep.)
Cement	2610	2610	2610	2610	2610
River Sand	8980	6740	4490	2250	0
POC sand	0	2420	4840	7260	9690
Water	1450	1450	1450	1450	1450

Table 1.
Proportion of mixtures.

Figure 6.
Mixing of materials.

another three minutes with water (**Figure 6**). Three 50x50x50 mm cubes specimens from each mix were moulded for density, porosity and compressive strength testing. Also, three 200 mm high cylinders specimens for each mixes were prepared for sound absorption test. Compaction done lightly to obtain good porosity by using the vibrating table. After demoulding of the specimens on the following day, they were all cured in water at room temperature.

The compressive test for all specimens was carried out for the concrete aged 7 and 28 days of moist curing in accordance with ASTM C109/C109M [24]. The porosity test was conducted on 50 mm diameter by 200 mm length cylindrical specimens representing all mixtures in 1st batch. Two types of porosity are measured using volume method; interconnected porosity $\varnothing_{int}$ and closed porosity, $\varnothing_{closed}$. Total porosity, $\varnothing_{total}$ then calculated by summing interconnected and closed porosity.

The interconnected porosity test was done by applying the water displacement method to measure the accessible pores in concrete specimens i.e. displacing the absorbed water in concrete. Water absorbed into the concrete by interconnected pores can be beneficial information related to pore structure, and sound absorption performance by concrete. Meanwhile, the structure of the concrete pores is very important for strength material. The interconnected porosity is determined by using Eq. (1) [25].

$$\varnothing_{int} = \left[1 - \frac{w_2 - w_1}{\rho_w v}\right] * 100 \tag{1}$$

where, w_1: submerged weight of the porous specimen underwater (kg), w_2 weight of dry porous concrete specimen (kg), ρ_w: density of water (kg/mm^3), v: volume of porous concrete specimen (mm^3).

The specimens were totally dried until no further reduction of weight. The closed porosity is determined by using Eq. (2) [25].

$$\varnothing_{closed} = \left[1 - \frac{w_3 - w_1}{\rho_w v}\right] * 100 - \varnothing_{int} \tag{2}$$

where, w_3: totally dried weight of the porous specimen (kg),

where, w: weight of dry porous concrete sample (kg), w_2: submerged weight of the porous sample underwater (kg), ρ_w: density of water (kg/mm^3), v: volume of porous concrete sample (mm^3).

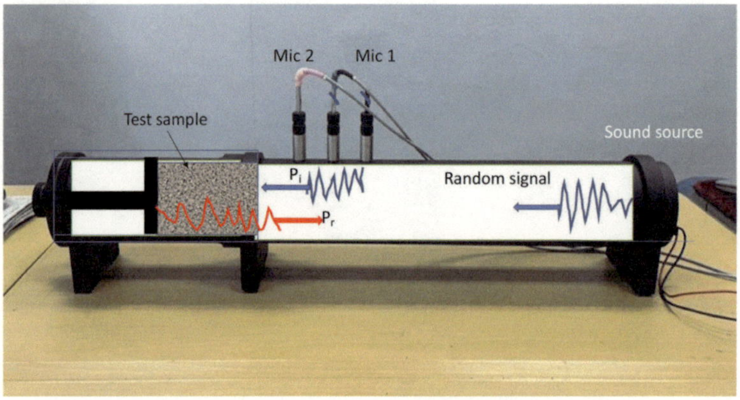

Figure 7.
Impedance tube set up for measuring specimen's sound absorption coefficient.

Sound absorption coefficient (SAC) or α of specimens were obtained by using impedance tube Type 4206-A, **Figure 7**, which was in accordance with ASTM E1050–98 [26]. SAC was determined using transfer-function method in a two-microphone method by placing the specimen at one end of the tube; involving the decomposition of a broadband stationary random signal into incident sound, P_i and reflected sound, P_r. The transfer function compensates for the possible gain and phase mismatch of the two microphones, then the measurement is repeated by interchanging the two channels. The complex reflection coefficient R is calculated by:

$$R = \left[\frac{H_1 - H_I}{H_R - H_1}\right] e^{j2k(l+s)} \tag{3}$$

Where H_1 is the frequency response function; H_1 is the frequency response function associated with the incident component; H_R is the frequency response function associated with the reflected component; j is defined as $\sqrt{-\mathbf{1}}$, k is wave number, l is the distance to the first microphone location from the specimen and s is spacing between the microphones.

The normalised surface impedance ratio of specimen, $(\frac{z}{\rho c})$ and α can be calculated;

$$\frac{z}{\rho c} = \frac{1+R}{1-R} \tag{4}$$

$$\alpha = 1 - |R|^2 \tag{5}$$

z is the surface impedance modulus of specimen which is obtained by calculating the characteristic air impedance ρc. Surface impedance implies the resistance of specimen surface to the sound energy. In this study ρc for temperature 25°C is 409 Rayls. Using this technique specimens' SAC in Eq. (5) were measured, and this was carried out by inserting the specimens in the impedance tubes and measuring the SAC absorption of the whole system.

3. Result and discussions

3.1 Porosity

Porosity is an important parameter in determining the sound absorption properties of materials. **Figure 8** shows the effect of increasing the percentage of POC in mixtures. Interconnected pores, mainly due to capillary pores [27], form channels to the other end surface that allow sound propagation, the same principle for the water penetration. While closed porosity occurs due to; (i) compaction that cause the air trap between the aggregate, (ii) POC pores and (iii) pores caused by hydrated cement. POC sand and natural river sand are covered with cement paste, thus makes closed pores in all specimens are identical. Without replacement, the interconnected porosity of specimen greater than that of 100% replacement of sand. This is also due to higher surface area of POC sand and its rough surface that makes the cement paste stick to the surface and cover micro-pores resulting in a decrease in interconnected pores. For substitution of 25–75% of natural sand results in a linear increasing relationship as shown in **Figure 9**.

The trend of changes of interconnected porosity and total porosity of mixture with POC 25–75% have very good relationship with the increment of POC

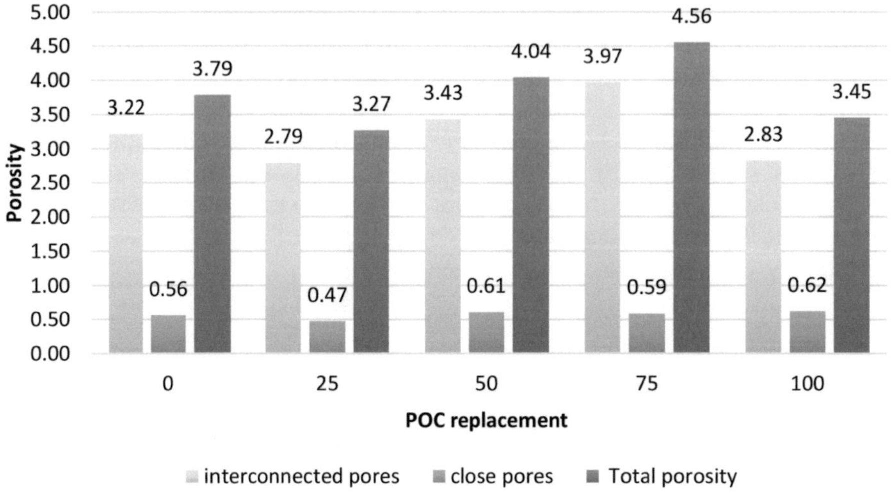

Figure 8.
Interconnected and closed porosities variation with the changes of POC percentage.

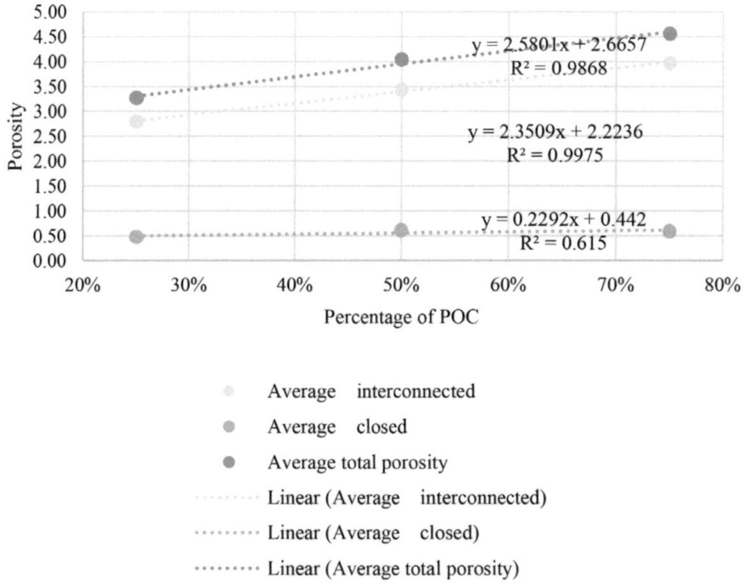

Figure 9.
Interconnected and closed porosities variation with the changes of POC percentage between 25–75%.

percentage with R^2 of 0.997 and 0.986, respectively. In this study, R^2 can be simplified as very good (>0.9), good (>0.8) [28], substantial (0.75), moderate (0.5), and weak (0.26) [29]. In summary, POC replacement between 50 and 75% increases interconnected and total porosity due to the angular shape and rough texture of POC sand, and the capillary porosity and connectivity of between capillary pores. **Figure 10** shows the irregular pores in both 100% sand and 50% POC replacement in samples. Based on these SEM micrographs, it is expected that the irregular pores for 100% sand has smaller diameters about 0.2 μm while 50% POC replacement with larger diameter of 0.6 μm. Larger pores was also created because of decrease of free water due to C-S-H bond formation and C–H gels crystallisation as surface area of POC is larger than natural sand.

(a) Natural river sand (b) 50% POC

Figure 10.
Morphology of sample 0% POC and 50% POC.

3.2 Sound absorption performance

Performance of SAC on samples tested using impedance tube test is shown in **Figure 11**. In general, all specimen curves have 2 peaks. The first peak is higher with a frequency around 300–400 Hz while the second peak is relatively low at a frequency of 1000 Hz. Anti-resonance occurs at 500 Hz with a SAC less than 0.1. For specimens containing 100% natural sand, the second resonance is somewhat unstable. However with POC replacement, the curves for all three specimens are almost the same. Also, there is no significant change in the SAC curve when the percentage of POC replacement is increased from 25–100%. However, close examination revealed that at 1000 Hz, the SAC curves for all three specimens produced almost identical SACs.

Figure 12 shows the average SAC for each sample containing 0%, 25%, 50%, 75%and 100%POC. Generally, as the % POC increase, the first dominant frequency shifts to a low frequency. It was obtained that the first dominant frequency and second dominant frequency can be described as follows;

$$f_1 = \frac{C}{\beta 4h}; \text{where } \beta = 1.25\text{–first dominant} \quad (6)$$

$$f_2 = \frac{3C}{\beta 4h}\text{f; where } \beta = 1.35. - \text{second dominant} \quad (7)$$

These findings is in opposite with the previous researches [18, 30–32], that an approximate relationship between the thickness, h, and dominant frequency f, is numerically by $f_1 = \frac{C}{4h}; \textit{ and } f_2 = \frac{3C}{4h};$ for first and second dominant frequency, respectively. C is wave velocity in the medium and h is thickness. This is showed that POC had changed the frequency for the 1st and 2nd peak. The maximum 1st peak of SAC occurs at a frequency of 315 Hz, which is good value of SAC of 0.41 to 0.52. The increase due to POC sand is about 5%. 315 Hz is a category of middle frequency range that usually causes problems in the elderly if the sound intensity exceeds the allowable threshold. The source of the noise that causes problems at 315 Hz including road traffic and train noise.

The results show the 2nd peak of maximum SAC occurs at a frequency of 1000 Hz with a good value of SAC of 0.36 when the POC is 50%. At 1000 Hz, POC sand yield in a 30% increase in SAC although POC sand has a porosity of 6% compared to the natural river sand of only 3%. This can be used to reduce the traffic noise from heavy traffic which it dominant frequency is between 800 to 1250 Hz

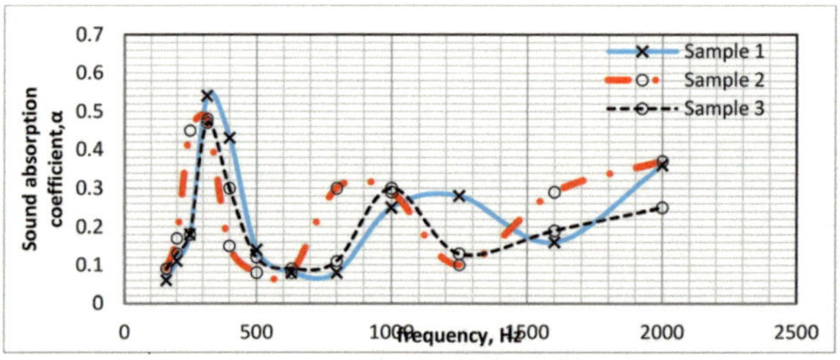

(i) Without POC

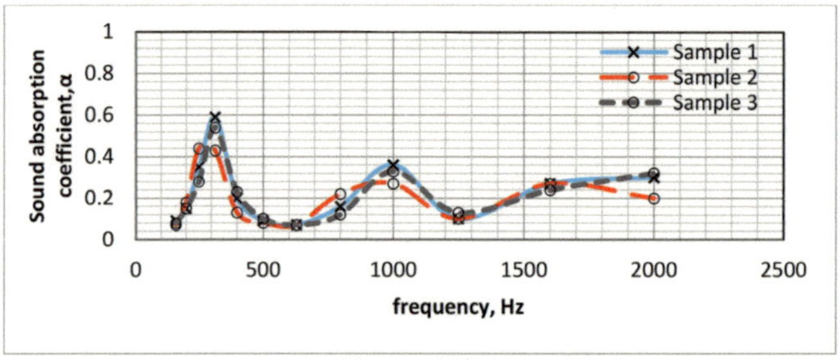

(ii) With 25% POC

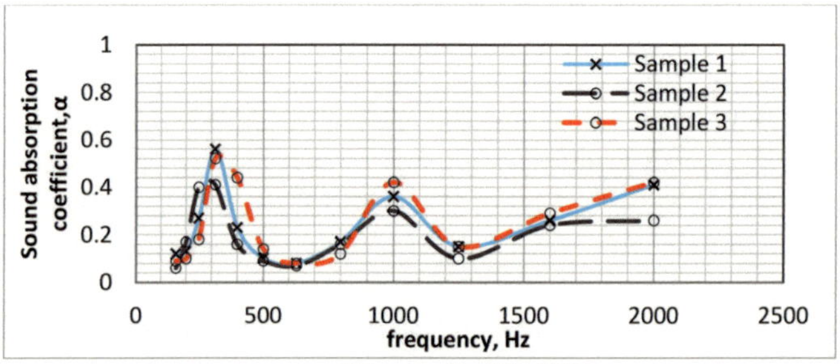

(iii) with 50% POC

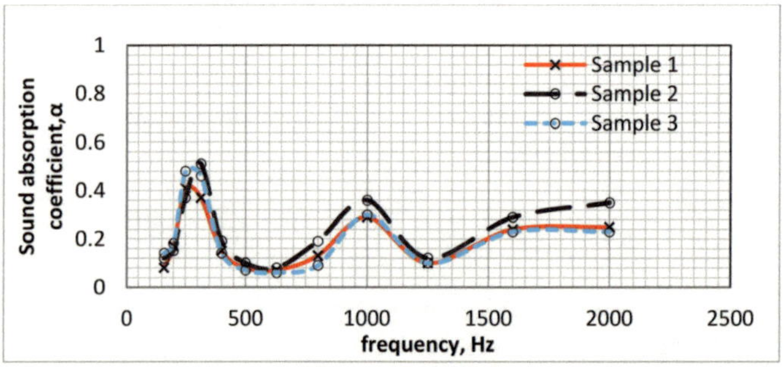

(iv) With 75% POC

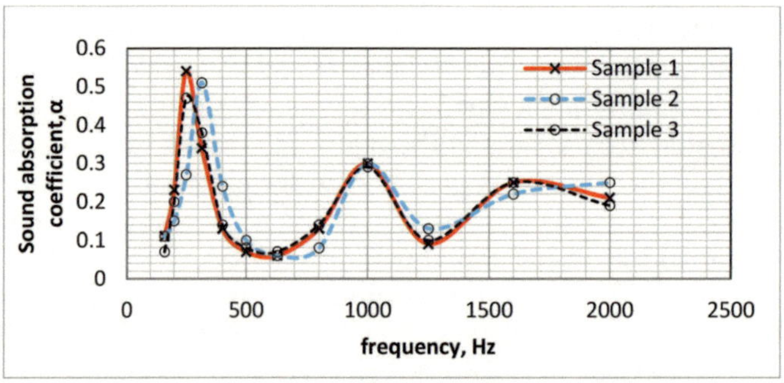

(v) With 100% POC

Figure 11.
Effect of POC percentage on SAC curve.

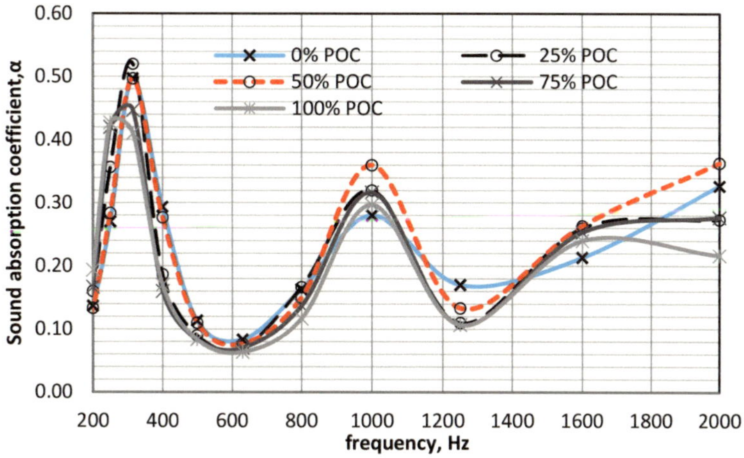

Figure 12.
Average sound absorption and POC percentage.

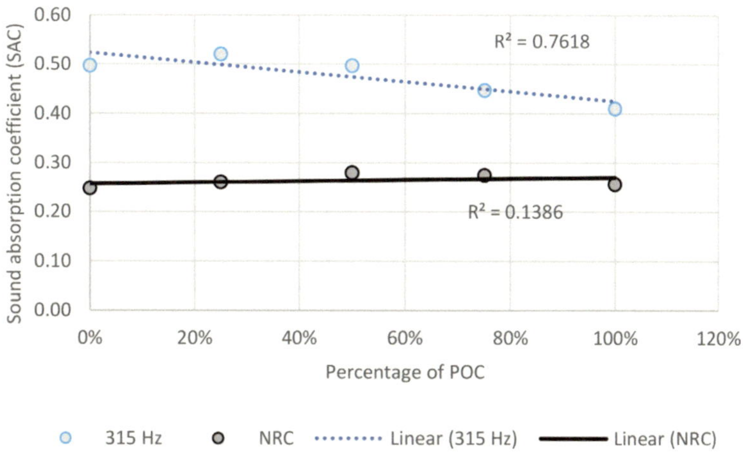

Figure 13.
Effect of POC percentage on NRC and 315 Hz.

[18]. Result from this study showed that all specimens have better result of SAC compared to that of mosaic tiles that have a very low SAC in the frequency range of 400 Hz and above of between 0.028 to 0.1 [33]. Overall increase of SAC is between 5 to 30% identical with previous studies using porous aggregate by [14–16].

3.2.1 Relationship between SAC and POC content

The average of SAC coefficient at 250 Hz, 500 Hz, 1000 Hz, 2000 Hz or noise reduction coefficient (NRC) is shown in **Figure 13**. NRC has weak linear relationship with increase of POC percentage with $R^2 = 0.14$ but surprisingly, SAC at 315 Hz has significant relationship with $R^2 = 0.78$, significant at 0.05 with the following expression;

$$\text{SAC at 315 Hz} = -0.523 - 0.098 * \text{POC} \left(R^2 = 0.761, 0.05\right) \quad (8)$$

3.2.2 Relationship between porosity and sound absorption coefficient

Figure 14 shows the relationship between interconnected porosity and the first peak, second peak SAC and NRC. All showed that interconnected porosity relatively

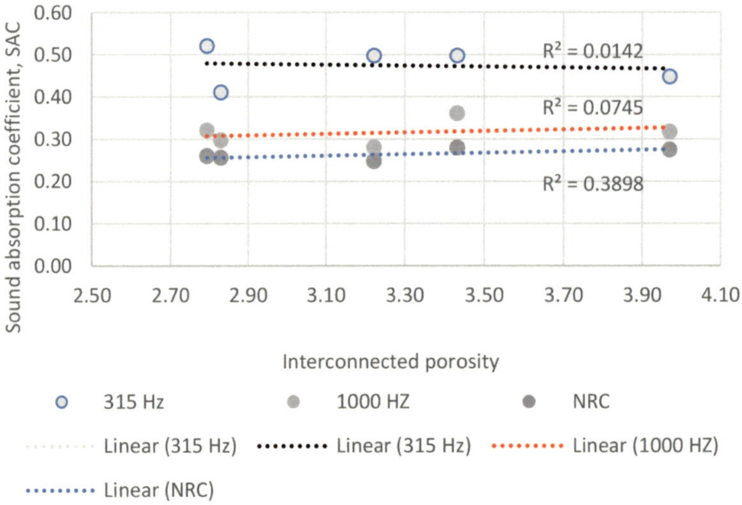

Figure 14.
Effect of interconnected porosity on NRC, 315 Hz and 1000 Hz.

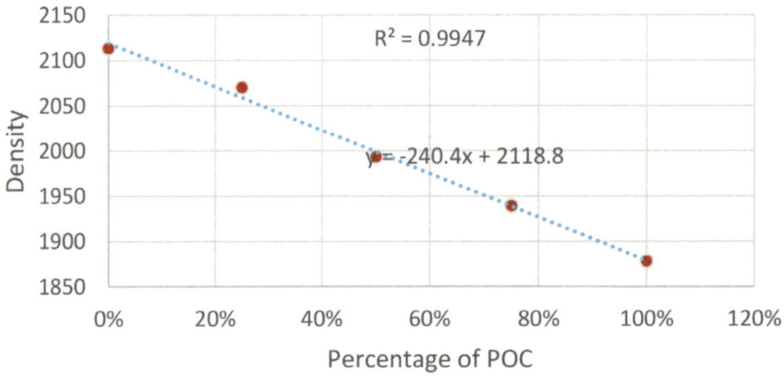

Figure 15.
Density of specimens with POC content.

has low relationship with SAC and NRC. This is indicated that interconnected porosity is not the factor influence the sound absorption. This finding is in disagreement with finding of Zhang et al. [16] that interconnected porosity has a significant relationship with the sound absorption properties of the concrete.

3.3 Density

As known, the specific gravity of the POC is lower due to micro-pores, and as a result, the replacement of natural sand with POC decreased the density of the specimens. The densities ranged from 1878 to 2070 kg/m^3 for replacement of POC percentage 25 to 100%. It can be seen that POC50–100% fell within the range of light weight concrete (between 900 to 2000 kg/m^3) while in previous work by Kanadasan et al. [34] 100% POC still resulted density more than 2000 kg/m^3. This could ideally fall under sustainable and energy efficient materials category [35].

Based on the density results, it can be observed that there is a direct relationship between the density and the percentage of POC as the density decreases linearly (**Figure 15**) as shown by Eq. (9). Such behaviour can be explained by taking into account the light weight properties of fine POC with high pore content [12], which reduces the mass per unit volume of mortar. It should also be noted that the POC itself is approximately 25% lighter than river sand [10], as mentioned in sect 1.

$$\text{Density} = -240.4 * \text{POC} + 2118.8 \left(R^2 = 0.994; p = 0.000\right) \tag{9}$$

Figure 16 shows the relationship between SAC at 315 Hz, 1000 Hz, NRC, and density of specimens containing POC of 0%, 25%, 50%, 75% and 100%. NRC has weak relation with density and this result opposite with findings from Gonzalez et al. [22] and Tzer el al. [21]. On the other hand, density has very good relationship with SAC at frequency 315 Hz with the following;

$$\text{SAC at 315 Hz} = -0.356 + 0.0004 * \text{density}\,(0.785, 0.045) \tag{10}$$

3.4 Compressive strength

Figure 17 shows the changes of compressive strength for 7 and 28 days. As can be seen, there is constant development of compressive strength within 7 and 28 days. At 28 days, the compressive strength of specimen decreases more

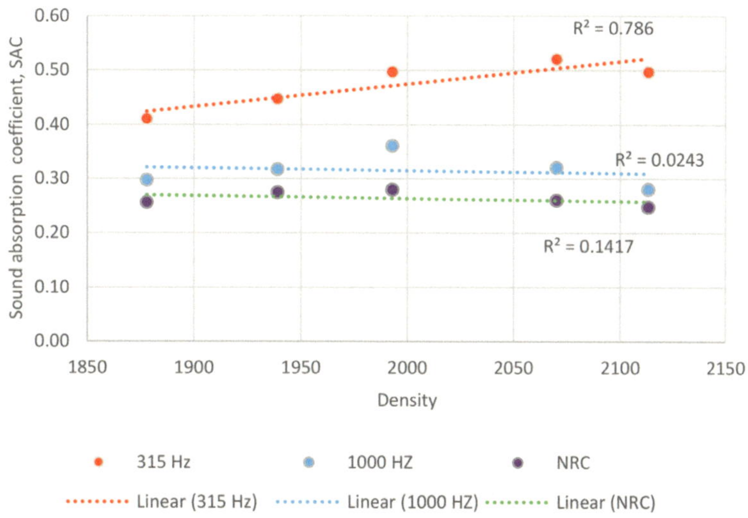

Figure 16.
Relationship between density and SAC.

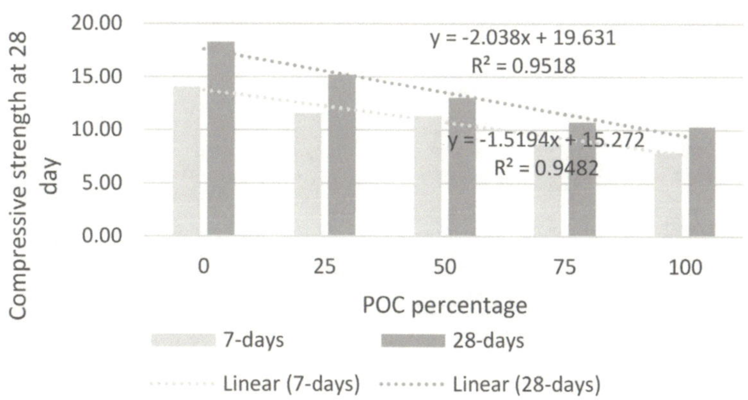

Figure 17.
Effect of POC percentage on compressive strength.

significant ($p = 0.001$) as the percentage of POC replacement increases. It is noted that all specimens meet the range for compressive strength of 5 N/mm^2 according to Specification for masonry units Part 2: Calcium silicate masonry units [36].

The relationship of compressive strength (f_c) and POC content at 28 days is as follows:

$$f_c = -0.081 * POC + 17.592\ (R^2 = 0.952; p = 0.004) \tag{11}$$

The reduction of compressive strength is due to reduction of density as explain in 3.3. The strength gradually reduced and almost 44% of strength was lost when replacement was 100%. This is due to fine POCs having micro pores in the internal structure have affected the strength capacity leading to a reduction in the strength of the mortar, this is also obtained by Kanadasan et al. [34]. Therefore, the higher the percentage of POC used then the more macro pores and this makes the mixture have even higher strength reduction. Regression analysis on compressive strength is statistically significant ($p = 0.006$) governed by density:

$$f_c = 0.034 * density - 53.57\ (R^2 = 0.938; p = 0.006) \tag{12}$$

Figure 18 shows specimen containing 50% POC failure in compression occurs quicker than specimen with 100% sand river due to porous POC contribute to lower density and lower strength which is in agreement with Eq. (11). Also, since the crushing value of aggregate (ACV) for POC is three times lower than that of ordinary [12, 34], this has given maximum effect of compressive strength compared to mixtures with river sand where the pores in POC allow greater crack spread than conventional aggregate. This type of failure similar with that of previous research [37–39].

3.5 POC concrete as noise control materials

Previous research show that reducing multiple sound reflections from building façades can be reduced by making them sound absorbing [40]. When considering NRC, the highest is given by 50% POC which dominated by SAC at 315 Hz and 1000 Hz of 0.5 and 0.36, respectively. Thus, based on this study the development of noise control materials for application on or as building façade can be based on SAC values exceeding 0.35 especially at 315 Hz and 1000 Hz. This is to address the problem of middle frequency range and dominant noise source from roads is

(a) Specimen with river sand

(b) Specimen with 50% POC sand replacement

Figure 18.
Compressive mode of failures of specimens.

Mixture	Cement = 2610 kg/m^3 River sand =4490 kg/m^3 POC sand = 4840 kg/m^3 Water = 1450 kg/m^3
Compressive strength	13.04 N/mm^2
NRC	0.3
Max SAC	0.5 (315 Hz) 0.4 (1000 Hz)

Table 2.
Noise control material suggestion.

sufficient apart from sufficient strength 5 N/mm^2. This means that concrete with a POC mixture of 50% replacing natural river sand has the potential to be used as masonry blocks that have sufficient strength and can absorb sound. This can replace application of conventional concrete having a SAC between 0.03 to 0.09 in the range of 400–4000 Hz as building façade [17]. A detailed summary of the properties and mixtures is shown in the **Table 2**.

4. Conclusion

This study focuses on the effect of the use of POC in mortar by substituting river sand by 25%, 50%, 75% and 100% on the sound absorption properties. Two important influences on sound absorption properties namely porosity types and density, and its relation with sound absorption properties have been studied. Compressive strength is also studied to obtain adequate strength. Morphology using SEM was also used to look at the microstructure of POC concrete. The following are the results obtained from this study and the conclusion:

Although POC contains micro-pores however inclusion of 100% POC reduces interconnected porosity and total porosity. The trend vice-versa when POC inclusion is between 50 and 75%. Generally, it is found that interconnected porosity has no relation with sound absorption coefficient (SAC) which contradict with previous research. However, it was found, sound absorption coefficient at 315 Hz has good relation with percentage of POC.

Further, interconnected porosity had no good association with density. Instead percentage of POC inclusion reduced density significantly and has a good linear relationship with sound absorption at 315 Hz. This study also proved that statistically there is no association between average of sound absorption at 250 Hz, 500 Hz, 1000 Hz and 2000 Hz or noise reduction coefficient with density which opposed to the previous research findings.

Finally, POC inclusion reduces compressive strength significantly however all specimens still poses good strength of more than 5 N/mm^2 fulfilling the specification standard for masonry units. It is suggested that inclusion of 50% POC produces concrete with good sound absorption at 315 Hz and 1000 Hz and may be used for alleviating the problem of noise from trains and roads. Thus, this mixture can be further suggested to be used for masonry block application for noise control materials.

Acknowledgements

This study was funded by the Research Management Centre, Universiti Teknologi Malaysia under the Research University Grant No. Q.J130000.2551.21H45.

Author details

Zaiton Haron[1*], Suhaida Ghalip[1], Khairulzan Yahya[1], Nadirah Darus[1], Herni Halim[2] and Roslli Noor Mohamed[1]

1 Department of Structures and Materials, School of Civil Engineering, Faculty of Engineering, Universiti Teknologi Malaysia, Skudai, Johor, Malaysia

2 Department of Environmental, School of Civil Engineering, Universiti Sains Malaysia, George Town, Penang, Malaysia

*Address all correspondence to: zaitonharon@utm.my

References

[1] Babalghaith AM, Koting S, Ramli Sulong NH, Karim MR, Mohammed SA, Ibrahim MR. Effect of palm oil clinker (POC) aggregate on the mechanical properties of stone mastic asphalt (SMA) mixtures. Sustainability (Switzerland). 2020;12(7).

[2] Index Mundi. No Title [Internet]. Indexmundi, 2020 Indexmundi Palm Oil Production by Country (2020). 2021 [cited 2021 Feb 20]. Available from: ttps://www.indexmundi.com/agriculture/?commodity=palm-oil,

[3] Shinoj S, Visvanathan R, Panigrahi S, Kochubabu M. Oil palm fiber (OPF) and its composites: A review. Vol. 33, Industrial Crops and Products. 2011.

[4] Aslam M, Shafigh P, Jumaat MZ. Oil-palm by-products as lightweight aggregate in concrete mixture: A review. Journal of Cleaner Production. 2016.

[5] Aslam M, Shafigh P, Jumaat MZ. Oil-palm by-products as lightweight aggregate in concrete mixture: A review. Journal of Cleaner Production. 2016.

[6] Karina M, Onggo H, Dawam Abdullah AH, Syampurwadi A. Effect of oil palm empty fruit bunch fiber on the physical and mechanical properties of fiber glass reinforced polyester resin. Journal of Biological Sciences. 2008;8(1).

[7] N.J. Van Eck LW. Manual for VOSviewer, Version 1.6.7. 2018.

[8] Kanadasan J, Razak HA. Mix design for self-compacting palm oil clinker concrete based on particle packing. Materials and Design. 2014;56:9–19.

[9] Haron Z, Yahya K, Galip NS, Zakaria MZ, Jahya Z, Darus N, et al. Acoustical Performance of Palm Oil Clinker Sand Sound Absorber. In: IOP Conference Series: Materials Science and Engineering. 2020.

[10] Mohammed BS, Foo WL, Abdullahi M. Flexural strength of palm oil clinker concrete beams. Materials and Design. 2014;53:325–331.

[11] Kanadasan J, Fauzi AFA, Razak HA, Selliah P, Subramaniam V, Yusoff S. Feasibility studies of palm oil mill waste aggregates for the construction industry. Materials. 2015;8(9):6508–6530.

[12] Abutaha F, Razak HA, Ibrahim HA. Effect of coating palm oil clinker aggregate on the engineering properties of normal grade concrete. Coatings. 2017;7(10):175.

[13] Kim HK, Lee HK. Influence of cement flow and aggregate type on the mechanical and acoustic characteristics of porous concrete. Applied Acoustics. 2010;

[14] Kim HK, Lee HK. Acoustic absorption modeling of porous concrete considering the gradation and shape of aggregates and void ratio. Journal of Sound and Vibration. 2010;

[15] Ngohpok C, Sata V, Satiennam T, Klungboonkrong P, Chindaprasirt P. Mechanical Properties, Thermal Conductivity, and Sound Absorption of Pervious Concrete Containing Recycled Concrete and Bottom Ash Aggregates. KSCE Journal of Civil Engineering. 2018;

[16] Zhang Y, Li H, Abdelhady A, Yang J. Effect of different factors on sound absorption property of porous concrete. Transportation Research Part D: Transport and Environment. 2020;

[17] Yu B, Ma H, Kang J. A hybrid model for investigating the effect of scattering from building façade on sound propagation in street canyons. Applied Sciences (Switzerland). 2019;9(2803): 1–17.

[18] Neithalath N. Development and Characterization of Acoustically Efficient Cementitious Materials. Journal of controlled release : official journal of the Controlled Release Society. 2004.

[19] Sato H, Ryu J, Kurakata K. Development of an on-site system for measuring the low-frequency noise and complainant's responses. Journal of Low Frequency Noise Vibration and Active Control. 2018;37(2).

[20] Attal E, Côté N, Shimizu T, Dubus B. Sound absorption by green walls at normal incidence: Physical analysis and optimization. Acta Acustica united with Acustica. 2019;

[21] Tie TS, Mo KH, Putra A, Loo SC, Alengaram UJ, Ling TC. Sound absorption performance of modified concrete: A review. Journal of Building Engineering. 2020.

[22] Stumpf González MA, Flach F, Reschke Pires J, Piva Kulakowski M. Acoustic absorption of mortar composites with waste material. Archives of Acoustics. 2013;38(3).

[23] ASTM. ASTM C33/C33M – 18: Standard Specification for Concrete Aggregates. ASTM. 2018.

[24] ASTM International. ASTM C109, Standard test method for compressive strength of hydraulic cement mortars (Using 2-in or [50 mm] cube specimen). 2002 p.

[25] Kim HK, Lee HK. Influence of cement flow and aggregate type on the mechanical and acoustic characteristics of porous concrete. Applied Acoustics. 2010;

[26] American society for testing and materials. ASTM E1050: Standard test method for impedance and absorption of acoustical materials using a tube, two microphones and a digital frequency analysis system. Astm international. 2019;

[27] Zhao H, Xiao Q, Huang D, Zhang S. Influence of pore structure on compressive strength of cement mortar. The Scientific World Journal. 2014; 2014.

[28] Badura M, Batog P, Drzeniecka-Osiadacz A, Modzel P. Regression methods in the calibration of low-cost sensors for ambient particulate matter measurements. SN Applied Sciences. 2019;

[29] Hussain S, Fangwei Z, Siddiqi AF, Ali Z, Shabbir MS. Structural Equation Model for evaluating factors affecting quality of social infrastructure projects. Sustainability (Switzerland). 2018;

[30] Kim HK, Lee HK. Influence of cement flow and aggregate type on the mechanical and acoustic characteristics of porous concrete. Applied Acoustics. 2010;

[31] Otaru AJ. Review on the Acoustical Properties and Characterisation Methods of Sound Absorbing Porous Structures: A Focus on Microcellular Structures Made by a Replication Casting Method. Metals and Materials International. 2020.

[32] Arenas C, Leiva C, Vilches LF, Cifuentes H. Use of co-combustion bottom ash to design an acoustic absorbing material for highway noise barriers. Waste Management. 2013;

[33] Li KM, Lai CYC. A note on noise propagation in street canyons. The Journal of the Acoustical Society of America. 2009;126(2):644–655.

[34] Kanadasan J, Razak HA, Subramaniam V. Properties of high flowable mortar containing high volume palm oil clinker (POC) fine for

eco-friendly construction. Journal of Cleaner Production. 2018;170.

[35] Darvish P, Johnson Alengaram U, Soon Poh Y, Ibrahim S, Yusoff S. Performance evaluation of palm oil clinker sand as replacement for conventional sand in geopolymer mortar. Construction and Building Materials. 2020;258.

[36] BS:EN:771-2:2011. Specification for masonry units Part 2: Calcium silicate masonry units. BSI Standards Publication. 2011;

[37] Alengaram UJ, Muhit BA Al, Jumaat MZ Bin. Utilization of oil palm kernel shell as lightweight aggregate in concrete - A review. Construction and Building Materials. 2013.

[38] Shafigh P, Jumaat MZ, Mahmud H Bin, Hamid NAA. Lightweight concrete made from crushed oil palm shell: Tensile strength and effect of initial curing on compressive strength. Construction and Building Materials. 2012;

[39] Hamada HM, Skariah Thomas B, Tayeh B, Yahaya FM, Muthusamy K, Yang J. Use of oil palm shell as an aggregate in cement concrete: A review. Construction and Building Materials. 2020.

[40] Szabó D, Paulina Š, Rychtáriková M. Impact of Building Façade Properties on Noise Levels in Street Canyons. 2018;967–970.

Chapter 7

Palm Oil Tocotrienols in Cancer Chemoprevention and Treatment

Azman Abdullah, Ahmed Atia, Nadia Salem Alrawaiq, Mohd Kamil Md Yusof and Muhammad Fadzli Rusli

Abstract

Cancer remains a worrying cause of fatality worldwide despite the advancement in medicine. Among the dietary phytonutrients, tocotrienols have been extensively studied for their bioactivity against cancer. Palm oil is a rich source of tocotrienols. The most common formulation of tocotrienols is the tocotrienol-rich fraction of palm oil (TRF). The anticancer activities of tocotrienols were once presumed due to their antioxidant and free radical scavenging properties. However, recent evidence suggested that tocotrienols are capable of demonstrating cancer-fighting properties through their influence in various signalling pathways. The selectivity of tocotrienols in killing cancer cells without affecting normal cells is indicative of their potential role in cancer treatment and prevention. Tocotrienols had proven to be particularly effective in the chemoprevention and treatment of breast, colorectal, pancreatic, prostate and liver cancers in many in vitro and in vivo animal experiments. However, the efficacy of tocotrienols in the management of human cancers are still questionable due to their poor bioavailability and lack of well-designed clinical trials. Nevertheless, due to their superb safety profiles, palm oil tocotrienols are still considered ideal candidates for future large scale clinical trials to prove their efficacy to treat or prevent cancers in humans.

Keywords: Palm oil, vitamin E, tocotrienols, TRF, chemoprevention, antioxidant

1. Introduction

Vitamin E is a fat-soluble antioxidant that comprises two major classes, the tocopherols and the tocotrienols. Structurally, tocotrienols are similar to tocopherols in which both isomers of vitamin E consist of chromanol core ring and polyprenyl side chain. However, the difference lies in the fact that the polyprenyl side chain of tocotrienols has three unsaturated bonds at 3′, 7′ and 11′ positions, which are connected to carbon number 2 of the chiral centre [1] (**Figure 1**). Similar to tocopherols, tocotrienols also have four different isomers, which is determined by the number and position of the methyl groups on the chromanol ring. They are α-tocotrienol, β-tocotrienol, γ-tocotrienol 7 and δ-tocotrienol [2] (**Figure 2**). Tocotrienol discovery was first annotated in 1961 and was described in greater detail in 1964, in which it was identified from the latex of the rubber plant, *Hevea brasiliensis*, in 1964 [3]. Tocotrienols first received critical attention in the early 1980s when it was first reported that they were able to lower cholesterol levels *in vitro* (hepatocytes of chicken and mouse) as well as *in vivo* (animal

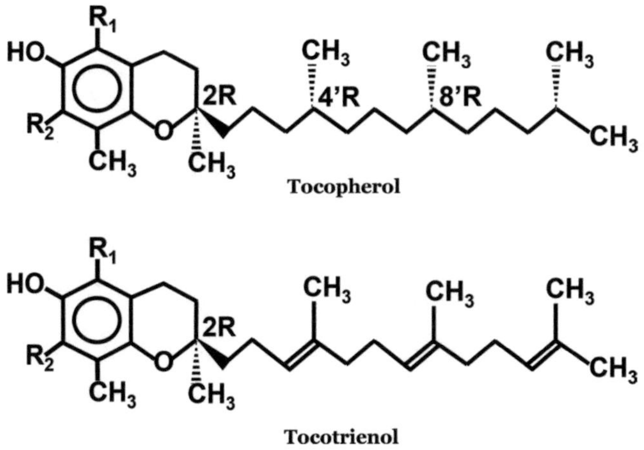

Figure 1.
The chemical structure of tocopherols and tocotrienols.

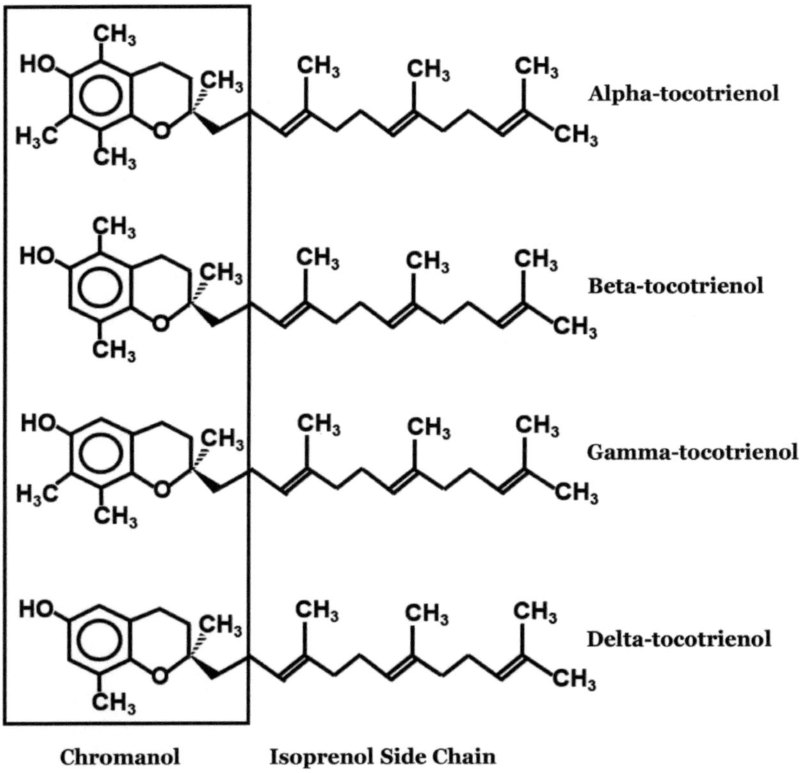

Figure 2.
Chemical structures of the four isoforms of tocotrienols.

experimentation using chicken and mouse). In the early 1990s, the anti-cancer properties of tocotrienols were also reported [3].

The concentrations of tocotrienols in various plants have been well-characterized and described. Palm oil, rice bran, wheat germ oil, coconut oil and annatto seeds have all been ascertained to contain tocotrienols in varying concentrations [4, 5]. Palm oil contains 0.66% α-tocotrienol, 0.019% β-tocotrienol, 0.71% γ-tocotrienol, 0.31% δ-tocotrienol. Rice bran contains 0.43% α-tocotrienol,

0.08% β-tocotrienol, 0.63% γ-tocotrienol, 0.04% δ-tocotrienol. Wheat germ oil 1.94% α-tocotrienol and 0.05% δ-tocotrienol, whereas in coconut oil there is 0.2% α-tocotrienol, 0.04% γ-tocotrienol, 0.76% δ-tocotrienol [6, 7]. In annatto seeds, most of the tocotrienol is δ-tocotrienol with a content ranging from 140 to 147 mg/100 g of dry seed [8]. Tocotrienols are the predominant vitamin E found in palm oil, rice bran and annatto seeds [9]. Out of all the various sources of tocotrienols, palm oil is the most sustainable, most easily available, and one of the richest sources of tocotrienols, i.e., the vitamin E isomer most extensively used in Asian countries as a health supplement. The various vitamin E components found in palm oil include α-tocopherol, α-tocotrienol, γ-tocotrienol, and δ-tocotrienol. The tocotrienol-rich fraction of palm oil (TRF) consists of 23.54% α-tocotrienol, 43.16% γ-tocotrienol, 9.83% δ-tocotrienol, and 23.5% α-tocopherol and is extracted from palm oil post-esterification and following distillation, crystallization, and chromatography processes [6, 7].

As a potent antioxidant, tocotrienols exert a variety of beneficial biological and health effects. Tocotrienols have been shown to have cardioprotective and anti-ageing effects [10]. Studies have also suggested that tocotrienols prevent atherosclerosis and improved blood vessel function in diabetics [10]. In an animal experimental model, the deleterious effects of the strong oxidant ferric nitrilotriacetate on bones were prevented by the administration of tocotrienols [11]. Previous studies had shown that tocotrienols prevented hepatotoxicity due to exposure to harmful chemicals and prevented alterations in the acidity of stomach content due to the increase in gastrin secretion [12, 13]. Tocotrienol, in combination with ascorbic acid, had been shown to increase the levels of neutrophils and lymphocytes, thus strengthening the immunity of the body [14]. In patients with rheumatoid arthritis, tocotrienols administration was associated with a reduction in joint inflammation [15].

Tocotrienols are thought to treat and prevent cancer by suppressing the secretion of angiogenic factors from carcinogenic cells, and by acting as adjuvant drugs to improve the efficacy of existing anticarcinogenic drugs due to their immune-modulating properties [16]. Delta-tocotrienol has specifically been suggested to be a viable alternative treatment for brain and lung malignancies [17]. There have been suggestions that the tocotrienols, either alone or in combination with other bioactive compounds, is beneficial for the prevention and treatment of cancer patients [18, 19]. This review will therefore focus on the role of tocotrienols in cancer chemoprevention and cancer therapeutic management. Many studies have associated the antioxidant properties of vitamin E with the alleviation of carcinogenesis [20–22]. Therefore, this review will also discuss the antioxidative properties of vitamin E (which includes the tocotrienols) in the treatment of cancers. The antioxidant activity associated with vitamin E (tocopherols and tocotrienols) are thought to be the most effective free radical scavengers, which could prevent the generation of reactive oxygen species (ROS) molecules. It has been suggested that tocopherols and tocotrienols prevented living cells from becoming malignant by quashing the attack initiated by free radicals [23, 24]. In addition to their antioxidant properties, the anticancer effects of tocotrienols had also been shown to be related to their interaction with different intracellular signalling pathways. Therefore, this review will also elaborate upon the various pathways affected by tocotrienols in different types of cancer.

2. Antioxidants and free radicals

Antioxidants can exist in either water-soluble or water-insoluble forms. Vitamin C, the prototypical representative of water-soluble antioxidant, can be found in cellular fluids. Vitamin E, which comprise the tocopherols and tocotrienols, is the typical water-insoluble antioxidant and can mostly be found in cellular membranes [1, 2]. Antioxidants can further be classified as enzymatic or non-enzymatic entities. Interaction between antioxidants and free radicals will result in the inactivation of free radicals damaging effects on our bodies. The human body defence mechanism against the actions of free radicals would usually involve the enzymatic actions of antioxidants which would result in the reduction of lipid peroxidation levels [25].

Free radicals are molecules with one or more unpaired electrons. Free radicals cause damage when they react with other molecules to find electrons to pair with their unpaired electrons. The other molecules then lost their electrons, causing them to become free radicals themselves, thus creating a chemical chain reaction of free radical production [26]. Oxidative and chemical stress in the body due to pollutants, xenobiotics and certain foods can expedite the formation of free radicals [22, 27, 28]. The free radical chain reaction may cause damage to cellular homeostasis due to its potential in causing alterations in the lipid, protein and DNA structure [27]. The damaged molecules may initiate mutation and growth of tumours. Several studies throughout the last few decades have suggested that oxidative stress plays a role in the development of many conditions, including cancer, cardiovascular disease, neurodegenerative disorders and inflammatory diseases such as arthritis [29].

The mitochondria is the chief target of free radical damage due to its preponderance to produce reactive oxygen species (ROS). The metabolic processes occurring within the mitochondria e.g., the electron transport chain may cause leakage of electrons. These electrons may in turn react with water to form ROS such as the superoxide radical, or via an indirect route the hydroxyl radical. These radicals then damage the mitochondria's DNA and proteins, and these damaged components in turn are more liable to produce ROS by-products. Thus a positive feedback loop of oxidative stress is established that, over time, can lead to the deterioration of cells and later organs and the entire body [30].

The most highly reactive free radical species are the hydroxyl radical (OH), hydrogen peroxide (H_2O_2), superoxide anion radical (O_2-), and peroxynitrite radical (ONOO-) [31–33]. The hydroxyl (OH) radical is the most active oxygen species amongst the others, whereby it could potentially cause serious biological perturbations and peroxidation of lipid-based cellular membranes. Superoxide radical (O_2-) inflicts severe damage after interaction with various biological cellular components [34]. H_2O_2 toxicity is due to the oxidation of proteins, membrane lipids and DNA by the peroxide ions [35]. Peroxynitrite radical is an oxidant and nitrating agent. Because of its oxidizing properties, peroxynitrite inflicts damage to various cellular components e.g. DNA and proteins. Peroxynitrite formation in the body has been attributed to the reaction of the superoxide free radicals with the nitric oxide free radicals [36, 37].

Lipid peroxidation is the oxidative damage caused by free radicals when they attack the polyunsaturated fatty acids (PUFAs) of the cell membranes. Lipid peroxides or lipid oxidation products are the results of this oxidation process. These fatty acid radicals are unstable molecules that react immediately with molecular oxygen, thus creating a peroxyl-fatty acid radical. This radical is also an unstable species and reacts towards other free fatty acids to produce more lipid peroxides, or cyclic peroxides if they reacted with each other. This cycle can linger on continuously as the new fatty acid radical reacts in the same way. Reactive aldehydes such as malondialdehyde (MDA) and 4-hydroxynonenal (HNE) would be the end products of lipid peroxidation, which are potentially mutagenic and carcinogenic. For example, MDA reacts with DNA to create DNA adducts [38–40].

When a free radical reacts with an inert molecule, a new radical molecule is formed, which is why the process is called a "chain reaction mechanism". The free radical reaction terminates when two radicals react and produce a non-radical species. Certain molecules within the cells can accelerate the termination of lipid peroxidation by neutralizing free radicals, thus protecting the cellular membrane from being damaged. Such molecules are known as antioxidants, of which the vitamin E tocotrienols are important examples [41].

3. Vitamin E as antioxidants and free radical scavengers

The number of methyl groups on the chroman ring influence the antioxidative effects of the various isomers of vitamin E. Delta-tocopherol and alpha-tocopherol each have one and three methyl groups respectively. The alpha isomers of vitamin E are thought to have the highest antioxidant activity, followed by the beta, gamma and delta isomers. It is generally believed that the antioxidative of tocopherol is greater than tocotrienol. However, more recent studies have indicated that tocotrienol has higher antioxidant properties [42, 43].

As a free radical scavenger, vitamin E reacts with ROS indirectly in vitro. As tocopherol has been extensively studied compared to tocotrienol, its pathway during reaction with ROS is well-documented. This reaction will eventually lead to the formation of α-tocopheroxyl radical (α-Toc•) [44]. The antioxidative reaction goes on and on, until the final product i.e., lipid peroxide is formed. α-tocopherol is able to deliver hydrogen ion to peroxyl radical (LOO•) during lipid hydroperoxide formation (LOOH). The α-tocopherol radical formed (α-Toc•) will undergo a series of reactions with the peroxyl radical (LOO•) to eventually produce inert adducts (**Figure 3**) [45, 46].

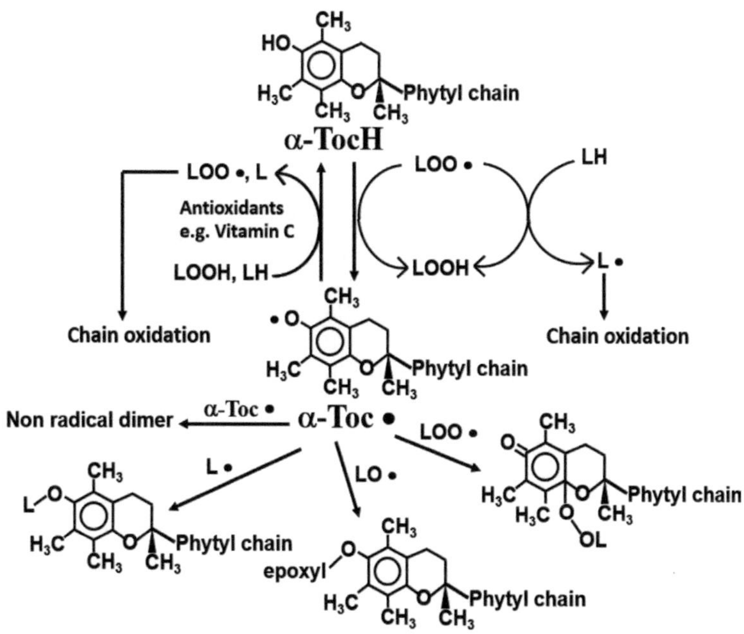

Figure 3.
Summary of whole antioxidative reaction of vitamin E. The figure was modified from Ref. [46]. The reactions involving tocotrienols are thought to be similar to that of tocopherols. Abbreviations: α-Toc•, α-tocopheroxyl radical; α-TocH, α-tocopherol; L•, lipid radical; L, low density lipoprotein; LH, unsaturated fatty acid; LO•, alkoxy radical; LOO•, lipid peroxyl radical; LOOH, lipid hydroperoxide.

Vitamin E is also known to react with other radicals as well. The reaction of vitamin E with alkoxy radical produces α-tocopherol radical (α-Toc•). A previous report had also stated that the reaction between two α-tocopherol radical (α-Toc•) will produce inert non-radical dimers [47]. In terms of inhibition of lipid peroxidation, the alpha isomer of vitamin E has the highest potency, followed by the beta and gamma isomers which are equipotent, and the least potent is the delta isomer [46].

In in vivo situations, a variety of hydrophilic and hydrophobic antioxidant components exists and reacts with various moieties. It has been shown that the hydrophilic antioxidant ascorbic acid is able to reduce α-tocopherol radical (α-Toc•) back to α-tocopherol (α-TocH) [48].

The reactions involving tocotrienols are thought to be similar to that of tocopherols. In fact, a previous study had found that consumption of tocotrienol-rich rice bran oil decreased levels of plasma phospholipid hydroperoxide (PLOOH) in rats [49]. It must be remembered however that the supramolecular structures derived from various derivatives are too complex to be simulated by in vitro models to be compared with in vivo conditions [46].

4. Antioxidant effects of tocotrienols

Similar to tocopherol, tocotrienol displays antioxidative effects by acting as a scavenger towards the chain-propagating peroxyl radical [50]. Alpha-tocotrienol is more potent as an antioxidant than alpha-tocopherol with regards to the scavenging of peroxyl radicals in liposomes [51]. Alpha-tocotrienol is disseminated more evenly in the phospholipid bilayer of the plasma membrane than alpha-tocopherol. Alpha-tocotrienol also collides with radicals in a more efficient manner, thus giving further credential to alpha-tocotrienol as a better antioxidant than alpha-tocopherol [50]. In the phospholipid bilayer of the plasma membrane, the distribution of alpha-tocotrienol is more evenly spread. The collision between free radical species and alpha-tocotrienol was also found to be more efficient. These two reasons contribute to the assumption that alpha-tocotrienol possess a better antioxidative effect compared to alpha-tocopherol [50].

Another study also concluded that alpha-tocotrienol is more potent than alpha-tocopherol, whereby tocotrienols were able to inhibit the free radical activation within monocytes [52]. In this experiment, the generation of free radicals was achieved by injecting rodents with a toxic chemical (ferric nitrilotriacetate) intraperitoneally, in which the rodents were supplemented with alpha-tocotrienol and alpha-tocopherol concurrently throughout the study. It was discovered that alpha-tocotrienol could inhibit free radical activation at a lower dose compared to alpha-tocopherol [52].

It had been suggested that many health benefits of tocotrienols have been linked to their antioxidative activity. A previous study had indicated that tocotrienols were able to decrease the oxidative stress level in subjects with hyperlipidaemia and carotid stenosis [53]. Tocotrienols have been shown to exert antiatherogenic effects through oxidative and non-oxidative pathways [54]. The reduction in DNA damage is brought about by the free radical scavenging activity of tocotrienols. Tocotrienol supplementation might provide benefit to healthy older adults by protecting their DNA from damage [55]. The reduction in DNA damage is postulated to prevent the onset of cancer [3].

5. Anticancer effects of tocotrienols

The results of various studies in cell lines and animals indicated that the damage incurred by free radicals could be prevented by antioxidants. However, the conclusions derived from observational studies in humans are mixed, due to the diverse difference in human genetics, general health and metabolic capability of our bodies [21]. Many diseases (including cancer) are strongly related to oxidative stress conditions [10]. Oxidative stress has always been associated with the process of carcinogenesis and the adverse effects of cancer management i.e., radiotherapy and chemotherapy [56]. Therefore, it was once postulated that similar to the tocopherols, the major actions of tocotrienols in preventing cancer is through their antioxidant and free radical scavenging activities [45].

Several important signalling pathways are activated by the presence of oxidative stress in the cell. These pathways could facilitate tumour development through deregulation of cellular proliferation, angiogenesis, and metastasis [56]. It was noted, however, that consumption of a diet rich in antioxidants e.g., vitamin E and vitamin C, has been beneficial in preventing cancer [57–60]. A study on the pattern of human food consumption had inferred that the intake of foods rich in antioxidants is correlated with decreased breast cancer incidence within the population [61].

A lot of research has been published since the 1980s on the anticarcinogenic effects of tocotrienols. This has led to an accumulation of numerous works of literature on the anticancer activity of tocotrienols, especially on breast cancer. Apart from breast malignancy, other cancers in which tocotrienols intervention had been postulated to be beneficial are liver, prostate, pancreatic, cervical, colorectal, and skin cancers. From the year 2000 onwards, it had become very clear that the anticancer mechanisms of tocotrienols goes way beyond their antioxidant and free radical scavenging properties. Apart from inhibiting the growth of cancer, tocotrienols are able to initiate apoptosis in cancerous cells. The anticancer effects of tocotrienols are manifested through the inhibition of angiogenesis and tumour cell progression [2, 62]. An isomer of tocotrienols i.e., delta-tocotrienol, was found to inhibit the expansion of cancerous cells and initiated the process of apoptosis significantly in an in vitro study using human colon cancer cell lines [63]. Using animal models, delta-tocotrienol was shown to reduce the incidence of colon cancer and initiated apoptosis in rats [63]. Delta-tocotrienol was able to halt the rate of proliferation of cancerous pancreatic cells without affecting normal pancreatic cells [63]. The growth of colorectal cancer cells was found to be prevented by administering tocotrienols supplementation to mice [63]. Studies using cancerous stem cells indicated that melanoma can be treated and prevented with delta-tocotrienol [64].

6. Mechanism of action of tocotrienols in cancer prevention and treatment

Tocotrienols possess several mechanisms which are deemed to be anticarcinogenic i.e., including anti-inflammatory, anti-proliferative, anti-survival, anti-angiogenic properties. These properties are mediated through the regulation of several signalling pathways which influence the process of carcinogenesis [24].

Tocotrienols affect anti-inflammatory activities by repressing COX-2, hypoxia-inducible factor-1 (HIF-1), inducible nitric oxide synthase (iNOS), prostaglandin E2 (PGE2), and interleukin (IL)-1, IL-6, and IL-8. There are convincing suggestions that the anticarcinogenic activity of tocotrienols is mediated mainly by the repression of two main transcription factors, NF-κB and STAT3 [24].

Tocotrienols were found to suppress the tumour necrosis factor-alpha (TNF-α)-induced NF-κB activation, resulting in the downregulation of its associated gene products which are responsible for the survival of cells, which includes inhibitors of apoptosis proteins (IAP)-1, −2, B-cell lymphoma 2 (Bcl-2), B- cell lymphoma-extra large (Bcl-xL), cellular FLICE-like inhibitory protein (cFLIP), X-linked inhibitor of apoptosis (XIAP), survivin etc. [65]. The regulators of STAT3 e.g., Src kinase, Janus kinase (JAK) 1and JAK2 are all inhibited by tocotrienols, which eventually resulted in the downregulation of STAT3. Tocotrienols suppress the growth of cancer cells by means of downregulation of 3- hydroxy-3-methylglutaryl coenzyme A reductase [24], cyclin-dependent kinases (CDK)-2, −4, −6, c-Jun, c-Myc, mitogen-activated protein kinases (MAPK), phosphoinositide 3-kinase (PI3K)/Akt, Wnt/β-catenin, amongst others [65]. Tocotrienols have also been reported to facilitate the arrest of cellular growth and apoptosis through activation of CDK inhibitors (p21, p27, p53), caspase-3, −8, 9, Bcl-2-associated X (Bax) and enhancing cleavage of Bid, Apaf-1, Fas, etc. [54, 66]. Furthermore, tocotrienols were found to inhibit angiogenesis in cancer through downregulation of angiopoietin-1 fibroblast growth factor (FGF), metalloproteinase-9 (MMP-9), TNF-α, and vascular endothelial growth factor (VEGF) [16, 24, 63].

7. Tocotrienols as chemopreventive and chemotherapeutic agents for various cancers

The anticarcinogenic potential of tocotrienols was first acknowledged in 1985, in which the a-tocotrienol isomer had been shown to lengthen the age of rats afflicted with tumour [3]. Subsequently, numerous studies have been performed to highlight the postulated anti-cancer properties of tocotrienols in breast, colorectal, liver, lung, pancreas, prostate and stomach carcinomas. The following paragraphs cover the details of various studies performed using tocotrienols concerning cancer chemoprevention and management (**Table 1**).

7.1 Tocotrienols in breast cancer

Tocotrienols were found to possess anti-proliferative and pro-apoptotic activities in breast cancer cells. Tocotrienols had particularly been shown to reduce the levels of cyclin D1 and Cyclin-Dependent Kinases (CDK) 2, 4 and 6 and to enhance the expression of CDK inhibitors [24]. Tocotrienols are also implicated in the suppression of the PI3K/Akt/mTOR and the Ras/Raf/MEK/ERK signalling pathways and to decrease c-Myc levels [70]. Tocotrienols were able to induce intrinsic apoptosis in breast cancer cells, accompanied by cytochrome *c* release, mitochondrial membrane depolarization, caspase activation, DNA fragmentation and poly(ADP-ribose) polymerase cleavage [72, 77]. In breast cancer cells, tocotrienols were also able to trigger the extrinsic apoptotic pathway by activation of caspase-8 [78].

Tocotrienols were also shown to exert potent anticarcinogenic effects in mammary carcinomas through reduction of proliferation by means of downregulation of the HMG-CoA reductase activity and inhibition of cholesterol synthesis, and also through the induction of oxidative stress-related mitochondrial apoptosis [68, 73].

The Human Epidermal Growth Factor Receptor 2 (HER-2) is a receptor tyrosine-protein kinase that induces breast cell proliferation. It is encoded by the erbB-2 oncogene. Overexpression of this oncogene is found in 30% of all breast tumours, thus representing an important biomarker and target for breast cancer management [118]. A previous study had shown that HER-2 receptors and tocotrienols specifically accumulate in breast cancer cell lipid raft microdomains. Furthermore, it had been discovered that tocotrienols markedly transform the composition of the lipid rafts, with subsequent disruption of their integrity, and the consequent inactivation of the associated HER-2 receptors and the downstream signalling pathways [67].

Types of Cancer	In vitro/ In vivo	Cell line Model	Mechanism of action	Reference
Breast Cancer	In vitro[c]	MDA-MB-231; T-47D; MCF-10A	↓Wnt/β- catenin	[65]
	In vitro[c]	SKBR3; BT474	↓HER2	[67]
	In vitro	MCF-7; T47D; MDA-MB-23	↓HMGR	[68]
	In vitro[c]	MCF-7 and MDA-MB-231	↑ER stress	[69]
	In vitro	+SA; MCF-7	↓c-Myc	[70]
	In vitro[c]	+SA; MCF-7; MDAMB-231	↓Akt/mTOR	[63]
	In vivo[h]	+SA mammary tumor growth in syngeneic mice	↓Akt/mTOR	[71]
	In vitro	MDA-MB-231; MCF-7	↓NF-κB	[72]
	In vitro	—	↓PI3K/Akt	[63]
	In vitro[d]	SKBR3	↓p38; ↓ERK1/2	[73]
	In vitro[d,e]	4 T1	↓IL-8; ↓VEGF	[63]
	In vitro[c]	MCF-7;MDA-MB 231	↑ER stress	[63]
	In vitro [c]	MCF-7; MDA-MB-435	↑JNK/CHOP/DR5	[63]
	In vitro[c]	MCF-7	↑NQO2	[24]
	In vitro[e]	MCF-7	↑ER-β	[74]
	In vitro[c]	+SA	↓G1-S progression	[24]
	In vitro[a,c,d]	MDA-MB-231; MCF-7	↑NRF2; ↓KEAP1	[24]
	In vitro	MDA-MB-231; MCF-7	↑JNK; ↑p38 MAPK; ↑DR5	[63]
	In vitro[d,e]	4 T1	↑IL-24; ↓VEGF; ↓IL-8	[63]
	In vivo	BALB/c mice inoculated with 4T1[i]	↑ IL-24	[63]
	In vitro	MDA-MB-231	↑ER-β	[75]
	In vitro	+SA	↑ER stress	[63]
	In vivo[e]	BALB/c mice inoculated with 4T14T1[i]	↓VEGF	[63]
	In vitro[c]	MCF-7	↓NF-κB	[65]
	In vitro[c]	+SA	↓PI3K/PDK-1/ Akt	[63]
	In vivo	Athymic mice inoculated with MCF-7[j]	↓c-Myc	[76]
	In vitro[c]	MDA-MB-231	↑Mitochondrial disruption	[77]
	In vitro	+SA	↑Caspase-8	[78]
	In vivo[e]	Sprague–Dawley rats administered with DMBA[k]	↓HMG-CoA	[79]
Colorectal Cancer	In vivo[c]	Athymic mice inoculated with HCT 116 [j]	↓NF-κB	[80]
	In vivo[e]	Balb/c nude mice inoculated with SW620 human colon cancer cell [j]	↓Wnt	[65]

Types of Cancer	In vitro/ In vivo	Cell line Model	Mechanism of action	Reference
	In vitro[d]	DLD-1	↑Caspases; ↓pAkt	[63]
	In vivo[f]	Athymic nude mice inoculated with DLD-1[j]	↑p21; ↑p27; ↑Caspase-3,-9; ↓pAkt	[63]
	In vitro[c]	SW620	↓Wnt	[65]
	In vitro	SW620	↓Wnt	[65]
	In vitro[c]	HT-29	↓β-catenin/Tcf	[81]
	In vitro[c]	HT-29	↑Bax/ Bcl-2 ratio	[82]
	In vitro[d]	DLD-1	↓HIF-1α	[16]
	In vitro[d]	DLD-1	↓PI3K/PDK/Akt	[83]
	In vivo[d]	Athymic Balb/c nude mice inoculated with DLD-1[j]	↓PI3K/PDK/Akt	[83]
	In vitro	DLD-1	↓Telomerase activity	[84]
	In vitro[e]	RKO	↑p53	[85]
Gastric Cancer	In vitro	SGC-7901	↓MMP-2; ↓MMP-9	[86]
	In vitro	SGC-7901	↓β-catenin	[87]
	In vitro[c]	SGC-7901	↑Caspase-3,-9	[88]
	In vitro	SGC-7901	↓MAPK	[89]
Lung Cancer	In vitro[b]	A549	↓Caspase-8	[90]
	In vitro	A549; U87MG	↑Caspase-8	[17]
	In vitro[d]	A549; H1650	↓Notch-1	[91]
	In vitro	A549	↓Akt; ↓HIF-2α	[92]
	In vitro[c]	H1299	↓NF-κB	[65]
	In vitro[g]	A549	↓Ras & RhoA prenylation	[93]
Pancreatic Cancer	In vitro[d]	L3.6pl; MIAPaCa-2	↓VEGF; ↓MMP-9	[63]
	In vivo[d]	L3.6pl orthotopic mice[m]	↓MMP-9	[63]
	In vitro[d]	MIA PaCa-2	↑Bax	[94]
	In vivo	LSL-Kras(G12D/+); Pdx-1-Cre[q] mice	↓MEK/ERK; ↓Akt; ↓NF-kB	[95]
	In vitro	Panc-1; MIA PaCa-2; BxPC-3	↓ErbB2	[65]
	In vitro[c]	BxPC-3; MIA PaCa-2; Panc-1	↓NF-κB	[96]
	In vivo	MIA PaCa-2 orthotopic mice[m]	↓NF-κB	[96]
	In vitro[c]	MIA PaCa-2; PC3	↓STAT3	[65]
	In vivo[c]	MIA PaCa-2 orthotopic mice[m]	↓STAT3	[65]
	In vitro	PANC-1; MIA PaCa-2; BxPC-3	↓HMGR	[97]
Prostate Cancer	In vitro	PC3	↓Fyn/HIF-1α	[98]
	In vitro	PC-3; LNCaP	↑DHS; ↑DHC	[99]
	In vivo	LNCaP mice xenograft[l]	↑DHS; ↑DHC	[99]
	In vitro	CHO-7; SRD-15	↓SREBP-2	[100]

Types of Cancer	In vitro/ In vivo	Cell line Model	Mechanism of action	Reference
	In vitro	PC-3;CRL-1435;LNCaP;CRL-1740	↓TGFβ2	[101]
	In vitro	PC-3; DU145	↓ CD133/CD44	[102]
	In vitro[d]	PCa	↓NF-κB; ↓EGFR	[103]
	In vitro[e]	LNCaP; DU145; PC-3	G0/G1 cell cycle arrest	[104]
Liver Cancer	In vitro[c]	HepG2	↑Peroxiredoxin-4	[105]
	In vivo	HCCLM3_Luc2[m]	↓Akt/ mTOR	[106]
	In vitro[c]	HepG2;C3A; SNU-38; PLC/ PRF5	↓STAT3	[107]
	In vitro	HepG2	↓TG;↓VLDL secretion	[108]
	In vitro	MH134	↑ PARP fragmentation	[109]
	In vivo	MH134[n]	↑ PARP fragmentation	[109]
	In vitro	Hep3B	↑Caspase-8, −9	[110]
	In vitro[d]	HepG2	↑S phase arrest	[111]
	In vivo	4NQO[o]	↑S phase arrest	[111]
	In vivo	DEN & AAF[p]	↓GSH-Px;↓GST; ↓GSSG-Rx	[112]
	In vivo	AAF[p]	↓GGT; ↓UDP-GT	[12]
Skin cancer	In vitro[d]	BLM; A375	↑ER-stress	[113]
	In vivo[d]	A375 xenograft[j]	↑ER-stress	[113]
	In vitro	B16	↑ERK	[114]
	In vitro	B16F10	↑Syntaxin7; ↑Vps16; ↑Vps3; ↑Vps41	[115]
	In vitro[d]	A2058; A375	↑Caspase-3; ↓CDK-4	[116]
	In vitro	C32; G361	↓NF-κB; ↑JNK	[117]

Abbreviations: ↑- Upregulation; ↓- Downregulation; AAF - 2-acetylaminofluorene; Bax - BCL2-Associated X; Bcl-2 - B- cell lymphoma 2; CDK - cyclin-dependent kinase; DEN - Diethylnitrosamine; DHC - Dihydroceramide; DHS - Dihydrosphingosine; DMBA - 7,12 Dimethylbenz(a)anthracene; DR - death-receptor; EGFR - epidermal growth factor receptor; ER - Endoplasmic reticulum; ER-β -Estrogen receptor beta; ErbB- erb-b2 receptor tyrosine kinase; ERK - extracellular signal–regulated kinase; GGT-gamma-glutamyltranspeptidase; GSSG-Rx-glutathione reductase; GSH-Px-glutathione peroxidase; GST- glutathione S-transferase; HER2 - Human epidermal growth factor receptor 2; HIF-Hypoxia- inducible factor; HMGR - HMG-CoA reductase; IL - Interleukin; JNK - Jun amino-terminal kinases; KEAP - Kelch Like ECH Associated Protein; MAPK - Mitogen-activated protein kinases; MMP - Matrix metalloproteinases; mTOR - mechanistic target of rapamycin; NF-κB - nuclear factor kappa-light-chain-enhancer of activated B cells; NQO2-NAD(P)H dehydrogenase quinone 2; pAkt - phosphorylated Akt; PARP - Poly (ADP-ribose) polymerase; PDK - 3-phosphoinositide dependent protein kinase; PI3K-Phosphatidylinositol-4,5-bisphosphate 3-kinase; Ras - rat sarcoma virus; pRb- phosphorylated retinoblastoma protein; SREBP-2 -sterol-regulatory-element-binding protein isoform 2; STAT3 - Signal transducer and activator of transcription 3; TGF-β - Transforming growth factor beta; TG - triglyceride; UDP-glucuronyltransferase; VEGF-Vascular endothelial growth factor; VPS - Vacuolar protein sorting-associated protein.
[a]*Alpha-tocotrienol.*
[b]*beta-tocotrienol.*
[c]*gamma-tocotrienol.*
[d]*delta-tocotrienol.*
[e]*Tocotrienol-rich fraction (TRF).*
[f]*Rice bran tocotrienol.*
[g]*6-O-carboxypropyl-alpha-tocotrienol.*
[h]*Oxazine derivatives of gamma- & delta-tocotrienol.*
[i]*BALB/c mice.*
[j]*athymic nude mice.*

[k]Sprague–Dawley rats.
[l]xenografts mice.
[m]Orthotopic mice.
[n]C3H/HeN mice.
[o]ddY mice.
[p]Rattus norvegicus.
[q]Mutant mice.

Table 1.
The Summary of Tocotrienols Mechanism of Action Against Different Cancers.

Almost 70% of human breast cancers are oestrogen-dependent and oestrogen receptor-positive. Tocotrienols have been postulated to facilitate the nuclear translocation of the anti-proliferative Oestrogen Receptor (ER) β and to decrease the tumorigenic ERα expression in breast cancer cells [74]. Additionally, a previous study had suggested that the delta isomers of the tocotrienols were able to induce cytotoxic effects in breast cancer cells independently of their ER status [68].

In ERβ positive breast cancer cells, tocotrienols potentially activated the ERβ signalling pathway and increased the expression of the estrogen-responsive genes such as MIC-1. This will subsequently trigger the caspase-dependent apoptosis pathway [75]. Past research has also shown that tocotrienols induce apoptosis in breast cancer cells by increasing endoplasmic reticulum (ER) stress and autophagy [69, 119]. Moreover, tocotrienols also induced apoptosis *in vitro* by inhibiting NF-κB, PI3K/Akt/mTOR, and subsequently downregulate FLIP, increase Bax/Bcl-2 ratio and PARP cleavage [63, 72].

Another finding revealed that the apoptotic property of gamma-tocotrienols in breast cancer cells is mediated by *de novo* ceramide synthesis-dependent activation of JNK/CHOP/DR5 signalling. The anti-angiogenic effect of delta-tocotrienol against breast cancer was also mediated by its ability to reduce the synthesis of pro-angiogenic markers such as IL-8 and VEGF [63]. Apart from these findings from cellular experiments, a number of studies performed in animals have also confirmed the anti-cancer activities of tocotrienols against breast cancer. For example, it has been strongly suggested that tocotrienols increased tumour latency and reduced the tumour incidence and number in 7,12 dimethylbenz(a)anthracene-induced breast cell carcinomas in rats [120]. A recent study had suggested that gamma-tocotrienol could potentially acts as proteasomes inhibitor which was able to conquer the deficiencies in growth-inhibitory or pro-apoptotic molecules in breast cancer cells. The inhibition of proteasome proteins was postulated to induce apoptosis in breast cancer cells [121].

Apart from the various tocotrienol isoforms, other formulations of tocotrienols have also been shown to be effective in the treatment and prevention of breast cancer. This is exemplified by tocotrienol oxazine derivatives, which is effective in suppressing breast carcinoma progression by inactivating the pAkt, NF-κB, COX-2, cyclin D1, CDK-2, −4, −6 pathways, and also by amplifying the levels of cell-cycle arrest proteins (p21 and p27) [71].

A relatively new formulation of tocotrienols i.e., the annatto tocotrienols (comprising 90% delta-tocotrienol and 10% gamma-tocotrienol), which is derived from the annatto fruit, has been found to suppress the progression of breast carcinoma by accelerating the process of apoptosis and senescent-like growth arrest in HER-2/neu transgenic mice [122].

The tocotrienol rich fraction derived from palm oil (TRF) has also been shown to be effective in experimental breast cancer management. Treatment of MCF7-injected athymic mice with tocotrienol-rich fraction (TRF) resulted in significant downregulation of c-Myc and CD59 glycoprotein precursor gene involved in the immune system, which positively contributed towards its anticarcinogenic effects [76]. Another tocotrienol product derived from palm oil i.e., tocomin, which consists of a mixture of naturally occurring tocotrienols and tocopherols, induced apoptosis in breast cancer cells [119].

A small-scale clinical trial was conducted in female patients with early breast cancer in order to investigate the efficacy of tocotrienols administration as a potential adjuvant with tamoxifen [123]. These patients were diagnosed with either stage I or II oestrogen receptor-positive breast cancer, and were divided into two groups. The treatment group was administered 400 mg/day tocotrienol-rich fraction (TRF) derived from palm oil, in addition to conventional tamoxifen chemotherapy. The control group was administered a placebo, together with tamoxifen. The result of the experiment indicated that the 5-year breast cancer-specific survival was 98.3% in the treatment group and 95% in the control group, while the 5-year disease-free survival was 86.7% and 83.3%, respectively. The mortality risk was 60% lower in the TRF group versus controls, however, it was not statistically significant. This was probably due to the small sample size of the experiment. Accordingly, these studies comprehensively summarize the possible benefits of tocotrienols (in particular the TRF) in the prevention and management of breast cancer.

7.2 Tocotrienols in colorectal cancer

The potential advantages of tocotrienols in the treatment of colorectal cancer have been investigated for the past several years. It was recently established that the delta isomer of tocotrienol could potentially suppress the progression of colon cancer cells by downregulating Wnt-1, β-catenin, c-jun, and cyclin D1 signals [124]. Correspondingly, the gamma isomer of tocotrienol was able to hinder the expansion of HT-29 colon carcinoma cells through the suppression of Bcl-2 and increment of Bax and caspase-3 signals [82]. The delta isomer of tocotrienol was proclaimed to decrease telomerase activity by inhibiting PKC activity in colorectal adenocarcinoma cells, which subsequently causes the downregulation of c-Myc and human telomerase reverse transcriptase (hTERT) expression, thus indicating tocotrienols' anticarcinogenic potential [82, 84].

Furthermore, the gamma isomer of tocotrienol also significantly induced apoptosis *in vitro* and *in vivo* by downregulating NF-κB and its regulated gene products [80, 82]. Additionally, previous experiments have also claimed that tocotrienols suppressed colon carcinoma growth and initiated apoptosis by inactivating pAkt, upregulating CDK inhibitors (p21, p27), caspase-3, and − 9, and deregulating β-catenin/Tcf (T-cell factor) signalling [66, 81].

Remarkably, the tocotrienol-rich fraction (TRF) derived from palm oil also initiated apoptosis in colon carcinoma cells through the activation of p53 and alteration of Bax/Bcl2 ratio [85]. Furthermore, an in vivo study has established that TRF reduced the growth of human colon cancer mice xenografts via the inhibition of Wnt pathway [125]. In addition, past research has also demonstrated that the delta isomer of tocotrienol prevented the hypoxia-induced VEGF and IL-8 overexpression and decreased the HIF-1α expression, which subsequently inhibited the angiogenesis process in colon cancer cells [83, 126]. In summary, these studies strongly validated tocotrienols as promising neutraceuticals for the management of colon carcinoma.

7.3 Tocotrienols in gastric cancer

The gamma isomer of tocotrienol was first discovered to induce intrinsic apoptosis in human gastric cancer cells through the repression of the MAPK signalling [88, 89]. The gamma isomer of tocotrienol was also found to exhibit potent anti-metastatic and anti-angiogenic activity in gastric carcinoma. Specifically, it inhibited cell migration and invasion potential, by lowering the expression of the matrix metalloproteinases MMP-2 and MMP-9 and by raising the levels of Tissue Inhibitor of Metalloproteinase-1 (TIMP-1) and TIMP-2 [86]. The gamma-tocotrienol also significantly prevented the over-expression of hypoxia-mediated HIF-1α and VEGF synthesis by alteration of the ERK signalling pathway [127]. Furthermore, gamma-tocotrienol significantly reduced the expression of VEGFR-2 in HUVEC cells grown in a conditioned medium of gastric adenocarcinoma cells [87]. Recently, a study had shown that the gamma-tocotrienol significantly inhibited human gastric cancer SGC-7901 and MGC-803 cellular growth in vitro as well as in xenograted mice through its effect on the NF-κB activity [128]. Additionally, gamma-tocotrienol was noted to intensify the anticarcinogenic activity of capecitabine in human gastric carcinoma cell lines, as well as in nude mice xenografted with human gastric carcinoma cells [129].

7.4 Tocotrienols in lung cancer

In pulmonary carcinomas, the various tocotrienol isoforms may potentially be acknowledged as the new interventional alternatives for the successful management of the disease. In order to substantiate this claim, a previous study had suggested that all of the tocotrienol isoforms convincingly induced apoptosis in lung carcinoma cells through the activation of caspase-8 and mitochondrial cytochrome c release [17, 90]. Anticarcinogenic activity towards pulmonary carcinomas exerted by delta-tocotrienol was also achieved by raising the levels of microRNA, miR-34a, which subsequently suppressed the levels of Notch-1 and its downstream targets i.e., Bcl-2, cyclin D1, and survivin [91]. 6-O-carboxypropyl-alpha-tocotrienol, which is an analogue of alpha-tocotrienol, has been found to exhibit more potent anticarcinogenic activity in A549 lung carcinomatous cells [92, 93]. In addition, past research had suggested that tocotrienols inhibited lung carcinoma cells adaptation of hypoxic conditions by inhibiting Src-induced Akt activation, as well as through reducing HIF-2a levels [92]. Tocotrienols had also been claimed to induce apoptosis in human pulmonary adenocarcinoma that contained Ras mutation [93]. Additionally, a recent study showed that delta-tocotrienol inhibited glutamine uptake in non-small-cell lung cancer (NSCLC) cell lines A549 and H1229, which subsequently triggered the inhibition of cellular proliferation and induction of apoptosis via downregulation of the mTOR pathway [130].

7.5 Tocotrienols in pancreatic cancer

The gamma-tocotrienol isomer halted pancreatic cancer progression, both in cellular and animal studies, through inhibition of NF-kB innate activation and consequently, inhibition of the gene products under NF surveillance e.g. as c-Myc, CXCR-4, cyclin D1, MMP-9 and VEGF, amongst others [96]. Besides, delta-tocotrienol exhibited its chemopreventive activity by inhibiting the growth of pancreatic intraepithelial neoplasm (PIN) in K-ras transgenic mouse model via suppression of the K-ras downstream gene products such as MEK/ERK, PI3K/Akt and NF-kB, and intensified Bax and caspase-3 activities [95].

Both cellular and animal work has ascertained that delta-tocotrienol initiated the G1 cell cycle arrest through increasing the nuclear accretion of p27 (Kip1). The tocotrienols also suppressed pancreatic carcinoma growth through the inhibition of ErbB2 and mevalonate signalling, as well as initiation of the caspase-dependent apoptotic pathway [79, 97, 131]. Delta-tocotrienol was found to trigger EGR-1 signalling, thus increased the expression of Bax and eventually turned-on the apoptotic process [94]. In a small-scale phase 1 clinical trial involving patients with pancreatic ductal neoplasia, gamma-tocotrienol administration was shown to be safe and effective in increasing the cleaved caspase-3 level in the cells of the cancerous tissue, which indicated apoptosis of the cancer cells [132]. In addition, it was recently discovered that delta-tocotrienol significantly suppressed the pancreatic ductal adenocarcinoma (PDAC) and PDAC stem-like cells metastatic process in vivo [133].

7.6 Tocotrienols in prostate cancer

Studies performed on human prostate carcinoma cells had shown that gamma-tocotrienol stimulated the process of apoptosis and autophagy in these cells. Gamma-tocotrienol supplementation in LNCaP-xenograft mouse significantly reduced tumour progression in these animals [99].

Additional studies further discovered that gamma-tocotrienol stimulated apoptosis of prostate carcinoma cells by causing under-expression of TGFβ2, NF-κB, EGFR, Fyn/HIF-1α and Id genes, concurrent with the stimulation of the JNK signalling pathway [98, 101, 103]. Fascinatingly, gamma-tocotrienol carried out its anti-invasive effect by inhibiting mesenchymal markers and significantly improving the expression of E-cadherin andβ -catenin [103]. It has also been published that gamma-tocotrienol increasingly attenuated the expression of prostate cancer stem cells (CSC) markers CD133/CD44, which are thought to be the main cause of remission in androgen-independent prostate cancer [102]. Moreover, tocotrienols were found to degrade the sterol regulatory element-binding protein-2 (SREBP-2) in prostate cancer cells, thereby reducing their viability through reduced cholesterol level [100].

Correspondingly, TRF also illustrated its chemopreventive and therapeutic activities against prostate cancer by inducing cell cycle arrest and apoptosis studies involving cell lines [104]. Furthermore, daily supplementation of TRF to transgenic adenocarcinoma mouse prostate mouse model (TRAMP) significantly suppressed tumour formation and high-grade neoplastic lesions due to the overexpression of pro-apoptotic proteins Bcl-2 antagonist of cell death (BAD), caspase-3 and CDK inhibitors (p21 and p27) [134].

7.7 Tocotrienols in liver cancer

Previous studies have substantiated the anticarcinogenic effects of tocotrienols against hepatocellular carcinoma (HCC) through its effect on a number of important signalling pathways [42, 107, 109–111]. The potency of tocotrienol isoforms against HCC cells were regarded as delta-tocotrienol > beta-tocotrienol > alpha-tocotrienol > gamma-tocotrienol [111, 120]. Gamma-tocotrienol was thought to suppress the proliferation of HCC cells by overexpression of peroxiredoxin-4 (Prx4) [105]. Moreover, in HCC, gamma-tocotrienol intervention resulted in increased expression of Bax, caspase-8 and caspase-9; raising Bid fragmentation; suppressing STAT3 and its associated proteins such as Bcl-2, Bcl-xL, survivin, cyclin D1, Mcl-1, and VEGF. All these aforementioned mechanisms gave rise to the anti-proliferative and apoptotic effects in HCC. [107, 110].

Interestingly, palm oil tocotrienols (palmvitee) supplementation surprisingly diminished the progression of hepatocarcinogenesis initiated by chemical carcinogens such as diethylnitrosamine (DEN)/2-acetylaminofluorene (AAF) in rats [12, 112].

Specific tocotrienol isomers such as gamma and delta-tocotrienols also substantially decreased the progression of HCC in mouse xenograft and orthoptic models through the nullification of the Akt/mTOR signaling pathway [106, 109]. It was pertinent to note that both the gamma and delta-tocotrienols were sequestered only in tumour sites and not in normal tissues, which implied that these tocotrienol isomers act specifically on the tumours [109]. In HCC, gamma-tocotrienols displayed lipid-lowering function, which might facilitate its anticarcinogenic activity [108, 135]. The mechanism in which tocotrienols exert its lipid-lowering ability includes the regulation of various lipogenic enzymes e.g., fatty acid synthase (FAS), sterol regulatory element-binding transcription factor 1(SREBF1), stearoyl CoA desaturase 1(SCD1) and carnitine palmitoyl transferase 1A (CPT-1A) [135].

In addition, the tocotrienols were also able to inhibit the upstream regulators of lipid homeostasis genes e.g., diacylglycerol O-acyltransferase 2 (DGAT2), apolipoprotein B (ApoB)-100, SREBP1/2 and HMGCR. These reactions will cause a reduction of TGs, cholesterol and very low-density lipoprotein (VLDL) secretion in HepG2 HCC cells [108]. Accordingly, these studies conjointly exposed the potential of tocotrienols in the prevention and treatment of HCC.

7.8 Tocotrienols in skin cancer

One of the earliest application of tocotrienols is in the protection of skin against ultraviolet rays damage. In recent years, the beneficial effects of tocotrienols against skin cancer had been acknowledged. For example, delta-tocotrienol and its peroxy dimer were discovered to be able to prevent the expansion of melanoma cells [136]. The delta isomer of tocotrienols was demonstrated to trigger G1 cell cycle arrest in A2058 and A375 human melanoma cells. This was achieved through suppression of CDK-4 and activation of caspase-3 pathway [116]. Additional studies had demonstrated that gamma-tocotrienol initiated apoptosis in melanoma cells due to inhibition of NF-κB, EGFR, Id family proteins and JNK signalling pathway [117]. The anticarcinogenic effect of δ-T3 has also been deducted by its capability to induce apoptosis via ER stress in melanoma conditions, either in cellular and animal model research [64, 113].

An animal study had found that intravenous administration of alpha-tocotrienol encapsulated within transferrin-bearing vesicles resulted in exhaustive tumour eradication in 60% of B16-F10 melanoma tumour and lengthened the durability of the mice [137]. The gamma-tocotrienol treatment prevented cell invasion in human melanoma through blockade of mesenchymal markers and reclamation of E-cadherin and γ-catenin [117]. In addition, delta-tocotrienols prevented the synthesis of melanin in B16 melanoma cells through stimulation of ERK signalling pathway, which resulted in the under-expression of melanogenesis-related proteins e.g., microphthalmia-associated transcription factor (MITF), tyrosinase and tyrosinase-related proteins, TYRP-1 and TYRP-2 [138].

Interestingly, it was also discovered that tocomin (a tocotrienol preparation derived from palm oil) facilitated the degradation of melanosomes in melanoma cells by upregulating endosome docking/fusion proteins including syntaxin7, vacuolar protein sorting-associated protein Vps16, Vps33, and Vps41 [114]. Mutually, these studies had indicated the undeniable potential of tocotrienols for the management of skin cancer.

7.9 Other cancers

Besides the types of cancers mentioned above, tocotrienols were also postulated to be able to prevent and treat other forms of carcinomas. Previous research has indicated that the gamma isomer of tocotrienols was able to initiate apoptosis in neuroblastoma SH-SY5Y cells through several postulated mechanisms [115]. Another previous research had concluded that gamma-tocotrienol negated the effect of NF-κB in squamous cell carcinoma of the tongue, which resulted in apoptosis [139]. It was also postulated that delta-tocotrienol triggered apoptosis by means of repression of STAT3 signalling in human bladder cancer [140]. Additionally, it has been concluded that alpha- and gamma-tocotrienol triggered apoptosis in HeLa cervical cancer cells through induction of Bax and IL-6 gene expression, promoting cytochrome c release, and suppression of Bcl-2, cyclin D3, CDK6, and p16 gene expression machinery [141, 142]. Likewise, gamma- and delta-tocotrienol also initiated apoptosis in cervical cancer by stimulating ER stress [143]. In multiple myeloma, gamma-tocotrienol suppressed NF-κB and STAT3 signalling and their associated proteins such as Bcl-2 and cyclin D1, thus indicating its postulated role as a potential alternative therapy for multiple myeloma sufferers [139, 144]. A recent result from a small-scale phase II clinical trial involving patients with ovarian cancers showed that the combination of tocotrienol and bevacizumab increased the rate of disease control, progression free survival and overall survival significantly [145].

8. Conclusion

The tocotrienols have been acknowledged in preclinical studies as a unique, safe and effective natural compound that could be described as a multi-target anticarcinogenic agent. Many studies had confirmed the superiority of tocotrienols in terms of health benefits compared to their chemical siblings i.e., the tocopherols. Studies after studies had arduously confirmed the ability of tocotrienols to regulate various crucial signalling pathways related to the development and growth of cancer (**Table 1**). This makes tocotrienols prime candidates to be used as a mono-chemotherapeutic anticarcinogenic agent, as well as in combination therapy with other well-known chemotherapeutic agents, with regards to cancer management. No doubt, several studies in humans have confirmed the safety and efficacy aspects of tocotrienols. However, the purported superiority of tocotrienols as potent therapeutics is severely hampered by their insolubility and low bioavailability. In order to address these issues, tocotrienols have been formulated in the form of nanoemulsions, liposomes, micelles complexes, as well as lipid conjugates with established chemotherapeutic drugs in order to elevate their bioavailability and efficacy. Nevertheless, a further appraisal is still needed for these formulations with regards to their actual efficacy. Consequently, more studies are still needed to find suitably effective tocotrienols delivery systems and to further improve the bioavailability and efficacy of tocotrienols. The limited amount of tocotrienols found in nature, and the prohibitive financial means needed for their extraction, led to the implication that tocotrienols are quite expensive to produce. Therefore, more studies are on the cards to find cheaper extraction procedures that produces a high yield of tocotrienols-rich fraction, as well as individual tocotrienols isomers depending on the situation. Furthermore, a considerable amount of research is still needed to elucidate which isomers are effective against which types of carcinomas. It is quite surprising that no major clinical trial has been attempted to explore the potential of tocotrienol in cancer. Therefore, large-scale clinical trials and high-profile translational studies are indeed crucial to establish once and for all the efficacy of tocotrienols or their isomers in the treatment of various types of cancers.

Acknowledgements

The authors would like to thank all the staff in the Department of Pharmacology Faculty of Medicine, Universiti Kebangsaan Malaysia, for their support in various aspects of academic and research work. Mohd Kamil Md Yusof and Muhammad Fadzli Rusli were partly funded by Universiti Kebangsaan Malaysia research grant GGP-2017-079.

Conflict of interest

The authors declare no conflict of interest, financial or otherwise.

Author details

Azman Abdullah[1*], Ahmed Atia[2], Nadia Salem Alrawaiq[3], Mohd Kamil Md Yusof[1] and Muhammad Fadzli Rusli[1]

1 Department of Pharmacology, Faculty of Medicine, Universiti Kebangsaan Malaysia, Kuala Lumpur, Malaysia

2 Department of Anesthesia and Intensive Care, Faculty of Medical Technology, Tripoli University, Tripoli, Libya

3 Department of Pharmacology, Faculty of Pharmacy, Sebha University, Sebha, Libya

*Address all correspondence to: azman.abdullah@ppukm.ukm.edu.my

References

[1] Baharuddin NS, Nasir NS, Nik Mohamed Kamal NNS, Saifuddin SN, Mohamad S. Vitamin E: an antioxidant with anticancer properties? Malaysian Journal of Medicine and Health Sciences. 2019; 15(Suppl 9):183-193.

[2] Abraham, A., Kattoor, A. J., Saldeen, T., & Mehta, J. L. (2019). Vitamin E and its anticancer effects. Critical reviews in food science and nutrition, 59(17), 2831-2838.

[3] Lodu W, Karwur FF. Mode of action of tocotrienol as anticancer. Journal of the Medical Sciences (Berkala ilmu Kedokteran). 2019;51(2):164-180.

[4] Sen CK, Rink C, Khanna S. Palm oil–derived natural vitamin E α-tocotrienol in brain health and disease. Journal of the American College of Nutrition. 2010;29(sup3):314S-323S.

[5] Theriault A, Chao JT, Wang QI, Gapor A, Adeli K. Tocotrienol: a review of its therapeutic potential. Clinical biochemistry. 1999;32(5):309-319.

[6] Sundram K, Sambanthamurthi R, Tan Y.A. Palm fruit chemistry and nutrition. Asia Pacific Journal of Clinical Nutrition. 2003;12(3):355-362.

[7] Liu D, Shi J, Posada LR, Kakuda Y, Xue SJ. Separating tocotrienols from palm oil by molecular distillation. Food Reviews International. 2008;24(4):376-391.

[8] Frega N, Mozzon M, Bocci F. Identification and estimation of tocotrienols in the annatto lipid fraction by gas chromatography-mass spectrometry. Journal of the American Oil Chemists' Society. 1998;75(12):1723-1727.

[9] Ahsan H, Ahad A, Siddiqui WA. A review of characterization of tocotrienols from plant oils and foods. Journal of Chemical Biology. 2015;8(2):45-59.

[10] Wong SK, Kamisah Y, Mohamed N, Muhammad N, Masbah N, Fahami NA, Mohamed IN, Shuid AN, Saad QM, Abdullah A, Mohamad NV. Potential role of tocotrienols on non-communicable diseases: a review of current evidence. Nutrients. 2020;12(1):259.

[11] Nazrun AS, Norazlina M, Norliza M, Nirwana SI. Tocotrienols as an anti-osteoporotic agent: the progress so far. J Osteoporosis Metab Disorders 2011. 4(1):1-14.

[12] Ngah WZ, Jarien Z, San MM, Marzuki A, Top GM, Shamaan NA, *et al.* Effect of tocotrienols on hepatocarcinogenesis induced by 2-acetylaminofluorene in rats. Am J Clin Nutr 1990; 53 (4suppl):1076-1081.

[13] Nafeeza MI, Kang TT. Synergistic effects of tocopherol, tocotrienol, and ubiquinone in indomethacin-induced experimental gastric lesions. Int J Vitam Nutr Res 2005; 75(2):149-55.

[14] Lestariningrum N A, Karwur F F, Martosupono M. Pengaruh vitamin E tokotrienol dan gabungannya dengan asam askorbat terhadap jenis leukosit tikus putih (*Rattus norvegicus* L.). Sains Medika 2012; 4(1):46-56.

[15] Radhakhrisnan A, Tudawe D, Chakravarthi S, Chiew GS, Haleagrahara N. Effect of γ- tocotrienol in counteracting oxidative and joint in collagen-induced arthritis inrats. Exp Ther Med 2014; 7(5):1408-1414.

[16] Montagnani Marelli M, Marzagalli M, Fontana F, Raimondi M, Moretti RM, Limonta P. Anticancer properties of tocotrienols: A review of cellular mechanisms and molecular targets. Journal of cellular physiology. 2019 Feb;234(2):1147-1164.

[17] Lim SW, Loh HS, Ting KN, Bradshaw TD, Zeenathul NA. Cytotoxicity and apoptotic activities of α-, γ- and δ-tocotrienol isomers on human cancer cells. BMC complementary and alternative medicine 2014; 14(1):1.

[18] Subramaniam S, Selvaduray KR, Radhakrishnan AK. Bioactive Compounds: Natural Defense Against Cancer? *Biomolecules*. 2019; 9(12):758.

[19] Fontana F, Raimondi M, Marzagalli M, Moretti RM, Marelli MM, Limonta P. Tocotrienols and cancer: From the state of the art to promising novel patents. Recent patents on anti-cancer drug discovery. 2019;14(1):5-18.

[20] Valko M, Rhodes C, Moncol J, Izakovic MM, Mazur M. Free radicals, metals and antioxidants in oxidative stress-induced cancer. Chemico-biological interactions. 2006;160(1):1-40.

[21] Glasauer A, Chandel NS. Targeting antioxidants for cancer therapy. Biochemical pharmacology. 2014;92(1):90-101.

[22] Rahal A, Kumar A, Singh V, Yadav B, Tiwari R, Chakraborty S, Dhama K. Oxidative stress, prooxidants, and antioxidants: the interplay. BioMed Research International. 2014;2014.

[23] Hussain T, Tan B, Yin Y, Blachier F, Tossou MCB, Rahu N. Oxidative Stress and Inflammation: What Polyphenols Can Do for Us? Oxidative Medicine and Cellular Longevity. 2016.

[24] Aggarwal V, Kashyap D, Sak K, Tuli HS, Jain A, Chaudhary A, Garg VK, Sethi G, Yerer MB. Molecular mechanisms of action of tocotrienols in cancer: Recent trends and advancements. International Journal of Molecular Sciences. 2019;20(3):656.

[25] Koruk M, Taysi S, Savas MC, Yilmaz O, Akcay F, Karakok M. Oxidative stress and enzymatic antioxidant status in patients with nonalcoholic steatohepatitis. Annals of Clinical & Laboratory Science 2004;34(1):57-62.

[26] Cui H, Kong Y, Zhang H. Oxidative stress, mitochondrial dysfunction, and aging. Journal of Signal Transduction. 2012;2012.

[27] Lobo V, Patil A, Phatak A, Chandra N. Free radicals, antioxidants and functional foods: Impact on human health. Pharmacognosy Reviews. 2010;4(8):118-126.

[28] Carocho M, Ferreira ICFR. A review on antioxidants, prooxidants and related controversy: Natural and synthetic compounds, screening and analysis methodologies and future perspectives. Food and Chemical Toxicology. 2013;(51):15-25.

[29] Pourahmad J, Salimi A, Seydi E. Role of oxygen free radicals in cancer development and treatment. In Free radicals and diseases 2016. IntechOpen.

[30] Afanas'ev I. Signaling and Damaging Functions of Free Radicals in Aging-Free Radical Theory, Hormesis, and TOR. Aging and Disease. 2010;1(2): 75-88.

[31] Jambunathan N. Determination and detection of reactive oxygen species (ROS), lipid peroxidation, and electrolyte leakage in plants. Methods Mol Biol. 2010

[32] Blokhina O, Virolainen E, Fagerstedt K V. Antioxidants, oxidative damage and oxygen deprivation stress: A review. Annals of Botany. 2003;(91):179-94.

[33] Triantaphylides C, Krischke M, Hoeberichts FA, Ksas B, Gresser G, Havaux M, Van Breusegem F, Mueller MJ. Singlet oxygen is the major

reactive oxygen species involved in photooxidative damage to plants. Plant physiology. 2008;148(2):960-968.

[34] Hazra B, Biswas S, Mandal N. Antioxidant and free radical scavenging activity of *Spondias pinnata*. BMC Complementary and Alternative Medicine. 2008;8(1):63.

[35] Patergnani S, Bouhamida E, Leo S, Pinton P, Rimessi A. Mitochondrial Oxidative Stress and "Mito-Inflammation": Actors in the Diseases. Biomedicines. 2021;9(2):216.

[36] Pacher P, Beckman JS, Liaudet L. Nitric oxide and peroxynitrite in health and disease. Physiological reviews. 2007 Jan;87(1):315-424.

[37] Szabó C, Ischiropoulos H, Radi R. Peroxynitrite: biochemistry, pathophysiology and development of therapeutics. Nature Reviews Drug discovery. 2007;6(8):662-80.

[38] Marnett LJ. Lipid peroxidation—DNA damage by malondialdehyde. Mutation Research/Fundamental and Molecular Mechanisms of Mutagenesis. 1999;424(1-2):83-95.

[39] Nimse SB, Pal D. Free radicals, natural antioxidants, and their reaction mechanisms. RSC advances. 2015;5(35):27986-28006.

[40] Ayala A, Muñoz MF, Argüelles S. Lipid peroxidation: production, metabolism, and signaling mechanisms of malondialdehyde and 4-hydroxy-2-nonenal. Oxidative medicine and cellular longevity. 2014;2014.

[41] Yoshida Y, Niki E, Noguchi N. Comparative study on the action of tocopherols and tocotrienols as antioxidant: chemical and physical effects. *Chemistry* and *Physics* of *Lipids*. 2003;123(1):63-75.

[42] Aggarwal BB, Sundaram C, Prasad S, Kannappan R. Tocotrienols, the vitamin E of the 21st century: its potential against cancer and other chronic diseases. Biochemical pharmacology. 2010;80(11):1613-1631.

[43] Maniam S, Mohamed N, Shuid AN, Soelaiman IN. Palm tocotrienol exerted better antioxidant activities in bone than α-tocopherol. Basic & Clinical Pharmacology & Toxicology. 2008;103(1):55-60.

[44] Yamauchi R. Vitamin E: mechanism of its antioxidant activity. Food Science and Technology International, Tokyo. 1997;3(4):301-309.

[45] Kamal-Eldin A, Appelqvist LÅ. The chemistry and antioxidant properties of tocopherols and tocotrienols. Lipids. 1996;31(7):671-701.

[46] Miyazawa T, Burdeos GC, Itaya M, Nakagawa K, Miyazawa T. Vitamin E: regulatory redox interactions. IUBMB life. 2019;71(4):430-441.

[47] Maulucci G, Bačić G, Bridal L, Schmidt HH, Tavitian B, Viel T, Utsumi H, Yalçın AS, De Spirito M. Imaging reactive oxygen species-induced modifications in living systems. Antioxidants & redox signaling. 2016;24(16):939-958.

[48] Vrolijk MF, Opperhuizen A, Jansen EH, Godschalk RW, Van Schooten FJ, Bast A, Haenen GR. The shifting perception on antioxidants: The case of vitamin E and β-carotene. Redox biology. 2015;4:272-278.

[49] Burdeos GC, Nakagawa K, Kimura F, Miyazawa T. Tocotrienol attenuates triglyceride accumulation in HepG2 cells and F344 rats. Lipids. 2012;47(5):471-481.

[50] Packer L, Weber S, Rimbach G. Molecular aspects of a tocotrienol antioxidant action and cell signaling. J Nutr. 2001;131:S369–S373.

[51] Serbinova E, Kagan V, Han D, et al. Free radical recycling and intramembrane mobility in the antioxidant properties of alpha-tocopherol and alpha-tocotrienol. Free Radic Biol Med. 1991;10:263-275.

[52] Ahmad N, Khalid B, Luke D, Ima Nirwana S. Tocotrienol offers better protection than tocopherol from free radical-induced damage of rat bone. Clin Exp Pharmacol Physiol. 2005;32(9):761-70.

[53] Tomeo AC, Geller M, Watkins TR, et al. Antioxidant effects of tocotrienols in patients with hyperlipidemia and carotid stenosis. Lipids. 1995;30:1179-1183.

[54] Black TM, Wang P, Colemen RA. Palm tocotrienols protect ApoE +/− mice from diet-induced atheroma formation. J Nutr. 2000;130:2420-2426.

[55] Chin SF, Hamid NA, Latiff AA, et al. Reduction of DNA damage in older healthy adults by Tri ER Tocotrienol supplementation. Nutrition. 2008;24:1-10.

[56] Reuter S, Gupta SC, Chaturvedi MM, Aggarwal BB. Oxidative stress, inflammation, and cancer: how are they linked? Free Radic Biol Med. 2010;49(11):1603-16.

[57] Klein E a, Thompson IM, Tangen CM, Crowley JJ, Lucia MS, Goodman PJ, et al. Vitamin E and the risk of prostate cancer: the Selenium and Vitamin E Cancer Prevention Trial (SELECT). JAMA. 2011;306(14):1549-56.

[58] Kline K, Lawson KA, Yu W, Sanders BG. Vitamin E and Cancer. Vitamins and Hormones. 2007;(76):435-61.

[59] Duhem N, Danhier F, Préat V. Vitamin E-based nanomedicines for anti-cancer drug delivery. Journal of Controlled Release. 2014;(182):33-44.

[60] Yun J, Mullarky E, Lu C, Bosch KN, Kavalier A, Rivera K, et al. Vitamin C selectively kills KRAS and BRAF mutant colorectal cancer cells by targeting GAPDH. Science. 2015;350(6266):1391-6.

[61] Pantavos A, Ruiter R, Feskens EF, De Keyser CE, Hofman A, Stricker BH, et al. Total dietary antioxidant capacity, individual antioxidant intake and breast cancer risk: The Rotterdam study. Int J Cancer. 2015;136(9):2178-86.

[62] Tang K, Liu J, Russell P, Clements J, Ling M-T. Gamma-Tocotrienol Induces Apoptosis in Prostate Cancer Cells by Targeting the Ang-1/Tie-2 Signalling Pathway. Int J Mol Sci. 2019;20(5):1164.

[63] Constantinou C, Charalambous C, Kanakis D. Vitamin E and cancer: an update on the emerging role of γ and δ tocotrienols. European Journal of Nutrition. 2020;59(3):845-57.

[64] Marzagalli M, Moretti RM, Messi E, Marelli MM, Fontana F, Anastasia A, et al. Targeting melanoma stem cells with the Vitamin E derivative δ-tocotrienol. Sci Rep. 2018;8(1):587.

[65] Tham SY, Loh HS, Mai CW, Fu JY. Tocotrienols modulate a life or death decision in cancers. International journal of molecular sciences. 2019;20(2):372.

[66] Shibata A, Nakagawa K, Tsuduki T, Miyazawa T. δ-Tocotrienol treatment is more effective against hypoxic tumour cells than normoxic cells: potential implications for cancer therapy. J Nutr Biochem. 2015;26(8):832-40.

[67] Alawin OA, Ahmed RA, Ibrahim BA, Briski KP, Sylvester PW. Antiproliferative effects of γ-tocotrienol are associated with lipid raft disruption

in HER2-positive human breast cancer cells. J Nutr Biochem 2016; 27: 266-77.

[68] Tiwari F, de Medina P, Caze-Subra S, Bystricky K, Balaguer P, Poirot M, *et al.* Molecular and biochemical analysis of the estrogenic and proliferative properties of vitamin E compounds. Front Oncol 2016; 5: 287.

[69] Tiwari R.V., Parajuli P., Sylvester P.W. γ-Tocotrienol-induced endoplasmic reticulum stress and autophagy act concurrently to promote breast cancer cell death, Biochem Cell Biol. 93 (2015) 306-320

[70] Parajuli P, Tiwari RV, Sylvester PW. Anti-proliferative effects of γ-tocotrienol are associated with suppression of c-Myc expression in mammary tumour cells. Cell Prolif 2015; 48: 421-35.

[71] Ananthula S., Parajuli P., Behery F.A., Alayoubi A.Y., El Sayed K.A., Nazzal S., Sylvester P.W. Oxazine derivatives of γ- and δ-tocotrienol display enhanced anticancer activity in vivo. Anticancer Res. 34 (2014) 2715-2726.

[72] Loganathan R, Selvaduray KR, Nesaretnam K, Radhakrishnan AK. Tocotrienols promote apoptosis in human breast cancer cells by inducing poly(ADP-ribose) polymerase cleavage and inhibiting nuclear factor kappa-B activity. Cell Prolif 2013; 46: 203-13.

[73] Viola V, Ciffolilli S, Legnaioli S, Piroddi M, Betti M, Mazzini F, *et al.* Mitochondrial-dependent anticancer activity of δ-tocotrienol and its synthetic derivatives in HER-2/neu overexpressing breast adenocarcinoma cells. BioFactors 2013; 39: 485-93.

[74] Comitato R, Leoni G, Canali R, Ambra R, Nesaretnam K, Virgili F. Tocotrienols activity in MCF-7 breast cancer cells: Involvement of ERβ signal transduction. Mol Nutr Food Res 2010; 54: 669-78.

[75] Comitato R., Nesaretnam K., Leoni G., Ambra R., Canali R., Bolli A, Marino MVirgili, F. A novel mechanism of natural vitamin E tocotrienol activity: involvement of ERbeta signal transduction, Am J Physiol Endocrinol Metab. 297 (2009) E427-437.

[76] Nesaretnam K., Ambra R., Selvaduray K.R., Radhakrishnan A., Reimann K., Abdul Razak G., Virgili F. Tocotrienol-rich fraction from palm oil affects gene expression in tumors resulting from MCF-7 cell inoculation in athymic mice, Lipids. 39 (2004) 459-467.

[77] Takahashi K, Loo G. Disruption of mitochondria during tocotrienol-induced apoptosis in MDA-MB-231 human breast cancer cells. Biochem Pharmacol 2004; 67: 315-24.

[78] Shah S, Sylvester PW. Tocotrienol-induced caspase-8 activation is unrelated to death receptor apoptotic signaling in neoplastic mammary epithelial cells. Exp Biol Med 2004; 229: 745-55.

[79] Ahsan H., Ahad A., Iqbal J., Siddiqui W.A. Pharmacological potential of tocotrienols: a review, Nutr Metab (Lond). 11(2014) 52.

[80] Prasad S., Gupta S.C., Tyagi A.K., Aggarwal B.B. γ-Tocotrienol suppresses growth and sensitises human colorectal tumours to capecitabine in a nude mouse xenograft model by down-regulating multiple molecules, Br J Cancer. 115 (2016) 814-824.

[81] Xu W., Du M., Zhao Y., Wang Q., Sun W., Chen B. γ-Tocotrienol inhibits cell viability through suppression of β-catenin/Tcf signaling in human colon carcinoma HT-29 cells, J Nutr Biochem. 23 (2012) 800-807.

[82] Xu W.L., Liu J.R., Liu H.K., Qi G.Y., Sun X.R., Sun W.G., Chen B.Q. Inhibition of proliferation and induction of apoptosis by gamma-tocotrienol in human colon carcinoma HT-29 cells, Nutrition. 25 (2009) 555-566.

[83] Shibata A., Nakagawa K., Sookwong P., Tsuduki T., Oikawa S., Miyazawa T. Tumor anti-angiogenic effect and mechanism of action of delta-tocotrienol, Biochem Pharmacol. 76 (2008) 330-339.

[84] Eitsuka T., Nakagawa K., Miyazawa T. Down-regulation of telomerase activity in DLD-1 human colorectal adenocarcinoma cells by tocotrienol, Biochem Biophys Res Commun. 348 (2006) 170-175.

[85] Agarwal M.K., Agarwal M.L., Athar M., Gupta S. Tocotrienol-rich fraction of palm oil activates p53, modulates Bax/Bcl2 ratio and induces apoptosis independent of cell cycle association, Cell Cycle. 2 (2004) 205-211.

[86] Liu H-K, Wang Q, Li Y, Sun W-G, Liu J-R, Yang Y-M, *et al.* Inhibitory effects of γ-tocotrienol on invasion and metastasis of human gastric adenocarcinoma SGC-7901 cells. J Nutr Biochem 2010; 21: 206-13.

[87] Li Y, Sun W-G, Liu H-K, Qi G-Y, Wang Q, Sun X-R, *et al.* γ-Tocotrienol inhibits angiogenesis of human umbilical vein endothelial cell induced by cancer cell. J Nutr Biochem 2011; 22: 1127-36.

[88] Sun W, Xu W, Liu H, Liu J, Wang Q, Zhou J, *et al.* γ-Tocotrienol induces mitochondria-mediated apoptosis in human gastric adenocarcinoma SGC-7901 cells. J Nutr Biochem 2009; 20: 276-84.

[89] Sun W, Wang Q, Chen B, Liu J, Liu H, Xu W. Gamma-tocotrienol-induced apoptosis in human gastric cancer SGC-7901 cells is associated with a suppression in mitogen-activated protein kinase signalling. Br J Nutr 2008; 99: 1247-54.

[90] Lim S.W., Loh H.S., Ting K.N., Bradshaw T.D., Zeenathul N.A. Antiproliferation and induction of caspase-8-dependent mitochondria-mediated apoptosis by β tocotrienol in human lung and brain cancer cell lines. Biomed Pharmacother. 68 (2014) 1105-1115.

[91] Ji X., Wang Z., Geamanu A., Goja A., Sarkar F.H., Gupta S.V. Delta-tocotrienol suppresses Notch-1 pathway by upregulating miR-34a in non-small cell lung cancer cells. Int J Cancer. 131 (2012) 2668-2677.

[92] Kashiwagi K., Harada K., Yano Y., Kumadaki I., Hagiwara K., Takebayashi J., Kido W., Virgona N., Yano T. A redox-silent analogue of tocotrienol inhibits hypoxic adaptation of lung cancer cells, Biochem Biophys Res Commun. 365 (2008) 875- 881.

[93] Yano Y., Satoh H., Fukumoto K., Kumadaki I., Ichikawa T., Yamada K., Hagiwara K., Yano T. Induction of cytotoxicity in human lung adenocarcinoma cells by 6-O-carboxypropyl-alpha-tocotrienol, a redox-silent derivative of alpha-tocotrienol, Int J Cancer. 115 (2005) 839-846.

[94] Wang C., Husain K., Zhang A., Centeno B.A., Chen D.T., Tong Z., Sebti S.M., Malafa M.P. EGR-1/Bax pathway plays a role in vitamin E δ-tocotrienol-induced apoptosis in pancreatic cancer cells, J Nutr Biochem. 26 (2015) 797-807.

[95] Husain K., Centeno B.A., Chen D.T., Fulp W.J., Perez M., Zhang Lee G., Luetteke N., Hingorani S.R., Sebti S.M., Malafa M.P. Prolonged survival and delayed progression of pancreatic intraepithelial neoplasia in LSL

KrasG12D/+;Pdx-1-Cre mice by vitamin E δ-tocotrienol, Carcinogenesis. 34 (2013) 858-863.

[96] Kunnumakkara A.B., Sung B., Ravindran J., Diagaradjane P., Deorukhkar A., Dey S., Koca C., Yadav V.R., Tong Z., Gelovani J.G., Guha S., Krishnan S., Aggarwal B.B., {Gamma}-tocotrienol inhibits pancreatic tumors and sensitizes them to gemcitabine treatment by modulating the inflammatory microenvironment, Cancer Res. 70 (2010) 8695-8705.

[97] Hussein D., Mo H. d-δ-Tocotrienol-mediated suppression of the proliferation of human PANC-1, MIA PaCa-2, and BxPC-3 pancreatic carcinoma cells, Pancreas. 38 (2009) e124-e136.

[98] Shiozawa N., Sugahara R., Namiki K., Sato C., Ando A., Sato A., Virgona N., Yano T., Inhibitory effect of a redox-silent analogue of tocotrienol on hypoxia adaptation in prostate cancer cells, Anticancer Drugs. 28 (2017) 289-297.

[99] Jiang Q., Rao X., Kim C.Y., Freiser H., Zhang Q., Jiang Z., Li G. Gamma tocotrienol induces apoptosis and autophagy in prostate cancer cells by increasing intracellular dihydrosphingosine and dihydroceramide, Int J Cancer. 130 (2012) 685-693.

[100] Krycer J.R., Phan L., Brown A.J. A key regulator of cholesterol homoeostasis, SREBP-2, can be targeted in prostate cancer cells with natural products, Biochem J. 446 (2012) 191-201.

[101] Campbell S.E., Rudder B., Phillips R.B., Whaley S.G., Stimmel J.B., Leesnitzer L.M., Lightner J., Dessus-Babus S., Duffourc M., Stone W.L., Menter D.G., Newman R.A., Yang P., Aggarwal B.B., Krishnan K. γ-Tocotrienol induces growth arrest through a novel pathway with TGFβ2 in prostate cancer, Free Radic Biol Med.; 50(2011) 1344-1354.

[102] Luk S.U., Yap W.N., Chiu Y.T., Lee D.T., Ma S., Lee T.K., Vasireddy R.S., Wong Y.C., Ching Y.P., Nelson C., Yap Y.L., Ling M.T. Gamma-tocotrienol as an effective agent in targeting prostate cancer stem cell-like population, Int J Cancer. 128 (2011) 2182-2191.

[103] Yap W.N., Chang P.N., Han H.Y., Lee D.T., Ling M.T., Wong Y.C., Yap Y.L. Gamma-tocotrienol suppresses prostate cancer cell proliferation and invasion through multiple-signalling pathways. Br J Cancer. 99 (2008) 1832-1841.

[104] Srivastava J.K., Gupta S. Tocotrienol-rich fraction of palm oil induces cell cycle arrest and apoptosis selectively in human prostate cancer cells, Biochem Biophys Res Commun. 346 (2006) 447-453.

[105] Abdul Rahman Sazli, Jubri Z., Abdul Rahman M., Karsani S.A., Md Top A.G., Wan Ngah W.Z. Gamma-tocotrienol treatment increased peroxiredoxin-4 expression in HepG2 liver cancer cell line, BMC Complement Altern Med. (2015) 15-64.

[106] Siveen K.S., Ahn K.S., Ong T.H., Shanmugam M.K., Li F., Yap W.N., Kumar A.P., Fong C.W., Tergaonkar V., Hui K.M., Sethi G. Y-tocotrienol inhibits angiogenesis-dependent growth of human hepatocellular carcinoma through abrogation of AKT/mTOR pathway in an orthotopic mouse model, Oncotarget. 5 (2014) 1897-1911.

[107] Rajendran P., Li F., Manu K.A., Shanmugam M.K., Loo S.Y., Kumar A.P., Sethi G. γ-Tocotrienol is a novel inhibitor of constitutive and inducible STAT3 signalling pathway in human hepatocellular carcinoma: potential role as an antiproliferative, pro-apoptotic and chemosensitizing agent, Br J Pharmacol. 163 (2011) 283-298.

[108] Zaiden N., Yap W.N., Ong S., Xu C.H., Teo V.H., Chang C.P., Zhang X.W., Nesaretnam K., Shiba S., Yap Y.L. Gamma delta tocotrienols reduce hepatic triglyceride synthesis and VLDL secretion, J Atheroscler Thromb. 17 (2010) 1019- 1032.

[109] Hiura Y., Tachibana H., Arakawa R., Aoyama N., Okabe M., Sakai M., Yamada K. Specific accumulation of gamma- and delta-tocotrienols in tumor and their antitumor effect in vivo, J Nutr Biochem. 20 (2009) 607-613.

[110] Sakai M., Okabe M., Tachibana H., Yamada K. Apoptosis induction by gamma- tocotrienol in human hepatoma Hep3B cells, J Nutr Biochem. 17 (2006) 672-676.

[111] Wada S., Satomi Y., Murakoshi M., Noguchi N., Yoshikawa T., Nishino H. Tumor suppressive effects of tocotrienol in vivo and in vitro, Cancer Lett. 229 (2005); (2):181-191.

[112] Rahmat A., Ngah W.Z., Shamaan N.A.,. Gapor A., Abdul Kadir K. Long-term administration of tocotrienols and tumor-marker enzyme activities during hepatocarcinogenesis in rats, Nutrition. 9 (1993) 229-232.

[113] Montagnani Marelli M., Marzagalli M., Moretti R.M., Beretta G., Casati L., Comitato R., Gravina G.L., Festuccia C., Limonta P. Vitamin E δ-tocotrienol triggers endoplasmic reticulum stress-mediated apoptosis in human melanoma cells, Sci Rep. 6 (2016) 30502.

[114] Choi B., Heo J.H., Kwon H.J., Lee E.S., Sohn S. Tocotrienols enhance melanosome degradation through endosome docking/fusion proteins in B16F10 melanoma cells, Food Funct. 4 (2013) 1481-1488.

[115] Tan J.K., Then S.M., Mazlan M., Raja Abdul Rahman R.N., Jamal R., Wan Ngah W.Z., Gamma-tocotrienol acts as a BH3 mimetic to induce apoptosis in neuroblastoma SH-SY5Y cells. J Nutr Biochem. 31(2016) 28-37.

[116] Fernandes N.V., Guntipalli P.K., Mo H. d-δ-Tocotrienol-mediated cell cycle arrest and apoptosis in human melanoma cells, Anticancer Res. 30 (2010) 4937-4944.

[117] Chang P.N., Yap W.N., Lee D.T., Ling M.T., Wong Y.C., Yap Y.L. Evidence of gamma-tocotrienol as an apoptosis-inducing, invasion-suppressing, and chemotherapy drug-sensitizing agent in human melanoma cells, Nutr Cancer. 61 (2009) 357-366.

[118] Asif HM, Sultana S, Ahmed S, Akhtar N, Tariq M. HER-2 positive breast cancer - A mini-review. Asian Pac J Cancer Prev 2016; 17: 1609-15.

[119] Tran A.T., Ramalinga M., Kedir H., Clarke R., Kumar D. Autophagy inhibitor 3- methyladenine potentiates apoptosis induced by dietary tocotrienols in breast cancer cells, Eur J Nutr. 54 (2015) 265-272.

[120] Ling M.T., Luk S.U., Al-Ejeh F., Khanna K.K., Tocotrienol as a potential anticancer agent. Carcinogenesis. 33 (2012) 233-239.

[121] Ramdas P, Radhakrishnan AK, Abdu Sani AA, Kumari M, Anandha Rao JS, Abdul-Rahman PS. Advancing the Role of Gamma-Tocotrienol as Proteasomes Inhibitor: A Quantitative Proteomic Analysis of MDA-MB-231 Human Breast Cancer Cells. Biomolecules. 2020;10(1):19.

[122] Shen M., Chan T.H., Dou Q.P. Targeting tumor ubiquitin-proteasome pathway with polyphenols for chemosensitization, Anticancer Agents Med Chem. 12 (2012) 891-901.

[123] Nesaretnam K, Selvaduray KR, Razak GA, Veerasenan SD, Gomez PA.

Effectiveness of tocotrienols -rich fraction combined with tamoxifen in the management of women with early breast cancer: a pilot clinical trial. Breast Cancer Research. 2010;12(5):R81.

[124] Yue MA, He N, Zhang J, Li D, Liu Y, Zhang J. Experimental study on delta- tocotrienol inhibits the Wnt pathway in the colon cancer cell SW620, Wei Sheng Yan Jiu. 41 (2012) 900-904.

[125] Zhang J.S., Zhang S.J., Li Q., Liu Y.H., He N., Zhang J., Zhou P.H., Li, M., Guan T., Liu J.R. Tocotrienol-rich fraction (TRF) suppresses the growth of human colon cancer xenografts in Balb/C nude mice by the Wnt pathway, PLoS One. 10 (2015) e0122175.

[126] Shibata A, Nakagawa K, Sookwong P, Tsuduki T,Tomita S, Shirakawa H. Tocotrienol inhibits secretion of angiogenic factors from human colorectal adenocarcinoma cells by suppressing hypoxia-inducible factor- 1α. J Nutr 2008; 138(11):2136-142.

[127] Bi S, Liu JR, Li Y, Wang Q, Liu HK, Yan YG, *et al.* γ-Tocotrienol modulates the paracrine secretion of VEGF induced by cobalt(II) chloride via ERK signaling pathway in gastric adenocarcinoma SGC-7901 cell line. Toxicology 2010; 274: 27-33.

[128] Sun WG, Song RP, Wang Y, Zhang YH, Wang HX, Ge S, Liu JR, Liu LX. γ-Tocotrienol-Inhibited Cell Proliferation of Human Gastric Cancer by Regulation of Nuclear Factor-κB Activity. Journal of agricultural and food chemistry. 2018;67(1):441-51.

[129] Manu KA, Shanmugam MK, Ramachandran L, Li F, Fong CW, Kumar AP, *et al.* First evidence that γ-tocotrienol inhibits the growth of human gastric cancer and chemosensitizes it to capecitabine in a xenograft mouse model through the modulation of NF-κB pathway. Clin Cancer Res 2012; 18: 2220-9.

[130] Rajasinghe LD, Hutchings M, Gupta SV. Delta-tocotrienol modulates glutamine dependence by inhibiting ASCT2 and LAT1 transporters in non-small cell lung cancer (NSCLC) cells: A metabolomic approach. Metabolites. 2019;9(3):50.

[131] Shin-Kang S., Ramsauer V.P., Lightner J., Chakraborty K., Stone W., Campbell S., Reddy S.A., Krishnan K. Tocotrienols inhibit AKT and ERK activation and suppress pancreatic cancer cell proliferation by suppressing the ErbB2 pathway, Free Radic Biol Med. 51 (2011) 1164-1174.

[132] Springett GM, Husain K, Neuger A, Centeno B, Chen DT, Hutchinson TZ, Lush RM, Sebti S, Malafa MP. A Phase I Safety, Pharmacokinetic, and Pharmacodynamic Presurgical Trial of Vitamin E delta-tocotrienol in Patients with Pancreatic Ductal Neoplasia. EBioMedicine. 2015;2(12):1987-1995.

[133] Husain K, Centeno BA, Coppola D, Trevino J, Sebti SM, Malafa MP. δ-Tocotrienol, a natural form of vitamin E, inhibits pancreatic cancer stem-like cells and prevents pancreatic cancer metastasis. Oncotarget. 2017;8(19):31554-67.

[134] Barve A., Khor T.O., Reuhl K., Reddy B., Newmark H., Kong A.N. Mixed tocotrienols inhibit prostate carcinogenesis in TRAMP mice, Nutr Cancer. 62 (2010) 789-794.

[135] Burdeos G.C., Nakagawa K., Watanabe A., Kimura F., Miyazawa T. γ-Tocotrienol attenuates triglyceride through effect on lipogenic gene expressions in mouse hepatocellular carcinoma Hepa 1-6, J Nutr Sci Vitaminol (Tokyo). 59 (2013) 148- 151.

[136] de Mesquita M.L., Araújo R.M., Bezerra D.P., Filho R.B., de Paula J.E., Silveira E.R., Pessoa C., de Moraes M.O., Costa Lotufo L.V., Espindola L.S. Cytotoxicity of δ-tocotrienols from

Kielmeyera coriacea against cancer cell lines, Bioorg Med Chem. 19 (2011) 623-630.

[137] Karim R., Somani S., Al Robaian M., Mullin M., Amor R., McConnell G., Dufès C. Tumor regression after intravenous administration of targeted vesicles entrapping the vitamin E α-tocotrienol, J Control Release. 246 (2017) 79-87.

[138] Ng L.T., Lin L.T., Chen C.L., Chen H.W., Wu S.J., Lin C.C. Anti-melanogenic effects of δ-tocotrienol are associated with tyrosinase-related proteins and MAPK signaling pathway in B16 melanoma cells, Phytomedicine. 21 (2014) 978-983.

[139] Ahn K.S., Sethi G., Krishnan K., Aggarwal B.B., Gamma-tocotrienol inhibits nuclear factor-kappaB signaling pathway through inhibition of receptor-interacting protein and TAK1 leading to suppression of antiapoptotic gene products and potentiation of apoptosis, J Biol Chem. 282(2007) 809-820.

[140] Ye C., Zhao W., Li M., Zhuang J., Yan X., Lu Q., Chang C., Huang X., Zhou J., Xie B., Zhang Z., Yao X., Yan J., Guo H. δ-tocotrienol induces human bladder cancer cell growth arrest, apoptosis and chemosensitization through inhibition of STAT3 pathway, PLoS One. 10 (2015) e0122712.

[141] Wu S.J., Ng L.T., Tocotrienols inhibited growth and induced apoptosis in human HeLa cells through the cell cycle signaling pathway, Integr Cancer Ther. 9 (2010) 66-72.

[142] Xu W., Mi Y., He P., He S., Niu L. γ-Tocotrienol Inhibits Proliferation and Induces Apoptosis Via the Mitochondrial Pathway in Human Cervical Cancer HeLa Cells, Molecules. 22 (2017) 1299.

[143] Comitato R., Guantario B., Leoni G., Nesaretnam K., Ronci M.B., Canali R., Virgili F., Tocotrienols induce endoplasmic reticulum stress and apoptosis in cervical cancer cells, Genes Nutr. 11 (2016) 32.

[144] Kannappan R., Gupta S.C., Kim J.H., Aggarwal B.B. Tocotrienols fight cancer by targeting multiple cell signaling pathways, Genes Nutr. 7 (2012) 43-52.

[145] Thomsen CB, Andersen RF, Steffensen KD, Adimi P, Jakobsen A. Delta tocotrienol in recurrent ovarian cancer. A phase II trial. Pharmacological research. 2019;141:392-396.

Section 2

Elaeis guineensis Update

Chapter 8

Oil Palm Fatal Yellowing (FY), a Disease with an Elusive Causal Agent

Cleiton Barroso Bittencourt, Philippe de Castro Lins, Alessandra de Jesus Boari, Betania Ferraz Quirino, Wenceslau Geraldes Teixeira and Manoel Teixeira Souza Junior

Abstract

Fatal yellowing disease (FY) is a bud rot-type disease that severely affects oil palm plantations in Latin America. Since 1974, when it was first reported in Brazil, this disorder has been responsible for severe economic losses in the oil palm industry; and, for nearly 50 years, several studies have tried to identify its causal agent, without success. The etiological studies regarding FY in oil palm explored either biotic and abiotic stress scenarios, in a single or combined manner. Most recently, the hypothesis in favor of one biotic cause has lost some grounds to the abiotic one, mainly due to new insights regarding deficient aeration in the soil, which reduces the potential for oxy-reduction, causing changes in the ionic composition of the soil solution. This review presents an overview of the history of this disease and the several efforts done to fulfill Koch's postulates over the last 40 years, besides discussing recent studies that revisited this subject using some omics technics. We conclude by discussing further uses of omics via a multi-omics integration (MOI) strategy to help finally find out what is really behind the genesis of FY. Finding this elusive causal agent of FY out will allow either the development of a more efficient diagnostic tool and the advance in studies trying to find out the source of the genetic resistance hidden in the genome of the American oil palm.

Keywords: *Elaeis guineensis*, palm oil, epidemiology, tropical diseases, etiology, abiotic stress, biotic stress

1. Introduction

The African oil palm (*Elaeis guineensis* Jacq.) is a palm tree originally from the West Coast of Africa and currently distributed in three regions of the equatorial tropics; Africa, Southeast Asia, and Central and South America, where it exists in the wild, semi-wild, and cultivated form [1]. Among the oilseeds, it is the one with the highest average yield, producing 4 to 6 tonnes per year of vegetable oil [2]. The fruit palm oil (mesocarp) and the palm kernel oil (almond) are the raw material for several products in the food industry, in the cosmetic and personal hygiene industry, and the biofuels industry [3, 4]. Palm oil provides 36% of the global supply of vegetable oils with a considerable increase, from crude palm oil (CPO) production from 5.3 million tonnes in 1981 to 71.45 million tonnes in 2018 [2].

The Asian continent concentrates most of the CPO production, led by Indonesia and Malaysia, which together accounts for 85% of the world's CPO production [4]. However, the limited availability of areas for cultivation in Southeast Asia has opened new frontiers for expansion, culminating in the growth of Latin America's share in the global production of oil palm [5]. Latin America has the largest suitable area for oil palm cultivation, notably led by Brazil (2,283,000 km^2), Peru (458,000 km^2), and Colombia (417,000 km^2) [5]. Among Latin countries, Colombia is the world's fourth-largest producer of CPO and the first in the Americas, with an estimated production of 1.67 million tonnes in 2020, followed by Guatemala with 852 thousand tonnes and Honduras with 580 thousand tonnes [4].

Unfortunately, oil palm plantations in this geographical area are affected by a wide variety of pests and diseases that negatively affects productivity and discourage investment in this sector [1]. Notably, "bud-rot type" diseases pose the greatest threat to oil palm plantations in Latin America [6]. Among them, *Pudrición del Cogollo* (PC) and Fatal Yellowing (FY) are the diseases that cause most of the damage, both presenting a common symptom: rotting of the spear leaf that evolves until reaching the apical meristem system leading to the death of individuals [6, 7]. By far, the FY exhibits the most dramatic scenario because, in contrast to PC, its causal agent remains unknown, hindering effective sanitary control practices [8].

Fatal Yellowing was first identified by Reiking in 1928 in oil palm plantations in Panama, with cases reported in Colombia, Ecuador, Peru, Costa Rica, Venezuela, Suriname, Nicaragua, and, reportedly, in Central Africa, after that [6]. In Brazil, it was only in 1974 that the first symptomatic individuals were identified and, from the epidemiological explosion that occurred in the 1980s, FY started to represent the greatest threat to oil palm in the country [9]. As a result, several studies began to search for the possible biotic causal agent behind it and its putative vectors [8]. However, the research efforts made for more than 30 years have not exactly pointed out organisms directly linked to FY's cause [10–16]. Some studies also looked for possible abiotic causes, with inconclusive results so far [17–19]. Recently, techniques such as metabolomics, proteomics, and metagenomics started to be applied to provide insights into the possible FY etiology, initiating a new phase in the process to solve this problem [20–22].

Although Brazil has more than 30 million hectares with an aptitude for oil palm production, it currently has less than 1% of this area destined for this purpose [23]. Fatal Yellowing is the main contributor to hinder the expansion of the oil palm industry in Brazil, and the attempts to control the emergence of sick plants have not been successful, and its nature remains a mystery [10]. This review intends to analyze descriptively the studies carried out to investigate the FY problem in Brazil, besides pointing out new strategies employed for understanding the development of the disease, confirm the real cause behind it, and develop tools for early diagnostics.

2. The oil palm industry: social and economic importance

2.1 In the world

Oil palm is originally from West Africa and adapted to the intertropical areas of Africa, Asia, South and Central America [1]. It is the most profitable oil crop, as it presents a higher yield with a lower production cost [24]. Its oil yield is of the order of 4-6 tonnes per hectare per year of CPO, much higher than that presented by other crops, such as rapeseed (0.69 t), sunflower (0.69 t), and soybeans (0.44 t) [3]. Another positive point is that this crop uses only 6% of the area to produce 36% of the global oil supply, while soy, for example, occupies 40% of the land to generate 26% [4, 24]. Because of that, oil palm stands out as a player fundamental to help the world meet the growing global demand for vegetable oil in 2050 that will be around 240 million tonnes [25, 26].

The expansion of the oil palm industry has been strongly encouraged by governments and private sectors in Southeast Asia [27]. It is by far the most productive region in the world, supplying 85% of the CPO produced, reflecting the rapid expansion of the cultivated area that started in the middle of the last century [25]. The commercial oil palm plantations in Indonesia, for instance, went from 70 thousand hectares in 1961 to 6.78 million hectares in 2018, with a considerable increase of 9.582% during this period [2]. As a result, Southeast Asia production rose to 63.26 million tonnes in 2018, or a 22,378% increase in the period [2, 3].

Africa has not seen an expansion of the oil palm industry as significant as Southeast Asia in the last 60 years [3, 28]. The area occupied by oil palm increased from 3.55 million hectares in 1961 to 4.55 million hectares in 2018 in the African continent, representing an increase of only 33% (**Figure 2**) [2]. Meanwhile, the Americas now occupy 6% of the international market, producing around 4.89 million tonnes of palm oil in 2018, a 273% increase in the last two decades [2].

The considerable increase in oil palm production was supported mainly by the advances in genetic breeding programs that increased oil productivity more than 2 folds since 1960 [1].

Most of the CPO and its derivatives produced stays in the Asian markets that absorb 51% of the total, led by India, which imports 19.4%, and China 13.0% [29]. The European markets, which import 26%, have the Netherlands (6.1%) and Italy (4.3%) as the leading importers [23]. Africa (12%), the Middle East (4%), and Latin and North America (7%) also have a consumer market for vegetable oils, and palm oil from Southeast Asia helps to supply the demand [29]. The global vegetable oil market allocates 70% of total production to food and 30% to non-food industrial purposes, such as, for example, the production of cosmetics and personal hygiene products (24%) and as a raw material for the production of biofuels (5%) [26].

2.2 In Latin America

The increase in global palm oil production in the 21st century is due mainly to new plantations in producing countries, especially in Malaysia and Indonesia [27]. However, due to a reduction in the areas available for expanding cultivation in Southeast Asia, an opportunity opened up to expand to new frontiers to meet the growing global demand for palm oil [5]. As a result, Latin America became one of the most promising regions for oil palm cultivation, as it has one of the largest areas suitable for cultivation, notably represented by Brazil, Peru, and Colombia [5].

Brazil, Colombia, Costa Rica, Ecuador, Guatemala, Honduras, Mexico, Peru, and Venezuela produce together 4.65 million tonnes of palm oil, representing 6% of world production in 2020 [2]. Colombia is the largest oil palm producer in this region and ranks 4th in the World, with 1.61 million tonnes produced in 2018, or 33% from the annual palm oil of Latin America (**Figure 1**) [2, 30]. Guatemala produced a total of 875 thousand tonnes in 2018 what places the country in the 2nd position in Latin America [2, 31]. Honduras is in the 3rd, followed by Ecuador, Brazil, Costa Rica, and Venezuela [2].

2.3 In Brazil

The first oil palm plants arrived in Brazil in the 16th century, adapting very well to the Northeast region of the country [32]. The oil palm industry in Brazil stayed as

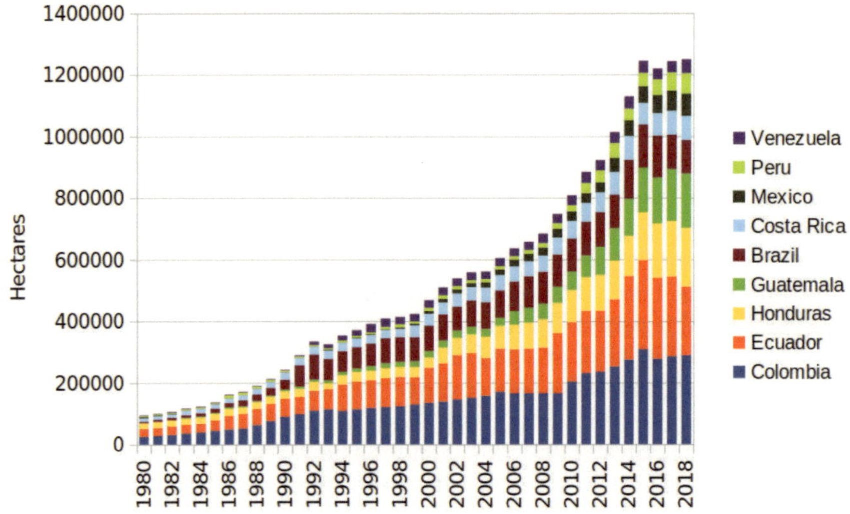

Figure 1.
Land use for oil palm cultivation in central and South America since 1980, in hectares. Source: adapted from our wold in data [2].

Figure 2.
Fatal yellowing (FY) disease in oil palm. (a) Oil palm plantation affected by FY; (b) individual showing signs of yellowing and necrosis of the intermediate leaves; (c, d, e) evolution of yellowing and dryness of the spear leaf with the presence of necrotic tissue, and (f) root section of an individual with signs of rot. Source: by authors.

a small industry until 1960, when, due to increasing demand for oil for cooling steel sheets in the national steel park, it started to experience significant growth [33]. In 1967, the oil palm cultivation expanded to the Pará State, in the North region of Brazil, with the first commercial-scale plantations comprising about 3.000 hectares [32].

Driven by technical advances and growth in demand for vegetable oils, there was a significant increase in the cultivated area of oil palm in Brazil, which went from 11 thousand hectares in 1980 to more than 236 thousand hectares in 2008 [3]. Brazil has large areas with the aptitude for oil palm production, favored by climatic conditions similar to the most productive regions in the world [1]. However, until 2014, less than 1% of this area was occupied by commercial plantations [34, 35]. Brazil's position as the 13th, and 23rd, in palm oil production and on the productivity scale, respectively, in the world, is due mainly to this under-utilization of available areas [3, 32].

Oil palm production is concentrated in Pará state, which accounts for 97.19% of the cultivated area and 97.99% of the national palm oil production, followed by Bahia (1.98%), Roraima, and Amazonas [36]. The expansion of cultivation to already deforested areas in the Amazon and other regions in Brazil is an opportunity to reduce pressure on forests and supply the palm oil demand from the food and energy sectors [35]. To make the plantations more environmentally sustainable, the Brazilian Government launched the agro-ecological zoning (ZAE) program in 2010, a legal mechanism to delimit the oil palm cultivation area [37]. This area include Acre, Amapá, Amazonas, Mato Grosso, Pará, Rondônia, Roraima and Tocantins, part of Maranhão and five municipalities in Goiás state, comprising about 59% of the Brazilian territory [35].

3. The fatal yellowing (FY) disease

3.1 History in the world

Several fatal syndromes of bud-rot severely affect plantations of oil palm in South and Central America [6, 38]. Bud-rot type disease was reported for the first time on oil palm plantations in Suriname in the 1920s, followed by another incidence in Panama reported by Reinking in 1927 [6]. In general, symptoms of bud-rot type diseases initiate with chlorosis of the youngest leaves and later necrosis that rapidly reach immature tissues of the meristem, causing a collapse of the spear leaf and plant death [9]. Bud rot diseases can take two forms: a lethal form found in Ecuador, Brazil, and in certain zones of Colombia and Suriname, and a non-lethal one, with a high recovery rate, found mainly in the Colombian Llanos [6]. The disease is synonym to a few other names such as "pudrición del cogollo" (PC) in most Spanish speaking countries, "PC típica" (PCt) or "PC diversa" (PCd) in the plantation Palmeras del Ecuador (PDE) in Eastern Ecuador, "amarelecimento fatal" (AF) in Brazil, "spear rot "in Suriname [1, 6, 7].

The first large-scale bud rot damage on oil palm plantations in Latin America was due to the PC disease in northern Colombia, where a field of 2,800 hectares located in the Turbo region was virtually devastated by PC in 1965 [9]. In Suriname, the spear rot was first registered in the Victoria region in 1976 on four-year-old oil palms in a plantation of 1,700 hectares. Despite the phytosanitary practices applied to control the disease, an exponential progression reduced the original area by 85% [39]. In Ecuador, the first PC cases happened in 1976 on four-year-old oil palms on the Pacific slopes of the Ecuadorian cordillera [1], and, like other regions, the plantation was decimated by the disease in a few years [6]. Recently Martinez et al. [7] carried out a study in Colombia to isolate microorganisms and reproduce PC in healthy oil palm plants and, in conclusion, they postulate that the oomycete *Phytophthora palmivora* is associated with the emergence of PC.

Fatal yellowing exhibits, by far, the most dramatic scenario among the bud-rot type diseases of oil palm in the Americas. The factors linked to the emergence of this disease in some countries remain unknown after experiencing more than 50 years of outbreaks in Brazil, Ecuador, Panama, Suriname, Costa Rica, Nicaragua, Honduras, Peru, and Venezuela [6, 9, 10].

3.2 History in Brazil

The FY disease first appeared in Brazil in 1974, with sporadic occurrences in a field established in 1967 near Benevides, a city in the Pará State [8, 12]. The disease progressed slowly in the following years, from 25 symptomatic plants in 1978 to 125 in 1981. In 1984, ten years after the first report, the number of plants diagnosed with FY was 465 [11]. In the first ten years after its first appearance, the disease progressed in a linear model, and the numbers of affected plants remained more or less the same per unit of time. This mode of progress indicated that the contamination did not occur from plant to plant. However, the numbers of affected plants rose to 9,968 in 1986 and 32,673 in 1987, starting a period of exponential increase [11]. In the first two decades after its first occurrence in Brazil, approximately 100 thousand oil palm trees died from this disease, resulting in the loss of entire plantations [11, 40].

Roguing was then put in place to maintain the source of the inoculum of a possible pathogen to a minimum, eliminating all plants showing symptoms up to one month after the discovery [40, 41]. The oil palm industry promoted training on the fast and precise recognition of FY symptoms to guarantee the effectiveness of this phytosanitary measure [42]. Despite it, the disease kept on occurring in plants between the 15th and the 16th year after planting, making FY one of the main problems of this crop in Brazil. Not surprisingly, this discouraged the expansion of oil palm cultivation in the affected regions [11]. As the inability to identify the causal agent and promote effective control of FY persists, the oil palm industry remains in a state of insecurity to expand in the regions affected by FY [42].

3.3 Symptomatology and diagnosis

Proper and early disease diagnosis is vital for applying control practices at the right moment. Without an efficient and effective early diagnosis of the disease and the disease-causing agent, any control measures will be inefficient [43]. Until the FY etiology is known and diagnostic systems developed, the only way to find out that a plant has this disease is by checking for characteristic symptoms and signs. Once a plant is diagnosed with FY, it must undergo roguing. In Brazil, symptoms identification in the field is still the only diagnostic system used for FY [8, 12].

An oil palm plant affected by FY shows necrosis or dryness of the spear leaf that evolves towards the base, then the region of the meristem rots, and a foul odor is felt in some cases (**Figure 2**) [12, 44]. The process of rotting of the meristem region, frequently observed in rainy seasons, motivated the initial designation of the disease as spear leaf rot [8, 40]. After losing the spear leaf, there is a general decline leading to plant death; however, some individuals during this process may temporarily re-issue a new one [12, 18]. In plants affected by FY, chlorosis appears in leaflets at the base of the intermediate leaves, which advances towards the extremity, followed by necroses frequently observed in younger leaves [6]. There is no synchronism between the spear leaf necrosis and the chlorosis of the leaflets. The FY symptoms always begin with leaflets chlorosis, which led to the Fatal

Yellowing disease name [1]. In Brazil, the oil palm tree usually dies 7 to 10 months after the onset of the first symptoms, but it can vary depending on the region [41].

Once the oil palm plant gets affected by FY, the developed bunches can reach the maturation stage and are not affected. However, the immature ones rot, and the inflorescence abort [40, 41]. The root system is visibly affected, and emission of new primary roots reduced, leading to a total cease of roots growth. FY kills the tips of the roots generating new false primary ones. In addition, the root tissue is usually necrotic at the beginning of the appearance of symptoms in the aerial part [45, 46]. On the other hand, no apparent internal symptoms are observed, such as rot or necrosis of the stipe and vascular system, a characteristic that is also seen in PC [41].

4. A genetic source of resistance to FY

The causal agent of FY is still unknown, but a possible genetic solution for this problem exists. This genetic solution resides upon the fact that the American oil palm (*Elaeis oleifera* (Kunth) Cortés) and the interspecific hybrids between this species and the African oil palm are considered resistant to this disease [47].

The genus *Elaeis* (from the Greek *Elaion* that means oil) belongs to the class Liliopsida (Monocotyledones), order Arecales (Palmales), family Arecaceae (Palmae), subfamily Arecoideae, tribe Cocoseae (Cocoinaea) and, subtribe Elaeidinae [48, 49]. This genus consists of two species, *E. guineensis* and *E. oleifera*, with a pantropical distribution and two distinct diversity centers, Nigeria and South America, respectively [50–52]. The former is the African oil palm, the predominant species in commercial plantations Worldwide, and known in Brazil as "Dendê"; and the latter is the American oil palm, which originated from Central and South America, and is known as "Caiaué" [53].

The American oil palm is endemic to Equatorial America, with natural populations distributed from Central America to northern South America, including the countries of Brazil, Colombia, Costa Rica, Ecuador, French Guiana, Honduras, Mexico, Nicaragua, Panama, Peru, Suriname, and Venezuela [1]. In Surinam, there are dense stands on poor, sandy soil, while in Colombia, it can grow in damp or even swampy situations near or on the banks of rivers [1].

The American oil palm also has a history of use as a source of vegetable oils and other products, but its most important value to the oil palm industry is its capacity to hybridize with the African oil palm [1]. The interest in the germplasm of this species is due to valuable characteristics for breeding programs of the African oil palm, such as slow growth, oil quality (mainly unsaturated oil) [54], and disease resistance, including FY [47].

These two species can sexually cross and generate fertile interspecific hybrids with intermediate characteristics to the two parental species [55]. Some interspecific hybrids between these species are already commercially available, and the Brazilian genetic group of *E. oleifera* is parental to most of them — Manicoré (BRS Manicoré from Embrapa, and [Mangenot × Manicoré] × La Mé from PalmElit SAS), Manaus (Amazon from ASD Costa Rica), and Coari (Coari × La Mé, Coari × Yangambi) [47].

Independent whether the origin of FY is biotic or abiotic, or a combination of both, once it is finally known, new studies will be necessary to confirm this genetic resistance and gain insights on possible strategies to transfer this resistance to the African oil palm more efficiently and effectively, besides the use of interspecific crosses followed by backcrosses.

5. The search for the causal agent

5.1 Biotic stress

5.1.1 Insects

After the epidemiological explosion of FY in 1986, Embrapa (the Brazilian Agricultural Research Corporation) started conducting studies on insects as a possible vector of the FY causal agent [8]. As the spread of the disease followed the direction of the prevailing winds, while natural barriers - such as roads, rivers, and glades - were not sufficient to prevent it supported this hypothesis [8, 56]. This hypothesis on a possible entomological role in the spread of FY also resided in the fact that this disease has similarities with the lethal yellowing-type disease that affects other palms. This disease that affects several other palms is due to insect-transmitted phytoplasmas [57]. Initially, from inventory obtained in plantations affected by FY in the municipalities of Alvaraes, in the Amazonas State, and Benevides, in the Pará State, the main insects suspected of being responsible for the transmission corresponded to *Persis* sp. and *Myndus crudus* because they are commonly found in oil palm plantations and depend on palm oil for nutrition [15].

Initially, an inventory of insects captured directly on the oil palm plantations located inside and outside areas with FY occurrence was generated. Healthy oil palm plants, isolated in cages made of wood and nylon canvas, received populations of the inventoried insects, and the plants monitored for symptoms appearance. After using almost one million insects in the FY transmission test, no symptomatic plant appeared, and there was no relationship between the affected areas with the collected insect fauna [15, 58]. Additional studies have attempted to establish a link between the insects *Contigucephalus* sp., *Omolicna* sp., and *Myndus crudus* and this disease, but they all gave negative results. Consequently, the authors discarded a Homoptera as the FY vector and suggested new studies on possible very active and rare insect species [8, 56].

Another study attempted to investigate the relationship between the presence of homopterans in the vegetation cover in oil palm plantations and the occurrence of FY [12]. No relation between the vegetation cover and FY occurrence appeared as the disease manifested itself either in an area covered with *Pueraria* spp. as in areas where there were grasses, especially *Brachiaria* spp. [25]. Studies using a series of chemicals in areas where FY occurs - such as insecticides, fungicides, and bactericides - did not reduce the appearance and development of FY [40].

5.1.2 Phytoplasmas

Phytoplasmas are prokaryotes of the Class Mollicutes that cause diseases in several plant species, including several economically important ones [59]. As biotrophic parasites, they colonize the elements riddled with the phloem and can also be found inside the vectors [60]. These organisms are responsible for Lethal Yellowing (LF), a fatal disease that affects the coconut (*Cocos nucifera* L.) and at least 36 other palm species in the Americas [61, 62].

Insects from the Homoptera order, popularly known as leafhoppers, are the vectors for most phytoplasmas causing disease in plants [63]. Biological characteristics, symptoms, and specificity of the insect vector were the focus of the first studies aiming to associate phytoplasmas with plant diseases [64, 65]. Later, new and more accurate DNA-based techniques started to dominate these studies, leading to the production of specific oligonucleotides for diagnosis [65].

Transmission electron microscopy was, for many years, the tool used for the detection and study of the cytological interaction between phytoplasmas and the hosts [66]. Studies using this tool were not successful in associating phytoplasma with FY, been replaced by new molecular techniques for the same purpose [8]. Studies carried out by Brioso et al. [67, 68] using nested-PCR in oil palm plants symptomatic for FY found just a very few samples positives for the presence of phytoplasmas from the SrI and 16SrI groups, which do not allow to associate these phytoplasmas to FY. An attempt to reproduce the disease was carried out by grafting intermediate leaf tips with FY into healthy seedling petioles and, during the period of two years, healthy individuals did not show symptoms characteristic of FY and, thus, the hypothesis proposing phytoplasma as the causal agent was discarded [12].

5.1.3 Fungi, bacteria, and nematodes

In the attempt to establish a causal relationship between plant pathogenic fungi, bacteria, and nematodes with FY, some studies tried to reproduce the symptoms in healthy young and adult oil palm plants inoculated with some of these microorganisms previously isolated from symptomatic plants [69, 70].

A pathogenicity test focused on studying the growth, reproductive and developmental habits of microorganisms, included one-year-old nursery plants with individual inoculations and a mixture of three fungi (*Fusarium* sp., *Pythium* sp., and *Coprinus* sp.) isolated from symptomatic plants; and again, the inoculum was unable to reproduce the disease in healthy oil palm trees [69]. The possibility of mechanical transmission between symptomatic and asymptomatic individuals by some microorganisms was also tested, with no success [69]. The chemical control attempts using fungicides or antibiotics failed to link fungi and bacteria to FY in oil palm [11].

Interestingly, some authors have observed similarities between the disease PC in Colombia and FY in Brazil. Furthermore, the oomycete *Phythophtora palmivora* was reported to be the PC causal agent [7]. The strategy used by Martinez et al. [7] was to remove tissue from oil palm plants exhibiting early symptoms of PC disease to inoculate fruit traps. Once microbial growth was observed in the fruits, tissue was transferred to culture media and pure cultures were obtained. Using the DNA isolated from the pure culture, amplification of the ITS region was performed and sequence analysis showed 99.9% homology to *P. palmivora*. The same study reported pathogenicity tests where sporangia were inoculated into the base of the spear of 150 oil palm nursery plants. After 3 to 4 days, the first symptoms of PC were observed in 85% of the plants [7]. However, full PC symptom development occurred in 15% of inoculated oil palm plants, and depended on environmental conditions. In another experiment, 20 immature spear leaves were inoculated with *P. palmivora,* and 3 days later all tissues were disintegrated, displaying a characteristic odor. Microscopy experiments showed the presence of *P. palmivora* in these tissues, and it was re-isolated using the fruit trap technique.

Nematodes are typically wormlike invertebrates able to live in the soil or inside plant structures such as roots, stems, leaves, and flowers and can cause morphological and developmental changes in their hosts [71]. The hypothesis of a nematode as a causative agent of FY came from observations of FY and the red ring disease - caused by the nematode *Bursaphelenchus cocophilus* - in the same area. Ferraz [72] did not observe this nematode in necrotic tissues or young leaves. Some studies found nematodes in the spear leaf rake and young leaves of symptomatic plants and the soil of oil palm plantations with a history of FY but were unable to link it to the appearance of this disease [24, 72].

5.1.4 Viruses and viroids

Other plant pathogens studied as potential causal agents of FY in oil palm were viruses and viroids. Several methods, including mechanical transmission, grafting, pollen-mediated dispersion, transmission electron microscopy, nested RT-PCR, RCA - rolling circle amplification, and electrophoresis, were used to test the hypothesis of a virus or a viroid as the causal agent of FY, without success [8, 10].

Lin et al. [73] evaluated extracts from plants with and without FY using the polyacrylamide gel electrophoresis technique, and the band patterns generated in both samples did not reveal any apparent difference. The same author also carried out a study to purify virus particles via separation with a fractional density gradient with no success [74]. Kitajima [75] evaluated ultrafine tissues from roots, leaves, and spear leaf of symptomatic and asymptomatic individuals by transmission electron microscopy, but no pathogen could be associated with FY.

Other studies have directed their efforts towards viroids, which are the smallest known phytopathogens, consisting basically of a single-stranded, circular RNA molecule not encapsulated [76, 77]. Beuther et al. [13] searched for viroids and viroid-like RNAs in oil palm plants using two-dimensional gel electrophoresis and return gel electrophoresis of nucleic acid extracts, with no success in showing a link between this type of pathogen and FY.

5.2 Abiotic stress

The initial pieces of evidence of a possible abiotic cause for FY came from observations made about the indefinite dissemination pattern in affected areas, with an exponential growth form not observed in the case of biotic stresses [78, 79]. Among the possible abiotic causes linked to the appearance of FY, there are lower and higher amounts of water, high or low temperature, high content of soluble salts in the soil, soil pH unsuitable for oil palm, nutritional deficiencies or excesses, presence of toxic organic compounds and intensity and balance of nutrients [78].

The regions with oil palm plantations and FY occurrence located in the North region of Brazil have soils with patches of quartz sand interspersed with patches of lateritic concretions and are subject to prolonged floodings, 5 to 6 months per year [41]. Thus, studies started aiming to understand the composition of the soil and its influence on FY emergence.

The concentrations of Cu, Fe, Mn, and Zn in the leaves of healthy and symptomatic oil palm plants and resistant interspecific hybrids were determined and found out that their concentrations were below the ideal range, suggesting their involvement in the appearance of FY [80]. Compact soils that stay temporarily saturated by rainfall suffer oxidation by anoxia, making it impossible for plants to absorb Fe [80]. Based on these observations, applications of ferrous sulfate were carried out on plants under different stages of FY, but after 120 days of the experiment, there was no regression of the disease in the evaluated oil palms [80].

The physical properties of the soil from areas with the occurrence of FY revealed that they were naturally well-drained and deep but had a thickening or compacting between the depths of 30 cm and 60 cm, as well as the occurrence of speckles in this depth, which results in soil saturation in the superficial layer during the rainfall season [81]. Bernardes [82] carried out chemical analysis on roots of symptomatic plants, and the results did not allow to pinpoint any element imbalance that could be responsible for FY. Another fact that needs consideration as possibly linked to a potential cause for the disease is the fact that at the moment when the first symptoms appear in the aerial part, the root system is severely impaired, which explains the plants' lack of response to fertilization and other interventions [82].

A series of field observations made in the heart of the oil palm production area in Brazil led to new hypotheses for a possible abiotic cause for FY [83]. The main field observations taken into consideration were: a higher occurrence of flooding in oil palm plantations, in comparison to the previous level, observed under native vegetation cover; the layers close to the soil surface without vegetation cover or with oil palm tend to stay close to water saturation for periods much longer than in the native forest; the presence of mottled-iron reduction in the profile of the oil palm plantations, and the redox-potential values (Eh) below −200 mV; and the presence of reduced iron ions on the soil surface in oil palm plantations during periods of intense rain [83].

The new hypotheses were brought together and summarized as: Deficient aeration reduces the potential for oxy-reduction in the soil, causing changes in the ionic composition of the soil solution (reduction of Fe^{3+} ions; NO^{3+}; Mn^{3+}). The soil solution with a high concentration of reduced ions initially causes damage to the root system (**Figure 3**) predisposing the oil palm plant to physiological disturbances (passive poisoning and attacks of secondary pathogens) whose symptoms are known as FY [84].

To gain insights into the idea of oxygen deficiency (hypoxia) in the origin of FY, a study by Encinas [85] evaluate the influence of land use and temporal variations on the dynamics of nutrients in the solution of soil and water at an oil palm plantation and a nearby area still with primary forest. Another by Muniz [83] compared the changes in water flow at an oil palm plantation and a nearby area still with native vegetation cover and evaluated its effects on iron dynamics and the structure of the soil. These two studies gathered additional shreds of evidence to further support this hypothesis, such as the electrical conductivity increased during a long flooding period (95 days), indicating that ions from the aggregates migrate to the solution; the soil pH increases after the initial flooding period, reaching values close to neutrality, with a subsequent reduction, but above the values found in aerated soil; the soil redox potential decreases during the flooding period, forming a highly reducing environment; the total carbon contained in the macroaggregates reduced

Figure 3.
Oil palm plant showing reduction of the root system in hypoxia conditions (A), and soil clouds showing the typical reductimorphic or oximorphic color mottles caused by stagnating soil environment (B). Source: Wenceslau Teixeira.

after flooding for a period of 11 days; the iron contained in the aggregates of Yellow Latosols with medium texture migrates to the soil solution under flooding conditions; there is a high negative correlation between the iron in the flooding solution and the DMG of the aggregates in the Yellow Latosols, and flooding for a period of 11 days promotes the destabilization of aggregates of Yellow Latosols with medium goethite texture.

6. New technologies to gain insights on the FY causal agent

The so-called 'omics' techniques (**Figure 4**) provide new opportunities to study oil palm FY. To get insights on FY possible causal agent, different research groups in Brazil have used metagenomics, metabolomics, and proteomics analysis [20–22]. To our knowledge, no work focusing on transcriptomics and FY has been published yet. The most commonly used approach in these studies is to compare healthy plants (without symptoms of FY) to those showing disease symptoms at different stages of progression. In contrast to more traditional non-molecular studies of FY, these techniques provide a global glimpse of the organism by looking at the associated microbiota (metagenomics), the complete protein content (proteomics), or metabolite content (metabolomics) of cells.

6.1 Metagenomics

Koch's postulate was fundamental to the identification of disease-causing microorganisms [86]. In short, the strategy of isolating and cultivating the potential pathogen, and inoculating it into a healthy organism to confirm the symptoms of the disease, brought many advances to the study of infectious diseases [87]. More recently, due mainly to the advent of next-generation sequencing (NGS) technologies, the frontiers of microbiology expanded to those microorganisms that we cannot cultivate by classical microbiology techniques. That has opened the possibility to test the hypothesis that a microorganism not grown in vitro easily is the cause of FY [88]. If this is the case, metagenomics would be the technique to study FY.

Metagenomics is a culture-independent approach to study microbial communities. A metagenomics strategy allows one to skip the step of isolation and cultivation of microbial species. Metagenomics studies can contribute to elucidate the identity

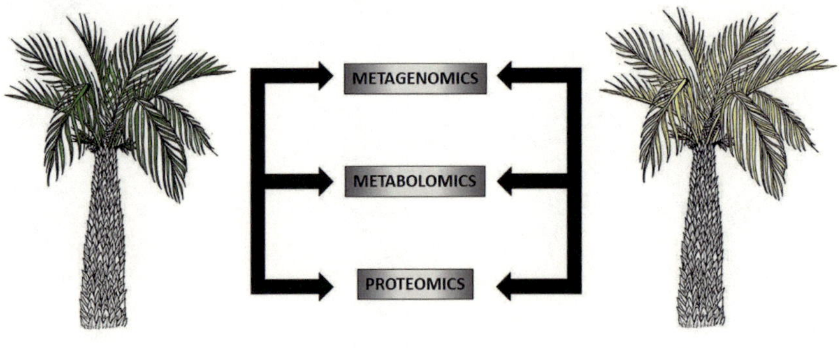

Figure 4.
Schematic showing a healthy oil palm tree (green leaves) and another one (yellow leaves) showing fatal yellowing (FY) symptoms. Different molecular techniques such as metagenomics, metabolomics and proteomics can be used to compare these contrasting biological situations. Metagenomics is a culture-independent technique that can be used to identify the microorganisms present. Metabolomics can used to identify and quantify cellular metabolites. Proteomics allows the identification of differentially expressed proteins. These 'omics' techniques are important high throughput tools that have been used to understand the biology of oil palm when challenged by FY disease. (credit: Clarissa Kruger).

and/or the genetic and metabolic capabilities of the microorganisms present in a sample, including any that are potentially pathogenic [89].

In this sense, metagenomics complements the classic techniques of isolation and cultivation of microorganisms, and one can apply it to study different classes of microorganisms (e.g., viruses, bacteria, fungi, archaea) [22, 90–92]. Metagenomics protocols begin with the extraction of total DNA from the sample of interest, which contains microorganisms. Samples can be many different ones, such as soil or plant parts with FY disease symptoms. There are distinct ways to study the microbial community from this DNA. Many studies in different plants use the ribosomal RNA (rRNA) gene or ITS amplification approach (i.e., PCR amplification with specific primers) to identify the microorganisms present, including a potential pathogen [93–95].

16S rRNA gene-specific primers amplify bacterial and archaeal sequences (16S rDNA). Similarly, the 18S rRNA gene and the ITS-specific primers amplify fungal sequences. The ITS refers to the internal transcribed spacer, the DNA situated between the small-subunit ribosomal RNA and large-subunit rRNA genes. The 16S rDNA, 18S rDNA, and the ITS regions are highly polymorphic, thus allowing taxonomical identification of the microorganisms present in a sample. The PCR-amplified DNA is then sequenced and submitted to bioinformatics analysis to compare the obtained sequences with sequence databanks, leading to a putative microorganism. In summary, this metagenomics approach that combines PCR amplification with NGS allows identifying microorganisms present in the community [96].

The first metagenomics work to use ITS amplification and high throughput NGS to study FY in Brazil was performed by Costa et al. [22], who evaluated fungal communities associated with leaves of oil palm plants, with and without symptoms of FY. Leaves from health plants and from plants showing FY symptoms in three different disease stages (stages 2, 5, and 8) were obtained. Because of the similarities between PC and FY, using primers specific to the genus *Phythophtora,* the authors attempted PCR-amplification of oil palm leaf samples showing symptoms of FY. Weak amplification was obtained in only one sample. Thus, this study provided preliminary evidence that DNA of the genus *Phytophtora* may not be commonly present in Brazilian FY, contrary to what has been reported in Colombia [7]. However, further experiments with more samples, and additional controls are needed to clarify the validity of this initial observation.

The Costa et al. [22] study reported the analyses of fungal diversity using the ITS region. Results showed that the fungal community in different healthy asymptomatic oil palm leaves are more similar to each other than those presenting FY disease symptoms. The fungal communities were not the same among all the symptomatic samples, and were not consistent even between samples at the same stage of FY disease. Importantly, no fungal taxon had its relative proportion increased in leaves across all the FY diseased oil palm plants. It was hypothesized that the changes observed in the fungal community composition could be a secondary effect of FY disease. Similar metagenomic studies to analyze the viral, bacterial and archaeal communities associated with FY are needed.

A less common metagenomic approach that can also be used to study plant disease is to assemble genomes from the metagenome obtained from plants showing symptoms of disease. In this case, instead of using PCR to amplify a specific gene, one can completely sequence the DNA extracted from the samples of interest, and use bioinformatics tools to assemble genomes (metagenome-assembled genomes) of the microorganisms present. This type of methodology allows, in addition to identifying microorganisms present, access to their genomes. This creates the possibility of studying the genetic relationship among the species present, and predicting metabolic capabilities as well as the interactions between the organisms of the community [97]. One limitation to this method, however, is that the plant host genome sequence needs to be available and subtracted *in silico* from microbial community sequences. If possible, it is useful to find a way to selectively extract microbial DNA from the samples before sequencing to avoid or reduce the presence of the plant host DNA [98]. It should be noted that if the complexity of the microbial community is high or if a lot of host DNA is present in the sequenced samples, inadequate sequencing depth might be an important limitation to this method. To our knowledge this approach has not been used yet to search for the causal agent of FY.

6.2 Proteomics and metabolomics

Proteome designates the set of proteins expressed by a cell, tissue, or organism at any given time [99]. Proteomic tools make it possible to obtain a protein profile with precision and sensitivity with the aid of electrophoresis, chromatography, mass spectrometry, and bioinformatics [99]. Proteomics is more and more used nowadays to understand plant responses to different biotic and abiotic stress conditions [100, 101].

In this context, and based on the hypothesis that the primary stress behind FY was abiotic and present in the soil, proteomics was applied to study this disease [21]. This hypothesis is based on observations regarding symptoms seen in the root system before they appeared in the aerial part [83]. Soil compaction, which hinders drainage and subject the roots to long periods of flooding in a hypoxia condition, would be in the origin of the stress [83].

Nascimento et al. [21] carried out a proteomic analysis to compare the protein profiles from symptomatic and asymptomatic oil palm plants, employing the mass spectrometry technique. The study looked for proteins linked to tolerance induction to relate the different areas collected and the distinct stages of the disease, analyzing the roots of symptomatic plants in early, intermediate, and final stages.

Proteins involved in the metabolism of phenylpropanoids and lignins, with a recognized role in reducing the effects of biotic and abiotic stress, were negatively regulated in symptomatic individuals, aggravating FY symptoms. In asymptomatic plants, enzymes such as S-adenosylmethionine - with a crucial role in methionine's biosynthetic metabolism - showed a recognized action in response to the stress. Plants with FY symptoms showed some pathogen-related proteins positively regulated, implying a progression of infection by biotic agents [21].

The hypothesis of a possible physiological dysfunction caused by factors present in the soil was reinforced by the large accumulation of antioxidant proteins in asymptomatic individuals [21]. The participation of the antioxidant system may indicate some level of resistance, considering that this system is vital for plants in conditions of soil flooding [102]. In addition, the accumulation of aldehyde dehydrogenase may indicate that the root system is under an anaerobic condition as it converts the acetaldehyde, promoting plant survival in this condition [21, 103]. Thus, these results indicate that plants affected by FY are in abiotic stress conditions and, with the damages done to the roots, it becomes a gateway for several opportunistic organisms [21].

In contrast to proteomics, metabolomics refers to a comprehensive analysis to identify the set of metabolites present in a sample with the aid of analytical techniques, such as liquid chromatographies or liquid–gas, associated or not with mass spectrometry, among others [104].

Rodrigues-Neto et al. [20] performed the first metabolomics work to study FY in Brazil using an untargeted metabolomics strategy to prospect metabolites differentially expressed in the leaves of FY symptomatic and asymptomatic plants. A high throughput method based on metabolic fingerprinting MS, using UHPLC coupled to high-resolution mass spectrometry (HRMS), was employed, and chemometric analysis, PCA and PLS-DA, were used to evaluate metabolic differences. This study aimed at prospecting a biomarker for FY early diagnosis, besides gaining insights on pathways responsive to this disease valuable for future improvement studies.

Nine secondary metabolites were detected in a higher concentration in the healthy plants in comparison to the FY affected ones: Glycerophosphorylcholine, arginine, asparagine, paniculatin or apigenin 6,8-di-C-hexose, tyramine, Chlorophyllide, 1,2-dihexanoyl-sn-glycero-3-phosphoethanolamine, proline, malvidin 3-glucoside-5-(6″-malonylglucoside) or kaempferol 7-methyl ether 3-[3-hydroxy-3-methylglutaryl-(1–> 6)]-[apiosyl-(1–> 2)-galactoside]. These metabolites made possible to identify different metabolic pathways that have been affected by the FY, such as the glycerophospholipid metabolism, the isoquinoline alkaloid biosynthesis, the flavonoid biosynthesis, the tetrapyrrole biosynthesis and citrate cycle derivatives pathways.

Unfortunately, due to the fact that these metabolites are already described in the literature as linked to other types of stress, they are not good candidate for biomarkers; except for two of them, glycerophosphorylcholine and 1,2-dihexanoyl-sn-glycero-3-phosphoethanolamine [20].

7. Final considerations

Fatal yellowing disease represents a threat of great magnitude to the Brazilian oil palm industry. For decades, several studies attempted to identify its causal agent without success. As a result, no measures used today can effectively reduce the economic loss for the oil palm industry due to this disease. The only glimpse of hope in solving this problem still resides in the genetic resistance found in the American oil palm. However, the road to transfer this resistance through interspecific crosses and backcrosses is very long and has many uncertainties.

The search for the primary stress leading to FY must go on, whether it is of biotic or abiotic origin - or the combination of both. Only then might be able to block its occurrence, or, if not possible to do that, develop early diagnostic tools to reduce its spread to a minimum.

Recent studies using single omics analysis have shown that these new technics can take the etiological studies regarding FY in oil palm to another level. We postulate that transcriptomics should be the next step in using omics to gain further insights regarding this disease. Even more, we believe that it should be done under the scope of a multi-omics integration (MOI) strategy, together with metabolomics, proteomics, and ionomics, at least.

Acknowledgements

The author(s) disclosed receipt of the following financial support for the research, authorship, and/or publication of this article: The grant (01.13.0315.00 - DendêPalm Project) for this study was awarded by the Brazilian Innovation Agency - FINEP.

Conflict of interest

The authors declare no conflict of interest.

Author details

Cleiton Barroso Bittencourt[1], Philippe de Castro Lins[2], Alessandra de Jesus Boari[3], Betania Ferraz Quirino[4], Wenceslau Geraldes Teixeira[5] and Manoel Teixeira Souza Junior[4*]

1 Universidade Federal de Lavras (UFLA), Lavras, MG, Brazil

2 Universidade de Brasília (UnB), Brasília, DF, Brazil

3 Embrapa Amazônia Oriental, Belém, PA, Brazil

4 Embrapa Agroenergia, Brasília, DF, Brazil

5 Embrapa Solos, Rio de Janeiro, RJ, Brazil

*Address all correspondence to: manoel.souza@embrapa.br

References

[1] Corley RHV, Tinker PB. The oil palm. 5th ed. Chichester: Wiley Blacwell; 2016. 1299 p. DOI: 10.1002/9781118953297

[2] Our World in Data. Oil palm. [Internet]. 2020. Available from: https://ourworldindata.org/palm-oil [Accessed: 2021-03-02]

[3] FAOSTAT. Food and agriculture organization of the United Nations. 2018. [Accessed: 2021-03-01]

[4] IndexMundi. 2020. Palm Oil Production by Country in 1000 MT. 2020. Available from: https://www.indexmundi.com/agriculture/?commodity=palm-oil. [Accessed: 2021-03-01]

[5] Furumo PR, Aide TM. Characterizing commercial oil palm expansion in Latin America: land use change and trade. Environmental Research Letters. 2017; 12(2):024008. DOI: 10.1088/1748-9326/aa5892

[6] Franqueville H. Oil palm bud rot in Latin America. Experimental Agriculture. 2003; 39(3): 225-240. DOI: 10.1017/S0014479703001315

[7] Martínez G, Sarria GA, Torres GA, Aya HA, Ariza JG, Rodríguez J. *Phytophthora* sp. es el responsable de las lesiones iniciales de la Pudrición del cogollo (PC) de la Palma de aceite en Colombia. Revista Palma. 2008; 29(3):31-41

[8] Boari ADJ. Estudos realizados sobre o amarelecimento fatal do dendezeiro (*Elaeis guineensis* Jacq.) no Brasil. Embrapa Amazônia Oriental-Documentos (INFOTECA-E). 2008

[9] Benítez É, García C. The history of research on oil palm bud rot (*Elaeis guineensis* Jacq.) in Colombia. Agronomía Colombiana. 2014; 32(3):390-398. DOI: 10.15446/agron.colomb.v32n3.46240

[10] Boari AJ, Teixeira WG, Venturieri A, Martorano L, Tremacoldi CR, Carvalho KB. Avanços nos estudos sobre o amarelecimento fatal da palma de óleo (*Elaeis guinnensis* Jacq.). In Embrapa Solos-Artigo em anais de congresso (ALICE). Tropical Plant Pathology, Brasília, DF, v. 37, ago. 2012. Suplemento. Edição dos Resumos do 45 Congresso Brasileiro de Fitopatologia, Manaus, 2012

[11] Van Slobbe WG. Amarelecimento Fatal (A.F.) at the oil palm estate Denpasa, Brazil. In: International Seminar on the Identification and Control of the Organism(s) and/or Other Factor(s) Causing the Spear Rot Syndrome in Oil Palm, Paramaribo (Suriname) 8-12 Mar 1988; Paramaribo (Suriname). Ministry of Agriculture; 1991. p. 75-80

[12] Trindade DR, Poltronieri LS, Furlan Júnior J. Abordagem sobre o estado atual das pesquisas para a identificação do agente causal do amarelecimento fatal do dendezeiro. In: Poltronieri LS, Trindade DR, Santos IP. (Ed.). Pragas e doenças de cultivos amazônicos. Belém, PA: Embrapa Amazônia Oriental; 2005. p. 439-450

[13] Beuther E, Wiese U, Lukáacs N, Van Slobbe WG, Riesner D. Fatal Yellowing of Oil Palms: Search for Viroids and Double-Stranded RNA. Journal of Phytopathology. 1992;136(4);297-311. DOI: 10.1111/j.1439-0434.1992.tb01312.x

[14] Brioso PST, Montano HG, Figueiredo DV, Poltronieri LS, Furlan Junior J. Amarelecimento fatal do dendezeiro: sequenciamento parcial do fitoplasma associado. Summa Phytopathologica. 2006; 32;50

[15] Celestino FP, Louise C, Lucchini F. Estudos de transmissão do amarelecimento fatal do dendezeiro (*Elaeis guineensis*, Jacq), com insetos suspeitos. In: Congresso Brasileiro de Entomologia, 14., 1993, Piracicaba, SP. Resumos. Piracicaba: SEB; 1993. 807 p. p.194

[16] Louise C. Inventory of Homoptera and Heteroptera in relation to the amarelecimento fatal disease. Spear rot of oil palm in tropical America. Proceedings.1990; 36-46

[17] Viégas I, Frazão D, Furlan Júnior J, Trindade D, Thomaz M. Teores de micronutrientes em folhas de dendenzeiros sadios e com sintomas de amarelecimento fatal. In: Embrapa Amazônia Oriental-Artigo em anais de congresso (ALICE). In: XXV Reunião brasileira de fertilidade do solo e nutrição de plantas; VIII Reunião Brasileira Sobre Micorrizas; VI Simpósio Brasileiro De Microbiologia Do Solo; III Reunião Brasileira De Biologia Do Solo. 24-26, october 2000; Santa Maria, RS. Fertbio; 2000.

[18] Silveira RI, Veiga AS, Ramos EJA, Parente JR. Evolução da sintomatologia do amarelecimento fatal a adubações com omissão de macro e micronutrientes. Belém, PA: Denpasa; 2000

[19] Baena ARC. Propriedades físicas do solo em áreas de ocorrência do amarelecimento fatal do dendezeiro. Embrapa Amazônia Oriental-Séries anteriores (INFOTECA-E). 1999; 1-3

[20] Rodrigues-Neto JC, Correia MV, Souto AL, Ribeiro JAA, Vieira LR, Souza MT, Abdelnur PV. Metabolic fingerprinting analysis of oil palm reveals a set of differentially expressed metabolites in fatal yellowing symptomatic and non-symptomatic plants. Metabolomics. 2018;14(10):1-16. DOI: https://doi.org/10.1007/s11306-018-1436-7

[21] Nascimento SVD, Magalhães MM, Cunha RL, Costa PHDO, Alves RCDO, Oliveira GCD, Valadares RBDS. Differential accumulation of proteins in oil palms affected by fatal yellowing disease. PloS one. 2018;13(4):e0195538. DOI: 10.1371/journal.pone.0195538

[22] Costa OYA, Tupinambá DD, Bergmann JC, Barreto CC, Quirino BF. Fungal diversity in oil palm leaves showing symptoms of Fatal Yellowing disease. PloS one. 2018;13(1):e0191884. DOI: 10.1371/journal.pone.0191884

[23] Kuss VV, Kuss AV, Rosa RG, Aranda DA, Cruz YR. Potential of biodiesel production from palm oil at Brazilian Amazon. Renewable and Sustainable Energy Reviews. 2015; 50:1013-1020. DOI: 10.1016/j.rser.2015.05.055

[24] Carter C, Finley W, Fry J, Jackson D, Willis L. Palm oil markets and future supply. European Journal of Lipid Science and Technology. 2007;109(4):307-3014. DOI: 10.1002/ejlt.200600256

[25] Barcelos E, Rios SDA, Cunha RN, Lopes R, Motoike SY, Babiychuk E, Skirycz A, Kushnir S. Oil palm natural diversity and the potential for yield improvement. Frontiers in plant Science. 1990;6. DOI: 10.3389/fpls.2015.00190

[26] Corley RHV. How much palm oil do we need? Environmental Science & Policy. 2009;12(2):134-139. DOI: 10.1016/j.envsci.2008.10.011

[27] Sheil D, Casson A, Meijaard E, Van Noordwijk M, Gaskell J, Sunderland-Groves J, Wertz K, Kanninen M. The impacts and opportunities of oil palm in Southeast Asia: What do we know and what do we need to know? Center for International Forestry Research. 2009; 51:1-80. DOI: 10.17528/cifor/002792

[28] Carrere, R. Oil palm in Africa: Past, present and future scenarios. WRM series on tree plantations. 2010. 5:111

[29] Pacheco P, Gnych S, Dermawan A, Komarudin H, Okarda B. The palm oil global value chain: Implications for economic growth and socialand environmental sustainability. Working Paper 220. Bogor, Indonésia. CIFOR; 2017. 55 p. DOI: 10.17528/cifor/006405

[30] Castiblanco C, Etter A, Aide TM. Oil palm plantations in Colombia: a model of future expansion. Environmental science & policy. 2013;27:172-83. DOI: 10.1016/j.envsci.2013.01.003

[31] GREPALMA. Palma of Guatemala. Available from: https://www.grepalma.org/en/development-for-guatemala/ [Accessed: 2021-03-02]

[32] Silva FL, Homma AK, Pena HWA. O cultivo do dendezeiro na Amazônia: promessa de um novo ciclo econômico na região. Embrapa Amazônia Oriental-Artigo em periódico indexado (ALICE). 2011; 158: 1-24

[33] Homma AKO, Furlan-Júnior J, Carvalho RAD, Ferreira CAP. Bases para uma Política de Desenvolvimento da Cultura do Dendezeiro na Amazônia. In: Viegas, I., Muller, A. (eds), A Cultura do Dendezeiro na Amazônia Brasileira, 1 ed., cap. 1. Belém, Pará, 2000

[34] Bentes EDS, Homma AKO. Importação e exportação de óleo e palmiste de dendezeiro no Brasil (2010-2015). In Embrapa Amazônia Oriental-Artigo em anais de congresso (ALICE). In: Congresso da Sociedade Brasileira de Economia, Administração e Sociologia Rural (SOBER); 14-17 Agosto 2016; Universidade Federal de Alagoas – UFAL. Maceió. 2016. p. 1-16

[35] Ramalho Filho A, Da Motta P, Freitas P, Teixeira W. Zoneamento agroecológico, produção e manejo para a cultura da palma de óleo na Amazônia. 1st ed. Embrapa Solos; 2010. 216 p

[36] Sistema IBGE de Recuperação Automática – SIDRA. Produção agrícola. Available from: https://sidra.ibge.gov.br/pesquisa/pam/tabelas/. 2019. [Accessed: 2021-03-02]

[37] Motta PEF, Naime UJ, Goncalves A, Baca J. Zoneamento agroecológico do dendezeiro para as áreas desmatadas do Estado de Rondônia. In Embrapa Solos-Artigo em anais de congresso (ALICE). In: XXXII CONGRESSO BRASILEIRO DE CIÊNCIA DO SOLO; 2-7 agosto 2009; Fortaleza: SBCS. 2009.

[38] Brazilio M, Bistachio NJ, Cillos Silva V, Nascimento DD. O Dendezeiro (*Elaeis guineensis* Jacq.) - Revisão. Bioenergia em Revista: Diálogos. 2012; 2(1): 27-45.

[39] Van de Lande HL. Studies on the Epidemiology of Spear Rot in Oil Palm (*Elaeis guineensis* Jacq.) in Suriname. Proefschrift Landbouwuniversiteit, Wageningen, the Netherlands. 1993

[40] Trindade DR, Furlan Júnior J. Amarelecimento fatal do dendezeiro. In: Muller AA; Furlan Júnior J. (Ed.). Agronegócio do dendê: uma alternativa social, econômica e ambiental para o desenvolvimento sustentável da Amazônia. Belém, PA: Embrapa Amazônia Oriental; 2001. p. 145-152

[41] Van Slobbe WG. Amarelecimento fatal: final report. Belém, PA: Denpasa, 1991. 100 p

[42] DENPASA. Amarelecimento Fatal. 2021. Available from: http://denpasa.com.br/pt-br/amarelecimento-fatal-af/ [Accessed: 2021-02-15]

[43] Riley MB, Williamson MR, Maloy O. Plant disease diagnosis. The Plant Health Instructor. 2002; 3. DOI: 10.1094/PHI-I-2002-1021-01

[44] Kastelein P, Van Slobbe WG, De Leeuw GTN. Symptomatological and histopathological observations on oil palms from Brazil and Ecuador affected by fatal yellowing. Netherlands Journal of Plant Pathology. 1990;96(2):113-117. DOI: 10.1007/BF02005135

[45] Ayala LS. Relatório de visita à Denpasa. In: DENPASA. Pesquisa sobre amarelecimento fatal do dendezeiro. Belém, PA. 2001. 319 p.

[46] Bernardes MSR. Relatório de visitas à plantação de Paricatuba, na Denpasa, visando à identificação das causas do AF (1999). In: DENPASA. Pesquisa sobre amarelecimento fatal. Belém, PA, 2001

[47] Rios SDA, da Cunha RNV, Lopes R, da Silva EB. Recursos genéticos de palma de óleo (*Elaeis guineensis* Jacq.) e caiuaé (*Elaeis oleifera* (HBK) Cortes). Embrapa Amazônia Ocidental-Documentos (INFOTECA-E);2012

[48] Dransfield J, Uhl NW, Asmussen CB, Baker WJ, Harley MM, Lewis CE. A new phylogenetic classification of the palm family, Arecaceae. Kew Bulletin. 2005;559-569

[49] Eiserhardt WL, Pintaud JC, Asmussen-Lange C, Hahn WJ; Bernal R; Balslev H; Borchsenius F. Phylogeny and divergence times of *Bactridinae* (Arecaceae, Palmae) based on plastid and nuclear DNA sequences. Taxon. 2011; 60(2):485-498. DOI:10.1002/tax.602016

[50] Bakoumé C, Wickneswari R, Siju S, Rajanaidu N, Kushairi A, Billotte N. Genetic diversity of the world's largest oil palm (*Elaeis guineensis* Jacq.) field genebank accessions using microsatellite markers. Genetic Resources and Crop Evolution. 2015;62(3):349-360. DOI:10.1007/s10722-014-0156-8

[51] Maizura I, Rajanaidu N, Zakri AH, Cheah SC. Assessment of Genetic Diversity in Oil Palm (*Elaeis guineensis* Jacq.) using Restriction Fragment Length Polymorphism (RFLP). 2006;53(1):187-195. DOI:10.1007/s10722-004-4004-0

[52] Ithnin M, The CK, Ratnam W. Genetic diversity of *Elaeis oleifera* (HBK) Cortes populations using cross species SSRs: implication's for germplasm utilization and conservation. BMC Genetics. 2017; 18(1):18-37. DOI:10.1186/s12863-017-0505-7

[53] Junior RAG, Lopes R, Cunha RNV, Abreu-Pina AJ, Quaresma CE, Campelo RD, Resende MDV. Ganhos de seleção para produção de cachos em híbridos interespecíficos entre caiaué e dendê. Pesquisa Agropecuária Brasileira. 2019;54(X): e00819-x. DOI: 10.1590/S1678-3921.pab2019.v54.00819

[54] España MD, Mendonça S, Carmona PAO, Guimarães MB, da Cunha RNV, Souza, M. T. Chemical characterization of the American oil palm from the Brazilian Amazon forest. Crop Science. 2018;58(5): 1982-1990. DOI: https://doi.org/10.2135/cropsci2018.04.0231

[55] Hormaza P, Fuquen EM, Romero HM. Phenology of the oil palm interspecific hybrid *Elaeis oleifera*×*Elaeis guineensis*. Scientia Agricola. 2012;69(4), 275-280. https://doi. org/10.1590/S0103-90162012000400007

[56] Trindade DR. Ações de pesquisas, objetivando a identificação do agente causal do amarelecimento fatal-AF do dendezeiro. Embrapa Amazônia Oriental-Outras publicações científicas (ALICE), 1995

[57] Harrison NA, Helmick, E. E., & Elliott, M. L. Lethal yellowing-type diseases of palms associated with phytoplasmas newly identified in Florida, USA. Annals of Applied Biology. 2008;153(1), 85-94 DOI: https://doi.org/10.1111/j.1744-7348.2008.00240.x

[58] Santos AF, Valois AC, Hartz JL. Workshop sobre a cultura do dendê. In: Workshop sobre a cultura do dendê; 24-27 october 1995. Manaus, Amazona. 1995

[59] Bertaccini A. Phytoplasmas: diversity, taxonomy, and epidemiology. Front Biosci. 2007;12(2):673-689. 10.2741/2092

[60] Bertaccini A, Duduk B. Phytoplasma and phytoplasma diseases: a review of recent research. Phytopathologia mediterranea. 2009;48(3):355-378. DOI: 10.14601/ Phytopathol_Mediterr-3300

[61] Harrison NA, Helmick EE, Elliott ML. Lethal yellowing-type diseases of palms associated with phytoplasmas newly identified in Florida, USA. Annals of Applied Biology. 2008;153(1):85-94. DOI: 10.1111/j.1744-7348.2008.00240.x

[62] Bertaccini A, Duduk B, Paltrinieri S, Contaldo N. Phytoplasmas and phytoplasma diseases: a severe threat to agriculture. American Journal of Plant Sciences. 2014;5, 1763-1788: DOI: 10.4236/ajps.2014.512191

[63] Trivellone V. An online global database of Hemiptera-Phytoplasma-Plant biological interactions. Biodiversity data jornal. 2019;7:e32910. DOI: 10.3897/BDJ.7.e32910

[64] Montano HG, Brioso PS, Pimentel JP. List of phytoplasma hosts in Brazil. Bulletin of Insectology. 2007; 60:129-130

[65] Montano HG. Fitoplasmas e fitoplasmoses no Brasil. Revisão Anual de Patologia de Plantas (RAPP). 2013;21:034-095

[66] Musetti R, Favali MA. Microscopy techniques applied to the study of phytoplasma diseases: traditional and innovative methods. Multidisciplinary Microscopy Research and Education. 2004. 2:72-80

[67] Brioso PST, Montano HG, Trindade DR, Poltronieri LS, Barcelos E, Veiga AS, Furlan-Júnior J. Fitoplasma do grupo 16S rRNA I associado ao amarelecimento fatal de Elaeis guineensis. In Congresso Paulista de Fitopatologia (Vol. 26); 2003

[68] Brioso PST, Montano HG, Figueiredo DV, Poltronieri LS, Furlan Junior J. Amarelecimento fatal do dendezeiro: sequenciamento parcial do fitoplasma associado. Summa Phytopathologica. 2006;32-35

[69] Silva HM. Relatório de atividades junto ao consultor em nematologia; 1995.

[70] Silva HM. Relatório de avaliação dos trabalhos com amarelecimento fatal. Belém, PA; 1989. 5 p

[71] Mendonça, 2016

[72] Ferraz LCCB. Relatório final - Apoio técnico na especialidade de nematologia de plantas. In: DENPASA. 2021

[73] Lin MT. Study on fatal yellowing of oil palms: two technical reports research contract Denpasa-Bioplanta.1990; 24 p.Typescripts (unpublisshed).

[74] Lin MT. Comparative analysis of oil palm tissues with and without fatal yellowing symptoms by centrifugation: technical report-research contract Denpasa-Bioplanta. 1989; 5 p. Typescripts (unpublisshed)

[75] Kitajima EW. Report to Uepae de Belém about E. M. observations on tissues of healthy and by AF affected palms from Denpasa. Brasilia, DF: Universidade de Brasília – Departamento de Biologia Celular. 1991. 2 p. Typescript (unpblished).

[76] Singh RPAG, Avila AC, Dusi AN. Boucher A, Trindade DR, Van

Slobbe WG, Ribeiro SG, Fonseca MEN, Association of viroid-like nucleic acids with the fatal yellowing diseases of oil palm. Fitopatologia Brasileira.1988; 13(4):392-394.

[77] Dollet M, Mazzolini L, Bernard V. Research on viroid-like molecules in oil palm. In: ACIAR PROCEEDINGS. Australian Centre for International Agricultural Research, 1993. p. 62-62.

[78] Bergamin-Filho A, Amorim L, Laranjeira FF, Berger RD, Hau B. Análise temporal do amarelecimento fatal, do dendezeiro como ferramenta para elucidar sua etiologia. Fitopatologia Brasileira. 1998; 23(3): 391-396.

[79] Laranjeira FF, Bergamin-Filho A, Amorim L, Berger RD, Hau B. Análise espacial do amarelecimento fatal do dendezeiro como ferramenta para elucidar sua etiologia. Fitopatologia Brasileira. 1998; 23(3): 397-403.

[80] Viégas IJM, Frazão DAC, Furlan-Júnior J, Trindade DR, Thomaz MAA. Teores de micronutrientes em folhas de dendezeiros sadios e com sintomas de amarelecimento fatal. In: XXV Reunião Brasileira de Fertilidade do Solo e Nutricão de Plantas, VIII Reunião Brasileira Sobra Micorrizas, VI Simpósio Brasileiro de Microbiologia do Solo, III Reunião Brasileira de Biologia do Solo. 22-26 October 2000; Santa Maria. Rio Grande do Sul. 2000. p. 1-4.

[81] Silveira RI, Veiga AS, Ramos EJA, Parente JR. Evolução da sintomatologia do amarelecimento fatal a adubações com omissão de macro e micronutrientes. Belém, PA: Denpasa, 2000. 35 p.

[82] Bernardes MSR. Relatório de visitas à plantação de Paricatuba, na Denpasa, visando à identificação das causas do AF (1999). In: DENPASA. Pesquisa sobre amarelecimento fatal. Belém, PA, 2001.

[83] Muniz RS. Alterações do fluxo hídrico e seus efeitos na dinâmica do ferro e na estrutura de um Latossolo Amarelo na Amazônia [thesis]. Rio de Janeiro: Universidade Federal do Rio de Janeiro. 2017.

[84] Teixeira W, Pina ADA, Boari ADJ, Martins GC, Lima WAA, Muniz RS, Gonçalves AO, Encinas OC, Araújo AC. A hipótese abiótica como agente causal do amarelecimento fatal (AF) da palma de óleo (*Elaeis guineensis* Jacq.) no Brasil. In XXXVI Congresso Brasileiro de Ciência do Solo. 30-04 august 2017; Belém. Pará; 2017.

[85] Encinas OC. Dinâmica da água e nutrientes na solução do solo em um dendezal (*Elaeis guineensis* Jacq.) na Amazônia Central [thesis]. Manaus: Universidade Federal do Amazonas; 2016.

[86] Hosainzadegan H, Khalilov R, Gholizadeh P. The necessity to revise Koch's postulates and its application to infectious and non-infectious diseases: a mini-review. European Journal of Clinical Microbiology & Infectious Diseases. 2020 Feb;39(2):215-218. DOI: 10.1007/s10096-019-03681-1

[87] Antonelli G, Cutler S. Evolution of the Koch postulates: towards a 21st-century understanding of microbial infection. Clinical Microbiology and Infection. 2016 Jul;22(7):583-584. DOI: 10.1016/j.cmi.2016.03.030

[88] Boolchandani M, D'Souza AW, Dantas G. Sequencing-based methods and resources to study antimicrobial resistance. Nature Reviews Genetics. 2019;20(6):356-370. DOI: 10.1038/s41576-019-0108-4

[89] Giwa AS, Ali N, Athar MA, Wang K. Dissecting microbial community structure in sewage treatment plant for pathogens' detection using metagenomic sequencing technology.

Archives of Microbiology. 2020 May;202(4):825-833. DOI: 10.1007/s00203-019-01793-y

[90] Garmaeva S, Sinha T, Kurilshikov A, Fu J, Wijmenga C, Zhernakova A. Studying the gut virome in the metagenomic era: challenges and perspectives. BMC Biology. 2019 Oct 28;17(1):84. DOI: 10.1186/s12915-019-0704-y

[91] Li X. Metagenomic screening of microbiomes identifies pathogen-enriched environments. Environmental Sciences Europe. 2019 Dec;31(1):37. DOI: 10.1186/s12302-019-0217-x

[92] Tupinambá DD, Cantão ME, Costa OYA, Bergmann JC, Kruger RH, Kyaw CM, et al. Archaeal Community Changes Associated with Cultivation of Amazon Forest Soil with Oil Palm. Archaea. 2016 Feb 24;2016:3762159. DOI: 10.1155/2016/3762159

[93] Alteio LV, Schulz F, Seshadri R, Varghese N, Rodriguez-Reillo W, Ryan E, et al. Complementary metagenomic approaches improve reconstruction of microbial diversity in a forest soil. mSystems. 2020 Mar 10;5(2). DOI: 10.1128/mSystems.00768-19

[94] Liu X-F, Liu C-J, Zeng X-Q, Zhang H-Y, Luo Y-Y, Li X-R. Metagenomic and metatranscriptomic analysis of the microbial community structure and metabolic potential of fermented soybean in Yunnan Province. fst. 2020 Mar;40(1):18-25. DOI: 10.1590/fst.01718

[95] Prodan A, Tremaroli V, Brolin H, Zwinderman AH, Nieuwdorp M, Levin E. Comparing bioinformatic pipelines for microbial 16S rRNA amplicon sequencing. PLoS One. 2020 Jan 16;15(1):e0227434. DOI: 10.1371/journal.pone.0227434

[96] Fadiji AE, Babalola OO. Metagenomics methods for the study of plant-associated microbial communities: A review. J Microbiol Methods. 2020 Mar;170:105860. DOI: 10.1016/j.mimet.2020.105860

[97] Bandla A, Pavagadhi S, Sridhar Sudarshan A, Poh MCH, Swarup S. 910 metagenome-assembled genomes from the phytobiomes of three urban-farmed leafy Asian greens. Scientific Data - Nature. 2020 Aug 25;7(1):278. DOI: 10.1038/s41597-020-00617-9

[98] Pascale A, Proietti S, Pantelides IS, Stringlis IA. Modulation of the root microbiome by plant molecules: the basis for targeted disease suppression and plant growth promotion. Frontiers in Plant Science. 2019;10:1741. DOI: 10.3389/fpls.2019.01741

[99] Chen S, Harmon A.C. Advances in plant proteomics. Proteomics. 2006; 6(20), 5504-5516. DOI:10.1002/pmic.200600143

[100] Kosová K, Vítámvás P, Prášil IT, Renaut J. Plant proteome changes under abiotic stress—contribution of proteomics studies to understanding plant stress response. Journal of proteomics. 2011;74(8), 1301-1322. DOI: 10.1016/j.jprot.2011.02.006

[101] Quirino BF, Candido ES, Campos PF, Franco OL, Krüger RH. Proteomic approaches to study plant–pathogen interactions. Phytochemistry. 2010;71(4), 351-362. DOI: https://doi.org/10.1016/j.phytochem.2009.11.005

[102] Alam I, Lee DG, Kim KH, Park CH, Sharmin SA, Lee H, ... & Lee, B. H. (2010). Proteome analysis of soybean roots under waterlogging stress at an early vegetative stage. Journal of Biosciences, 35(1), 49-62. DOI: 10.1007/s12038-010-0007-5

[103] Nakazono M, Tsuji H, Li Y, Saisho D, Arimura SI, Tsutsumi N, Hirai A. Expression of a gene encoding mitochondrial aldehyde dehydrogenase in rice increases under submerged

conditions. Plant Physiology. 2000;124(2), 587-598. DOI: 10.1104/pp.124.2.587

[104] Nakabayashi R, Saito K. Metabolomics for unknown plant metabolites. Analytical and bioanalytical chemistry. 2013;405(15), 5005-5011. DOI: 10.1007/s00216-013-6869-2

Chapter 9

Oil Palm (*Elaeis guineensis*) Cultivation and Food Security in the Tropical World

Famous Baa Adade

Abstract

The paper examined the nexus between oil palm cultivation in the tropics and food security. It was established that food security is real in oil palm producing nations of the tropical world and has grave consequences on people and the economies. Some of the identified drivers of food insecurity include oil palm production policies that do not support food production, problem of economies of scale among smallholder farmers, increased land use pressure from expanding industrial and urban areas as well as poor food distribution system, among others. From theoretical perspective, some authors expressed a direct relationship between palm oil production and food insecurity. However, a number of empirical studies from different regions of the tropics indicated that palm oil production promote food security. In addressing how oil palm cultivation mitigates or promote food insecurity, we looked at palm oil and the pillars of food security – food availability, access, stability and utilization. For this to be achieved there must be changes to the use of palm oil through private regulatory regimes like the Roundtable on Sustainable Palm Oil (RSPO), acquisition of only reserve agricultural land or unused land instead of grabbing land already in use by local communities, and ensuring food crop integration into plantations. Thus, policy makers willing to maintain the tropical rain forest, expand cultivation of oil palm must consider the drivers of food insecurity and how expansion of oil palm cultivation, especially under smallholdings, promotes or negatively affects food security in the tropical region.

Keywords: oil palm, food security, tropical world

1. Introduction

Palm oil, derived from *Elaeis guineensis*, is the world's most traded vegetable oil and 90% of it is used for food consumption while industrial consumption such as cosmetic production or fuel and diesel covers the remaining 10% [1]. Recent years have witnessed a massive expansion of oil palm monocultures in the tropics of Asia, Africa and South America, mainly to the detriment of rainforests, agroforests (timber and jungle rubber), and traditional arable crops [2, 3]. Tens of millions of farm households in the tropical areas continue to adopt the crop to enhance their livelihood [4]. This increasing adoption by smallholder farmers is carried out despite the crop's requirement of an expensive initial investment, managerial skills, and a switch to more capital-intensive farming practices [5]. The cultivated area of

IntechOpen

oil palm cultivation has increased greatly exceeding over 16 million hectares in 2017, with the yearly palm oil production increasing from one million tonnes in 1970 to 63 million tonnes by 2016 [6]. This was possibly because the crop is cheap to produce; it is more efficient as less land, fewer pesticides, and less fertilizers are used. For societies, especially in the tropics, having food security as their top priority, palm oil production could be one of the best approaches towards alleviating food insecurity.

The expansion in the production of palm oil has created concerns about the effect of such an activity on local food security and rural livelihoods. It has also produced diverse effects on the growers and the environment. Though some studies suggest that farm households' adoption of the crop contributes to alleviate poverty and improve households' income and living standards [5, 7–10], others have seen the activity as either promoting food security or aggravating the problem of food insecurity in some localities. For instance, [11, 12] have seen oil palm cultivation as an opportunity for fighting against rural poverty and food insecurity in several Southeast Asian countries, including in Indonesia and Malaysia.

However, [13] opined that the impact of oil palm expansion on food security is uncertain. This is possibly because, it is not always clear as to what extent, and under which prevailing circumstances, the production of the crop improves or compromises rural livelihoods and household food security [6, 14]. Some researchers [15] have argued that expansion of oil palm cultivation affects available land for food crop production and as a result food insecurity might be promoted. Others [16] have reasoned that waged employment in industrial crop plantations or smallholder oil palm cultivation can generate employment and income opportunities in rural areas of tropical region. Households can invest the obtained income to purchase food or improve farm productivity thereby increasing the yield of crops.

Available literature indicates that existing studies focused on the impact of oil palm expansion on some environmental and socio-economic issues rather than on food security; only few capture the interaction between oil palm cultivation and household food security. It is important to explore this area in order to understand the nature of the relationships between oil palm expansion in the tropics and household food security from a theoretical perspective and based on some empirical evidences. This paper, therefore, addresses the nexus between oil palm cultivation in the tropics and food security, and proffer solution on how land-use for oil palm can best be carried out to promote food security rather than food insecurity. It also aims to discuss identified several mechanisms through which oil palm production as an industrial crop interacts with the different pillars of food security – food availability, access, stability and utilization.

2. Global food security challenges

The Food and Agricultural Organization (FAO) projected that the agri-food sector would need to generate 50 percent more food by 2050 in order to meet the demand requirements, thereby making food security to be a global serious threat to millions of households in developing countries [17]. In the FAO's 2019 Summit on Food Security and Nutrition in the world, several levels of food insecurity were identified: moderate food insecurity and severe food insecurity. Moderate food insecurity occurs when people face uncertainties about their ability to obtain food and thereby forced to reduce, at times during the year, the quantity and/or quality of food they consume due to lack of money or other resources. Severe food insecurity, on the other hand, affects a community or a nation when people have likely run out of food, experienced hunger and, at the most extreme, gone for days without eating, putting their health and wellbeing at serious risk.

Despite significant progress in recent decades by most countries in the oil palm producing countries of the tropics, hunger and nutritional deficiencies still constitute serious challenges in farm households. Indonesia, the leading oil palm growing country in the world, which was once self-sufficient in rice and sugar failed to keep up with demand in the face of rising population and there are doubts about the future stability in the country's food system. This could be linked to the farm decisions of most oil palm farmers. Plantation farmers, for example, hardly cultivate food crops for their own consumption [5, 18]. Most plantation farmers heavily depend on agricultural cash income to purchase adequately diverse foods from such imperfect markets [18, 19], which consequently makes them vulnerable to substantial income and price shocks. Moreover, cultivating perennial and non-food commercial crops—that do not directly add to household dietary diversity through own consumption, are claimed to compete for resources (e.g., land) with other food crops that in turn negatively affects food availability and increase food prices [20]. This has significant implications in terms of food and nutrition security. According to the 2015 Global Hunger Index (GHI), Indonesia reduced its GHI score by about 25% and the rate was found to be higher in other oil palm growing nations like Thailand and Vietnam [21]. As a result, many of the nations have to resort to reliance on foreign imports to meet demand for key food products to ensure some level of food security. Concerning nutritional deficiencies, about 40% of the Indonesian population is affected by under-nutrition and micronutrient malnutrition, and majority of the affected group are farm households [22, 23]. It is not unlikely that similar situations may be prevalent in other oil palm growing regions of the tropical world.

For instance, [12] observed that the Jambi Province on the island of Sumatra, like other rural areas in South East Asia, has high levels of underweight and stunted children, poor household dietary diversity and pervasive micronutrient deficiency. The diets of farm households in the tropics in general are highly vulnerable to food prices and income shocks, most times due to global drop in prices of oil palm and rubber [24].

Food insecurity has grave consequences on people and the economy. Food insecurity could lead to stunting in children which is a significant health challenge. Between 2005 and 2015, the rate of stunting in children under the age of five in Indonesia increased from 28.6% to 36.4% According to the World Bank, children that experience stunting in their early development are less likely to graduate high school and are expected to earn ten percent less during their lifetime than their food secure peers.

In the Buvuma area of Uganda in East Africa food insecurity has been reported to be high. Poverty in the area made some farmers to sell their land cheaply to large oil palm companies. When they spent the money, several residents resorted to stealing food [25].

3. Drivers of food insecurity in the oil palm producing tropical land

The farming of oil palm, which is not a food crop, contributes to food insecurity. Piesse [21] opined that oil palm production does not further food security and makes importation of food more likely. The researcher also opined that the competition for land from industry and housing has pushed many farmers out of the market thereby further reducing the ability of palm oil producing nations like Indonesia to produce their own food.

Most food-crop farmers are smallholders and face problems of economies of scale. This has made it difficult to increase food production to meet self-sufficiency targets. Other factors that affect hunger and malnutrition include economic slowdowns and downturns, world price fluctuations for countries dependent on production of primary commodities and the ability to trade in free and open markets which leads to unemployment affects income and ultimately access to food. This is mainly affecting countries dependent on primary commodity exports, in South America, Asia and some countries in Africa.

Promoting policies of oil palm nucleus estates that is not balanced with sustained programmes that gradually increase agricultural food productivity and distribution, have the potential to reduce food security. The experience of Indonesia is a case in point. Though the country has the potential to produce enough food to feed its population, but it is prevented from doing so due to some policy issues. Large tracts of its agricultural land have been developed for oil palm plantations, which, while commercially successful, do nothing to bolster food security.

The food distribution network is one of the largest barriers to food security and increased consumption of domestically produced food. Increased land use pressure from expanding industrial and urban areas makes it difficult to find new agricultural land that is close to transport infrastructure. This is found to be partly responsible for Indonesia's poor domestic food production capacity. As the world's largest archipelagic state, Indonesia faces unique challenges that complicate its food distribution system. Transporting food throughout its 6,000 or so inhabited islands is a particularly difficult undertaking that the government has long grappled with.

Demographic transition also poses a potential problem for food supply, especially in Indonesia. Piesse [21] noted that the Indonesian middle class is currently the fourth largest in the world, after the US, India and China. By 2030, about 20 million households are likely to belong to the middle class. As the middle class grows to occupy a larger portion of the population, a shift in food preferences is likely to follow. Middle class consumers are more likely to purchase higher-cost food products, such as meat, dairy and processed foods, which Indonesia will struggle to supply through domestic production alone. Indonesia will continue to rely on foreign imports to meet domestic demand in key food products such as rice and beef. This is likely going to be the situation in other oil palm producing countries in South East Asia, Sub-Saharan Africa and in South America.

Low commodity prices are also likely to drive food insecurity. A consistent slide in the price of commodities over time may lead to depreciation and devaluation of currencies resulting in domestic price increases, including food prices. This then affects the ability of households to buy food as the cost of food relative to their incomes increases.

4. Nexus between oil palm and food security: theoretical perspectives and empirical evidences

Some researchers have expressed a direct relationship between palm oil production and food insecurity. For instance, [26], argues that palm oil production creates food insecurity in a direct way for local communities, especially rural and indigenous communities whenever government allocates to private plantation firms the land on which such communities depend on for their food and livelihood. Kimbowa [25] reported that Buvuma – Oil Palm Uganda Limited-BOPUL, a subsidiary of Oil Palm Uganda Limited and Bidco Uganda Limited in Kalangala arranged to acquire from the local communities 6, 500 hectares of land for oil palm cultivation while the out-growers of the companies will use 3, 500 hectares and this affected food production. Since their compensation in 2012, most of the residents have failed to secure alternative land for settlement and food production. In Nigeria, the Friends of the Earth also claimed that the allocation of agricultural land by the Cross River State Government to Wilmar International for the development of oil palm nucleus estates robbed the local communities of their land for arable crop production thereby promoting food insecurity in the affected communities.

Many plantation farmers are known to heavily depend on agricultural cash income to purchase adequately diverse foods [18, 19], thereby making them vulnerable to substantial income and price shocks. Furthermore, focusing on perennial and non-food commercial crops like oil palm that do not directly add to household dietary diversity through own consumption, is likely to encourage competition for land with other food crops which could negatively affect food availability and food prices [20].

A link between oil palm cultivation and food security has been established through some empirical studies. Using panel farm household data from Jambi province on the island of Sumatra, Indonesia, [24] examined the effects of oil palm adoption on dietary diversity, quantities of fruits and vegetables consumed calories, and food expenditure. Endogenous switching regression was applied to control for selection bias and to obtain counterfactual outcomes. Panel logit regression was also used to estimate the impact of oil palm adoption on dichotomous variables of household's diets, indicating whether or not the diet met the minimum adequacy level of fruits and vegetables consumed as well as intake of calorie, iron, zinc, vitamin A, and the average of the three micronutrients. Regardless of the diet indicator, oil palm adoption was found to have statistically significant and positive effects, indicating that oil palm cultivation leads to higher food and micronutrient adequacy in general. On an average, the adoption of oil palm increases the probability of consuming fruits and vegetables by 33.6%, calorie adequacy by 38.6%, iron adequacy by 36.4%, zinc adequacy by 54.9%, vitamin A adequacy by 33.1%, and average adequacy of the three micronutrients by 35%. Together, with the results from the endogenous switching regression, it was deduced that oil palm adoption improves the diets of farm households in the tropics, whether they belong to the migrants or the local communities. Hence, the nutritional impact might justify why farm households in the tropical region are rapidly expanding oil palm cultivation. Moreover, several socioeconomic, farm, and demographic factors impact oil palm adoption and, at the same time, shape the diversity and adequacy of diets in those households. The study therefore supports the idea that adopting a perennial and non-food commercial crop like oil palm does not worsen dietary quality and diversity in farm household. Rather land-use change through oil palm adoption significantly improves the diets of farm households in the tropics.

An exploratory study was also carried out by [6] to assess the food security outcomes of smallholder-based oil palm and rubber production at the household level in the forest region of Guinea using six standardized metrics of food security. The selected metrics covered different aspects of food security related to diet diversity, perceptions of hunger and coping behaviors in the face of food scarcity. Households involved in industrial crop production were compared with households that only grow food crops under subsistence conditions, using statistical tools like Propensity Score Matching (PSM) and Endogenous Treatment Effect Regression (ETER). The results obtained are mixed. Both oil palm and rubber have significantly lower levels of diet diversity (Food Consumption Score, FCS) than subsistence farmers. However, industrial crop smallholders have lower levels of perception of hunger compared to subsistence farmers, with oil palm farmers having significantly better values than other groups. Both the PSM and ETER analyses suggest that involvement in industrial crop production decreases food security in terms of diet diversity, but when it comes to perception of hunger, the involvement of oil palm production improves food security. The results of the ETER regarding involvement in oil palm production and coping strategy index were statistically significant and therefore suggest that involvement in industrial crop production improves coping strategy index thereby enhancing food security. Overall, results show that oil palm and rubber smallholders perform better than subsistence farmers on metrics that capture perceptions of hunger and coping behaviors. However, involvement in oil palm and rubber production reduces the levels of food security metrics that use shorter time scales and measure food diversity. This implies that involvement in industrial crop production does not enhance consistently food security across all metrics. This could be explained to arise from the strong sense of security that steady and higher income provides across time (food stability), that outweighs the shortcomings on diet diversity (food utilization).

Furthermore, [27] established a causal relationship between oil palm cultivation and farmers' household food security. Their study applied OLS and quantile regression models to household data in Indonesia to find the socio-economic factors that influence farmers' food expenditure and calorie intake, and to estimate the effect of oil palm expansion on food security across quantiles. The study indicated a statistically significant influence of the income from oil palm expansion on calorie intake. The study further showed that expansions of oil palm cultivated area, resulting in more crop income, could lead farmers to consume more nutritious food, but the food share in the household budget decreases, which is consistent with Engel's law. This is in consonant with the work of [5] who found that expansion of oil palm plantation by smallholder farmers positively affected nutritious food intake, particularly at the mid to upper tail of the expenditure distribution, implying that households in such categories spend their income to not only satisfy their basic calorie needs, but also consider the nutrient intake quantities in their daily diets.

5. Mechanism through which oil palm cultivation mitigates or promotes food insecurity

In addressing how oil palm mitigates or promotes food insecurity, we want to look at palm oil and the pillars of food security: availability, access, utilization and stability. There is the availability of palm oil in the tropical world – through production, distribution, and exchange. The crop is efficiently produced in comparison with other vegetable oil. Less land and other resources are required per hectare to produce palm oil and associated by-products. There is an efficient distribution system of the product through storage, processing, transport, packaging and marketing in most oil palm producing countries which eliminates waste in the value chain. This tends to make palm oil available in the countries that consume the product. A system of exchange or cash economy exists to acquire oil palm products in all seasons, thereby promoting food security. However, climate change affects the crop productivity and thus its products availability [28].

Oil palm products are accessible. They are affordable and preferable to individuals and households because of the nutrients in them. Palm oil, for instance, contains beta carotene, a precursor of vitamin A and also contains equal proportion of saturated and unsaturated fatty acids [29]. Oil palm products are sold in units that almost every household can access. Since the crop has been found to lift many of its growers out of poverty [30, 31], the revenue obtained from the sale of oil palm products can be used to access other food items not produced in the environment. Based on the crop income per hectare of average oil palm smallholder farmer estimated to be $2,200 per annum in Nigeria [31] and $1,400 per annum in Indonesia [13], most families in oil palm producing communities are likely have enough financial resources to purchase food at prevailing prices or have sufficient land and other resources to grow their own food.

Oil palm farmers enjoy both direct access to food and financial resources. Many of them, especially those who integrate food crops into their farms, produce food using human and material resources available to them while some others purchase food produced elsewhere, with financial resources obtained from oil palm farming. However, access to food must be available in socially acceptable ways, without, for example, resorting to emergency food supplies, scavenging, stealing or other forms of coping strategies as experienced in the Kalangala region of Uganda where large oil palm companies stripped the locals of their agricultural land and were left without land to farm, though compensated. The amount paid could not keep them for long.

Palm oil and associated products are safe for human and livestock consumption, thereby promoting food utilization as a pillar of food security. With the Roundtable on Sustainable Palm Oil (RSPO), emphasizing certification of oil palm fields and operations, palm oil is safe for ingestion and the nutrients therein are enough to meet the physiological requirements of each individual consumer. The Malaysian Oil Palm Council is doing a lot of sensitization about nutrition and palm oil preparation which can affect oil palm products' utilization and improve on food security.

Palm oil supply and prices can be considered to be relatively stable thereby making it possible to obtain the product supply over time. However, some forms of transitory palm oil food insecurity has existed in the past, thus making the products unavailable during certain periods of time. This has been due to drought resulting in crop failure and decreased food availability. There have also been cases of instability in markets resulting in food-price hikes causing transitory food insecurity. In virtually all oil palm producing nations, there are seasonal palm oil food insecurity resulting from the regular pattern of growing seasons of the crop. In Nigeria, for instance, the production level decreases during the rainy season which reduces the supply of the product to the market, thereby causing price rise. This limits the ability of some poor households to access the product or reduce the quantity bought.

6. How to govern palm oil production to militate against food insecurity

There must be changes to the use of palm oil if the direct and indirect food security contributions of the crop are to be maintained. One of such measures is through private regulatory regimes such as voluntary certification as practised under the Roundtable on Sustainable Soy. So, palm oil sustainability should be effectively promoted in order that plantation areas can be certified by the Roundtable on Sustainable Palm Oil (RSPO) established since 2014 under the Swiss Civil Code. Oil palm producing nations should adopt the food security criteria developed by ZEF in 2015. It is expected to be integrated in sustainability standards for different crops. This is likely to ensure that human rights to food at local level is not violated and the nations buying palm oil are complying with good purchasing practices required by international bodies like UN, OECD and EU.

In order to further reduce the direct consequences of large oil palm plantations on food and livelihood security of rural and indigenous communities, the government of oil palm producing areas regions should allow only acquisition of reserve agricultural land or unused land instead of "grabbing lands" that are already in use by local communities. This is because acquiring forested land or forest dwellings or lands that are lived on and farmed by local communities for generations is likely to cause conflict and deprive the local people from using the area for arable crop production most beneficial to their livelihood.

Another way to promote food security while encouraging oil palm cultivation is through food crop integration into plantations [32]. With crop integration as practised in Malaysia, Nigeria and other oil palm growing nations, food availability is promoted. Crops integrated include – pineapple, groundnut, banana, soya bean, sugar cane, sorghum, sweet corn, sweet potato, green pea, etc. In some cases livestock are reared and made to graze under the palms as being practised by Siat in their plantations in Nigeria, Ghana, and Gabon. In Eastern part of Nigeria, the farmers carry out spacing of their oil palm trees, large enough to prevent the tress from forming canopy, to promote arable farming in the midst of the palms. In this way, the smallholder farmers have adequate supply of vegetable oil and food crops like cassava, yam, vegetables, sweet corn and cocoyam planted in between the oil palms. Thus, food security is promoted.

7. Conclusion

The paper examined the nexus between oil palm cultivation in the tropics and food security. It was established that food security is real in oil palm producing nations of the tropical world and has grave consequences on people and the economies. Some of the identified drivers of food insecurity include oil palm production policies that do not support food production, problem of economies of scale among smallholder farmers, increased land use pressure from expanding industrial and urban areas as well as poor food distribution system, among others. From theoretical perspective, some authors expressed a direct relationship between palm oil production and food insecurity. However, a number of empirical studies from different regions of the tropics indicated that palm oil production promote food security. In addressing how oil palm cultivation mitigates or promote food insecurity, we looked at palm oil and the pillars of food security – food availability, access, stability and utilization. For this to be achieved there must be changes to the use of palm oil through private regulatory regimes like the Roundtable on Sustainable Palm Oil (RSPO), acquisition of only reserve agricultural land or unused land instead of grabbing land already in use by local communities, and ensuring food crop integration into plantations. Thus, policy makers willing to maintain the tropical rain forest, expand cultivation of oil palm must consider the drivers of food insecurity and how expansion of oil palm cultivation, especially under smallholdings, promotes or negatively affects food security in the tropical region.

Author details

Famous Baa Adade
Cambridge Assessment International Education Programme, Word of Faith Schools, G.R.A., Benin City, Nigeria

*Address all correspondence to: fb_adade@yahoo.com

References

[1] Shahbandeh M. Vegetable Oils: Global consumption by oil type 2013/14 to 2020/2021. www.staista.com> statistics>vegetableoil, January 26, 2021.

[2] Daulay A R, Intan K P, Barus B, Pramudya N B. Rice land conversion into plantation crop and challenges on sustainable land use system in the East Tanjug Jabung Regency. Proceeding of Society of Behavioural Sciences 2016. 227: 174-180. doi: 10.1016/j.sbspro.2016. 06.059

[3] Drescher J, Rembold K., Allen K, Beckschäfer P, Buchori D, Clough Y. Ecological and socio-economic functions across tropical land use systems after rainforest conversion. Philos. Trans. R. Soc. Lond. B Biol. Sci. 2016. 371:20150275. doi: 10.1098/ rstb.2015.0275

[4] Byerlee D, Falcon W P, Naylor R L. The Tropical Oil Crop Revolution: Food, Feed, Fuel, and Forests. New York, NY: Oxford University Press 2017.

[5] Euler M, Krishna V, Schwarze S, Siregar H, Qaim M. Oil palm adoption, household welfare, and nutrition among smallholder farmers in Indonesia. World Development 2017. 93, 219-235. doi: 10.1016/j.worlddev.2016. 12.019

[6] Boubacar S B, Mamady D, Cristiano M R, Alexandros G. Smallholder-Based Oil Palm and Rubber Production in the Forest Region of Guinea: An Exploratory Analysis of Household Food Security Outcomes. Agriculture 2019, 9, 41; doi:10.3390/agriculture 9020041 (accessed on 24 February 2021).

[7] Dewi S, Belcher B, Puntodewo A. Village economic opportunity, forest dependence, and rural livelihoods in East Kalimantan, Indonesia. World Development 2005. 33, 1419-1434. doi: 10.1016/j.worlddev.2004.10.006

[8] Sunderlin W D, Dewi S, Puntodewo A, Müller D, Angelsen A, Epprecht M. Why forests are important for global poverty alleviation: a spatial explanation. Ecological Society 2008; 13:24. doi: 10.5751/ES-02590-130224

[9] Tscharntke T, Leuschner C, Veldkamp E, Faust H, Guhardja E, Bidin A. Tropical Rainforests and Agroforests Under Global Change: Ecological and Socio-economic Valuations. Berlin, Heidelberg: Springer Berlin Heidelberg 2010.

[10] Euler M, Hoffmann M P, Fathoni Z, Schwarze S. Exploring yield gaps in smallholder oil palm production systems in eastern Sumatra, Indonesia. Agricultural System 2016. 146, 111-119. doi: 10.1016/j.agsy.2016.04.007

[11] Zen Z, Barlow C, Gondowarsito R. Oil Palm in Indonesian Socio-economic Improvement: A Review of Options. Working Papers in Trade and Development. Canberra: Research School of Pacific and Asian Studies, Australian National University 2005.

[12] FAO, WHO. National Nutrition Strategy paper of Indonesia: Second International Conference on Nutrition (ICN 2) 2014. Rome.

[13] Alwarritzi A, Nanseki T, Chomei Y. Impact of Oil Palm Expansion on Farmers' Crop Income and Poverty Reduction in Indonesia: An Application of Propensity Score Matching. Journal of Agricultural Science 2016; 8(1): 120-131.

[14] Lam, R.D.; Boafo, Y.A.; Degefa, S.; Gasparatos, A.; Saito, O. Assessing the food security outcomes of industrial crop expansion in smallholder settings: Insights from cotton production in

Northern Ghana and sugarcane production in Central Ethiopia. Sustain. Sci. 2017, 12, 677-693.

[15] Yengoh GT, Armah F A. Effects of large-scale acquisition on food insecurity in Sierra Leone. Sustainability 2015; 7, 9505-9539.

[16] Wiggins S, Henley G, Keats S. Competitive or Complementary? Industrial Crops and Food Security in Sub-Saharan Africa. Research Reports and Studies; Overseas Development Institute (ODI): London, UK, 2015.

[17] PalmOilToday. Palm oil holds promise for current and future food security: Health and nutrition. December 17, 2016

[18] Sibhatu K T, Krishna V V, Qaim M. Production diversity and dietary diversity in smallholder farm households. Proceeding, National Academy of Science, U.S.A 2015; 112: 10657-10662. doi: 10.1073/pnas.1510982112.

[19] Sibhatu K T, Qaim M. Farm production diversity and dietary quality: linkages and measurement issues. Food Security 2018; 10: 47-59. doi: 10.1007/s12571-017-0762-3. (accessed February 20, 2021).

[20] Li T M. Social impacts of oil palm in Indonesia: a gendered perspective from West Kalimantan in Occasional Paper 124 (Bogor: Center for International Forestry Research) 2015.

[21] Piesse M. Food security in Indonesia: A continued reliance on foreign markets. March 1 2016. www.futuredirections.org.au

[22] Isabelle M, Chan P. Seminar on young child nutrition: improving nutrition and health status of young children in Indonesia. Asia Pac. J. Clinical Nutrition 2011. 20, 141-147

[23] Ickowitz A, Rowland D, Powell B, Salim M A, Sunderland T. Forests, trees, and micronutrient-rich food consumption in Indonesia. PLoS ONE 2016. 11:e0154139. doi: 10.1371/journal.pone.0154139 (accessed February 20, 2021)

[24] Sibhatu K T. Oil Palm Boom and Farm Household Diets in the Tropics. Frontier Sustainable Food System 2019; 3(75): 1-14.

[25] Kimbowa I. Oil Palm growing threatening food security in Buvuma. https//ugandaradionetwork.net/stoey/oil-palm-threatening-food-security

[26] Nesadurai S E H. Food security, the palm oil–land conflict nexus, and sustainability: a governance role for a private multistakeholder regime like the RSPO?. The Pacific Review 2013; 26(5):505-529, DOI: 10.1080/09512748.2013.842311

[27] Alwarritzi A, Nanseki T, Chomei Y. The Effect of Oil Palm Expansion on Farmers' Household Food Security in Indonesia. Journal of Economics and Sustainable Development 2016; 7(24): 93-99. www.iiste.org ISSN 2222-2855 (accessed on 24 February 2021).

[28] Masipa T S. The impact of climate change on food securityin South Africa: Current realities and challenges ahead. Journal of Disaster Risk Studies 2017; 9(1): 44

[29] Kushairi A. Palm Oil and food security: Empowering change for high income, sustainability and inclusiveness. Malaysian Oil Palm Board International Conference on agricultural extension, Universiti Putra Malaysia, Serdang, Malaysia, 14-16 February, 2017.

[30] Rist L, Feintrenie L, Levang P. The Livelihood Impacts of Oil Palm: Smallholders in Indonesia. Biodiversity and Conservation 2010; 19 (4): 1009-1024.

[31] Adade B F, Okungbowa CI, Erhabor P O. Effect of Presco Oil Palm out-grower scheme on farmers' poverty status in Delta and Edo States, Nigeria. International Journal of Trend in Research and Development (IJTRD) 2018; 5(2): 427-435.

[32] Azhar B, Saadun N, Prideaux M, Lindenmayer D B. The global palm oil sector must change to save biodiversioty and improve food security in the tropics. Journal of Environmental Management 2017; 203(1) ; 457-466.

Chapter 10

Oil Palm Empty Fruit Bunches (OPEFB) – Alternative Fibre Source for Papermaking

Akpan Sunday Noah

Abstract

Elaeis guineensis (oil palm) is one of the most economical perennial oil crops for its valuable oil-producing fruits in tropical regions such as West Africa and South-East Asia. During oil extraction process, these fruits are usually stripped from the fruit bunches leaving behind empty bunches to be discarded as residues. Thus, empty fruit bunches (EFB) of *Elaeis guineensis* are usually considered as waste in the oil palm industry. The abundance of oil palm empty fruit bunches (OPEFB) has created enormous environmental issue, ranging from fouling, attraction of pests, greenhouse gas emissions to soil acidification, thus posing very serious threats to humans and the environment. Globally, in 2014 alone, over 22.4 million tons of EFB were estimated to have been produced. Therefore, exploring eco-friendly disposal methods and productive utilisation of oil palm EFB as alternative fibrous material for papermaking becomes imperative in converting waste to wealth, and initiating environmental wellness. *Elaeis guineensis* empty fruit bunch (EFB) fibre on the average measures 0.99 μm in length, while the fibre diameter and cell wall thickness are 19.1 μm and 3.38 μm respectively. Fibres of EFB are of ligno-cellulosic materials, consisting on the average of an estimated cellulosic content of 30–50%, 15–35% of hemicelluloses and the lignin constituting about 20–30% of extractive-free fibre. The rich cellulose base of EFB fibre makes *Elaeis guineensis* a good potential resource for papermaking furnish moreso that the pulp and paper industry is often referred to as the cellulose industry. Every 5 tons of EFB gives 1 ton of pulp for papermaking. This book chapter will therefore attempt to examine the fibre morphological characteristics of oil palm empty fruit bunch, the chemical properties of EFB fibre, papermaking potentials of empty fruit bunches and ultimately their impact on the environment.

Keywords: Empty fruit bunch (EFB), Oil palm, *Elaeis guineensis*, Fibre morphology, Chemical characterisation, Papermaking potentials, Environmental impact

1. Introduction

Oil palm, *Elaeis guineensis,* is cultivated on a vast scale as a source of oil in West and Central Africa, where it originated, and in Malaysia, Indonesia and Thailand, where it predominates and thrives luxuriantly, and has become well-established over years [1]. The explosive expansion of oil palm plantation in these regions and countries has generated enormous amount of vegetable waste, creating problems in re-planting operations and tremendous environmental concern. Thus, oil palm biomass refers to agricultural by-products generated from oil palm

IntechOpen

industries during re-planting, pruning and milling activities, which in most cases is left to decompose in the fields [2]. In essence, fibres of empty fruit bunches are either by-products of the process of extracting palm oil form the palm fruits, cultivation activities or remains of the trees at the end of their useful life [3]. The wastes are usually disposed indiscriminately or used by the locals as cooking fuel, both of which are not environmentally friendly During oil extraction process, the fruit bunches are left empty as residues after fruit extraction. The presence of these empty fruit bunches (EFB) at mill's gate is unavoidable, but the great headache these biomass residues cause mill management cannot be ignored in waste management and disposal strategies [4].

Millions of tons of oil palm empty fruit bunches on the average are generated annually in different countries across the globe. In Malaysia, for instance, over 5.2 million tons of EFB were generated in 2002 [5]. Thus, oil palm industry is the largest contributor of biomass in Malaysia. These biomass residues are continually generated in large quantities annually with only a small fraction being converted into value-added products while a large percentage are left underutilised [2]. Globally, in 2014, 22.4 million tons of EFB were estimated to have been produced [6], as waste from crude oil palm (COP) processing, the amount of which is abundantly high [7].

The abundance of oil palm empty fruit bunches (OPEFB) has created enormous environmental issue such as fouling and attraction of pests, thereby posing very serious threats to humans and environment. In the context of the afore-mentioned challenges, examining oil palm EFB as an alternative fibrous material to other known pulpable resources such as wood, bamboos, bagasse, straws and grasses, strikes an important concordant note in converting waste to wealth and enhancing environmental wellness.

In countries like Malaysia and Indonesia, oil palm is one of the non-woody plants that have shown great potential as papermaking raw materials. Therefore, this book chapter will attempt to examine the fibre morphological characteristics of OPEFB, the chemical nature of oil palm fibre, the papermaking potentials of empty fruit bunches (EFB) and ultimately their impact on the environment.

2. Fibre morphology and chemical characterisation of OPEFB

Empty fruit bunches are products from the oil palm processing industry. They have great potency as basic raw materials for fermentation because of their cellulose and hemicelluloses contents. The fibres of *Elaeis guineensis* empty fruit bunch (EFB) are of ligno-cellulosic materials, consisting on the average of an estimated cellulosic content of 30–50%, while the hemicelluloses and lignin constitute 15–35% and 20–30% respectively [8]. The OPEFB fibres, which are normally used as mulch for the palm oil mill, have been found to be a rich source of lignocellulosic material, especially cellulose, which can be 33.70–35.11% in composition for a press-shredded fibre [9]. According to reference [10], oil palm fibre exhibits the following percentage chemical composition as presented in **Table 1**.

A comparative broad varied value of the percentage chemical distribution of OPEFB fibres as reported by reference [11] is presented in **Table 2**.

As indicated in **Table 2**, cellulose is the main component in oil palm fibre, with the lignin content also relatively high. Hemicelluloses are of moderate quantity, and contain xylan as the main component. Extractives are of relative proportion, and can be found in traces and in considerable amount. According to reference [12], OPEFB has 50.9% cellulose, 29.6% hemicelluloses, 17.84% lignin, 3.4% ash and 3.21% extractives. Oil palm (OP) fibre contains comparatively high ash content, ranging from 1.6–6.69% [13]. This characteristic might contribute to an abnormal

Parameter	Mean value (%)
Holocellulose	59.6
Lignin	28.5
Ash	5.6
Protein	3.6
Lipid	1.0
Others	0.8
Source: Kobi and Isluzaki (2014).	

Table 1.
Chemical composition of oil palm fibre.

Cellulose	Hemicelluloses	Lignin	References
42.85	11.70	24.01	Rahman *et al.*, 2008
37.28	14.62	31.68	Sudiayani *et al.*, 2013
33.25	23.24	25.83	Mullati *et al.*, 2011
43–43.47	22.93–23.67	21–22.10	Mardawati *et al.*, 2014
Source: Kresnowati et al., 2015.			

Table 2.
Chemical composition of OPEFB fibre (dry basis).

Parameter	Mean value
Length-weighted fibre length (μm)	0.99
Fibre diameter (μm)	19.1
Cell wall thickness (μm)	3.38
Fibre coarseness (mg/m)	0.107
Fines (< 0.2 mm, %)	27.6
Rigidity index $(T/D)^3 \times 10^{-4}$	55.43
Lignin (%)	17.6
Holocellulose (%)	86.3
1% NaOH solubility (%	29.9
Hot water solubility (%	9.3
Alcohol-benzene solubility (%)	2.83
Source: Law and Jiang, 2001.	

Table 3.
Physical and chemical characterisation of OPEFB fibre.

mechanical wear of processing equipment. Thus, the potential build-up of silica in the black liquor recovery system might also be a source of concern in pulping oil palm material [14].

OP fibre is an important lignocellulosic raw material for the preparation of cost-effective and eco-friendly composite materials. One morphological peculiarity of oil palm fibres is that they have a much thicker cell wall when compared with those of wood, yielding substantially a high rigidity index. An electronic microscopic view of fibre cell wall layer reveals that oil palm fibres have structure similar to those of wood cell wall, with lignin distributed highest in the middle lamella in comparison to that of other cells [15]. Fibrous strands of oil palm EFB have unique structure characterised by several large vessel elements in the core region surrounded by vascular fibres [1].

According to reference [16], the average indices of the physical and chemical characterisation of OPEFB fibre can be presented as shown in **Table 3**.

3. Papermaking potential of OPEFB fibre

The utilisation of empty fruit bunches (EFB) of *E. guineensis* is more than an act of environmental friendliness. It is a means to create job and wealth. In order to transform the massive biomass waste generated during oil palm processing into a resource with industrial utility, a viable and sustainable area of utilisation is desirable. The pulp and paper industry becomes the inevitable destination, presenting itself as a veritable option for productive utilisation.

Moreover, substituting the lignocellulosic material of the fast diminishing wood resource with biomass of non-wood plant of various diversity, takes the burden off the forest, while at the same time, supporting natural biodiversity. Thus, using EFB of oil palm for papermaking ameliorates waste management challenges associated with its disposal. In the words of reference [17], every 5 tons of EFB gives 1 ton of pulp for papermaking. Therefore, the oil palm industry is now at the stage of seeking more value-added products, not only from oil and kernel, but also from its biomass. EFB of oil palm is now regarded as a potential feedstock to produce a variety of renewable and valuable biofuel and bio-based chemicals that can be derived from sugar, cellulose and lignocellulose, using furfural [18]. Hence, there is increasing ample opportunity to convert the available lignocellulosic biomass residues into pulp and paper, paperboard, medium density fibreboard (MDF) and other composites [19].

The global production of pulp and paper is expected to increase by 77% from 1995 to 2020 due to the increasing world population, in addition to improved literacy and quality of life worldwide [4]. Consequently, the high growth of paper consumption makes it more demanding to diversify the sources of papermaking fibres which are very much dependent on the forest for their supply. The continued high growth in paper consumption will surely lead to increased demand for papermaking fibres, creating additional pressure on the world dwindling forest resources. Therefore, exploring alternative fibre sources becomes imperative. Oil palm is one of the non-wood plants that show great potential as papermaking raw material.

Oil palm empty fruit bunches can be pulped by semi-chemical mechanical process. Clean pulp obtained by this process is suitable for making unbleached brown paper and moulded products. EFB of oil palm is also very adaptable to chemical pulping such as soda and kraft processes [20]. Soda EFB is claimed to be very suitable for manufacturing printing and writing papers, corrugated cartons, and other paper-based products [21].

Pulp produced from EFB responds favourably to mechanical treatment as reported by reference [22] in their work on the effect of beating time on fibre morphology and drainage time on soda pulp derived from oil palm empty fruit bunches. The beaten fibres were modified to become more shortened, swollen, flexible and collapsible into a smoother sheet with better formation and improved paper quality. The thick cell wall of oil palm fibres is likely to contribute to the production of sheet of high bulk and lower inter-fibre bonding potential in comparison with wood counterparts [23].

However, the paper quality obtained from EFB pulp is comparable to that of hardwood kraft pulp. And with total chlorine bleaching (TCF), pulps can be modified to make them much more conformable and suitable for papermaking.

According to reference [24], paper made from empty bunch pulps has good web characteristics and good printing properties. Consequently, empty fruit bunch of oil palm can serve as a sustainable alternative source of pulp and papermaking fibre.

4. Environmental impact of OPEFB

Oil palm biomass can generally be classified into oil palm fronds (OPF) and oil palm trunks (OPT), oil palm empty fruit bunches (OPEFB), palm kernel shells (PKS), mesocarp fibre (MF), and palm oil mill effluent (POME). In total, 44.85 MT of oil palm biomass is generated annually during the fresh fruit bunch processing, oil palm tree planting and pruning activities [12]. Enormous waste is often generated in the oil palm processing industry in form of empty fruit bunches after fruits extraction for palm oil and palm kernel production. These biomass residues are usually discarded indiscriminately to the detriment of environmental beauty. Where they are gathered away from the immediate vicinity of processing activities, this massive waste is often dumped around the periphery of factory or mill site to form heaps of unhygienic decaying biomass [25].

Proper management of this waste and its disposal is an ardent task and consequently create environmental hazards. These heaps of discarded empty fruit bunches become attractive sources for insects and pests, and a breeding ground for various infectious diseases [26]. The emission of foul-smelling odour at millsite is a constant reminder of the lurking health hazards with the potential for epidemic explosion within the environs of the oil palm processing factory.

Disposal of this massive solid waste causes pollution to the environment. Hence, success in converting this waste material into benefitting products would reduce cost of waste disposed and contribute towards cleaner environment [27]. As reported by reference [28], the use of biomass from the residues of African oil palm would reduce emissions from CO2 from 17.4 Tg p/year to 12.6 Tg p/year and from 3.0 PJ oil p/year to 23.0 PJ of oil p/year, corresponding to 72% and 67% of reduction respectively [28]. Nonetheless, some productive utilisation of these waste materials is not without its attendant effects on the environment. For instance, utilisation of EFB by fast pyrolysis has the potential environmental impact of SO_2 as the causes of acidification and C_2H_4 as the causes of photochemical oxidation process. Greenhouse emissions of CO_2 and CH_4 resulting from the burning of EFB, especially at landsite, are the major causes of Ozone layer depletion and the attendant accentuation of global warming [29].

Life cycle assessment (LCA) of the utilisation of EFB through recycling technologies for fuel, fibre and fertiliser products reveals that methane recovery and compositing are more environmentally friendly than other technologies as measured by reduction of greenhouse gas emissions. On the other hand, pulp and paper, and medium density fibreboard (MDF) production are favourable technologies for land use impacts. However, both recycling technologies for EFB utilisation require intense primary energy, high chemical uses and considerable emission from their waste treatment systems [30].

5. Conclusion

Empty fruit bunches of *Elaeis guineensis* are generated in enormous quantity as waste materials globally in oil palm processing industry. Fibres of *Elaeis guineensis* empty fruit bunch (EFB) are lignocellulosic materials, majorly consisting of an estimated cellulosic content of 30–50%, 15–35% of hemicelluloses and about

20–30% of lignin, based on extractive-free fibre. The rich cellulose base of EFB fibre makes *Elaeis guineensis* a good potential resource for papermaking furnish. In addition, the pulp and paper industry is often referred to as the cellulose industry. Notwithstanding these positive attributes, the abundance of empty fruit bunches as oil palm biomass residues poses serious challenges to humans and is of great consequence to the environment. Consequently, productive uses are needed for oil palm EFB, and research has shown that they can be utilised as an alternative, suitable and sustainable source of papermaking fibre.

Acknowledgements

I sincerely acknowledge the assistance of O. A Adegoke and O.E. Adegboyega, both of the Federal College of Forestry, Ibadan, Nigeria, in typesetting and formatting the manuscript in accordance with the template guidelines.

Conflict of interest

I declare that there is no conflict of interest in the concept, execution and outcome of the research work done towards the making of this book chapter.

Author details

Akpan Sunday Noah
Department of Wood and Paper Technology, Federal College of Forestry, Ibadan, Nigeria

*Address all correspondence to: alloynoah@gmail.com

References

[1] Law, K. N, Wan Rosli, W.D. and Ghazali, A. Oil palm empty fruit bunch, BioResources 2007; 2(3), 351-362.

[2] Onoja, E, Razak, F.I.B, Chadran, S,Mahat, N. A and Wahab, R.A. Oil palm (*Elaeis guineensis*) biomass in Malaysia: Present and Future Prospects, Waste and Biomass Volorisation 2019;10(28. DOI: 10.1007/s12649-018-0258-1.

[3] Momoh,E.O and Osofero, A.I . Recent deveopmentiin the application oil palm fibres in cement composites, Frontier of Structural and Civil Engineering2020; Vol.14, 94-105.

[4] Wan Rosli, W.D, and Law, K.N. Oil palm fibres as paper making materials: Potentials and challenges. BioRes. 2011; 6(1), 901-917.

[5] Tan aka, R and Yamamoto, K. Accumulation and availability of oil palm. In Proceedings of 3rd USM-JIRCAS Joint International Symposium 9-11, March, Penang, Malaysia, JIRCAS Working Report, 2014 No 39, P. 20-34

[6] Anyaoha, K.E, Sakrabani, R, Patehigolla, U and Mirozan, A.N. Evaluating oil palm fruit bunch processing in Nigeria. Waste Management and Research 2018; vol. 36 [3] 236-246. DOI. 10.1177/0734242x17751848.

[7] Sudradjat and Diansyah, A.N. International e-Conference o Sustainable Agriculture and Farming System. In Proceedings of IOP Conference Series:Earth and Environmental Science Vol.694, 24-25 September,Bojor, Indonesia 2021.

[8] Mardawati, E, Wermen, A, Bley, T, Kresnowati, M and Sediati, T. The enzymatic hydrolysis of oil palm EFB to xylose. Journal of Japan Institute of Energy 2014; 93, 973-978. https://dx.doi.org/103775/jie93.973

[9] Md Yunos, K.F.H, Samsu Bahamddin, A., Md Yunos, K.F, Hafid, H.S., Buzu, Z, Mokhtar, M.N, Sulaiman, A., Md Son, A. The Physico-chemical characteristics of residual oil and fibres from oil palm empty fruit bunches, BioRes. 2015; 10[1], 14-29.

[10] Kobi, Y and Isluzaki, A. Chemical composition of oil palm fibre and its feasibility as cellulosic raw material for sugar production. Agricultural and Biological Chemistry 2014; Vol.511, 1183-1187.

[11] Krisnowati, M. Mardawati, E. and Setiadi, R. Production of xylitol from oil palm EFB: A case study on bio refinery concept. Modern Applied Sciences 2015; Vol.9, No 7, 207-213.

[12] Padzil, F.N.M, Lee, S.H, Aimin, Z.M, Lee, C.H and Abdullah, L.C. Potential of oil palm EFB resources in Nano Cellulosic Hydrogel products for versatile application: A Review Material 2020; 13[5], 1245, doi:10.3390/ma13051245

[13] Bledzki, A.K. and Gassan, J. Composites reinforced with cellulose-based fibres. Progress in Polymer Science 1999; Vol 24, Issue 2, 221-274.

[14] Omar, F.N, Mohammed, M.A.P, and Bahadurridin, A.S. Microstructure modeling of silica bodies from oil palm empty fruit bunch (OPEFB) fibres. BioResources 2014; 9[1], 938-951. DOI:10.15376/diores9.1.938-951.

[15] Sreekala, M.S, Kumaran, M.G, and Thomas, S. Oil palm fibres: Morphology, Chemical composition, Surface modification and Mechnaical properties. Journal of Applied Polymer Science 1998; Vol 66, Issue 5, 821-835. https://doi:org/10.1002 (SYCI) 1097-4628 (10971031) 66:5<821:AD-APP253.0coj2x

[16] Law, K.N, and Jiang, X.P. Comparative papermaking properties

of oil palm empty fruit bunch. Tappi, J, 2001, 84 [1], 1-13.

[17] Ibrahim, R. Structural, mechanical and optical properties of recycled paper blended with EFB pulp. Journal of oil palm research 2003; Vol 15, No 2, 29-35

[18] Majesty, K. I and Herdiansyah.The empty palm oil fruit bunch as potential source of biomass in furfural production in Indonesia: Preliminary processing design and environmental perspective. J. Phys. 2019: Conf.Serv.1363012096.

[19] Suhaily, S.S, Jawaid, M, Abdulchalil, H.P.S, Mohammed, A.R and Ibrahim, P. A review of oil palm biocomposites for furniture design and appilications: Potentials and challenges. BioRes. 2012; 7[3], 4400-4423.

[20] Wan Rosli, W.D, Law, K.N, and Valade, J.L. Chemical pulping of oil palm empty fruit bunches, Cellulose Chem.Technol. 1998; 32,132-148

[21] Khoo, K.C and Tee, T.W. Pulp and paper from oil palm. Appita Journal. 1991; 44(6), 385-388

[22] Noah, A.S, Alao, O. J., Adeniyi, I. M., Abiola, J. K., Ayodele, E. B. beating effects of laboratory blender on pulp fibre characteristics of empty fruit bunches (EFB) of *Elaeis guineensis* Jacq. (oil palm) in papermaking. International Journal of Applied Research and Technology 2014; Vol. 3, No. 5, 111 – 117.

[23] Ghazali, A., WanRosli, W.D and Law, K.N. Alkali peroxide mechanical pulping (APMP) of oil palm linocellulosics: Part2. Empty fruit bunch (EFB) response to pretreatment. Appita Journal 2006; 59[1], 66-70

[24] Singh, P. Silaiman, O., Hashim, R, Peng, L.C. Using biomass residue from oil palm industry as a raw material for pulp and paper industry: Potential benefits and threats to the environment. Environmental Development &Sustainability 2012; 15[2], 367-383

[25] Yacob, S. Progress and challenges in the utilisation of oil palm biomass. Paper presented at the Asian Science and Technology Seminar, sponsored by the Japanese Science & Technology Agency, 8-9 March Jarkata Indonasia 2007. P.44-47

[26] Odenz Frias, E.J, Azama, A, Mata-Zey, L.E, Hermandez, S. Bioenergy potential and technical feasibility assessment of residues from oil palm processing. Biomass Energy 2020; 142, 1056-1068. DOI:10.1016/j.biombode 2020105668.

[27] Tao, H, Sheddon, J.L, Slade, E.M, Ludovic, H, Cul, J.R and Willis, K.J. et al. (2018). Application of oil palm empty fruit bunch effects on soil biota and functions: A case study in Sumatra, Indonesia. Agriculture, Ecosystem & Environment 2018; Vol156 105-113.

[28] Cabrales, H and Araque, O. The effect of moisture content, fibre length and compaction time on African oil palm empty fruit bunches briquette quality parameters. Heliyon 2020, Vol.6, Issue 12, e05607.

[29] Darajat, K., Hali, W. and Rahayi, D.E. Life cycle assessment [LCA], utilisation of oil palm empty fruit bunch as bioenergy. In the AIP Conference proceedings 2194.020019 P.1-7 https://doi.org//10/1063/1.5139751

[30] Cheen, Y.L and Shimada, S. Current state and environmental impact assessment for utilising oil palm empty fruit bunches for fuel, fibre and fertiliser: A case study of Malaysia. Biomass and Bioenergy 2013; 51, 109-124. DOI.10.1066/j.biombioe 2013. 01.012.

Chapter 11

An Overview of Oil Palm Cultivation via Tissue Culture Technique

Siti Khadijah A. Karim

Abstract

During the last three decades, plant cell, tissue, and organ culture have developed rapidly and become a major biotechnology tool in agriculture, horticulture, forestry, and industry. Many problems in conventional breeding techniques were solved via tissue culture techniques. Plant tissue culture technique permits the growing plants in test tube or closed container in vitro under controlled environment. This technique is devoted to solve two problems: 1) To keep the plant cells free from microbes. 2) To grow the desired plants by providing suitable nutrient medium and other environmental conditions. In this chapter, a review around plant tissue culture techniques that have been reported on oil palm breeding programme will be discussed. It is including the laboratory techniques, advantages and disadvantages of the technique, the problems to produce good and prolific oil palm tissue culture clones and mitigation measures that have been reported to overcome the problems. As a conclusion, this chapter reviews tissue culture techniques that could be used to propagate oil palm clones.

Keywords: oil palm, propagation, tissue culture, overview

1. Introduction

Plant tissue culture is a technique used to propagate plants in vitro (in test tube) under sterile conditions, often to produce clones of a plant. During the last three decades, plant cell, tissue, and organ culture have developed rapidly and become a major biotechnology tool in agriculture, horticulture, forestry, and industry. Many problems in conventional breeding techniques were solved via tissue culture techniques.

In nature, there are a variety forms of plants either seed plants for example, trees, herbs, grasses, or flowering plants for example, fruit-bearing trees. Plants exhibit the basic morphological for example, root, stem and leaves. However, they are vary with differences in cells and tissues, and their topography.

In tissue culture, the term 'culture' refers to the fragments of plants that are grown in nutrient media. There are several types of cultures, which are cell culture, tissue culture, organ culture, explant culture, and protoplast culture. Explant is the excised fragment of plants that is transferred into nutrient media. There are many types of explants, it can be roots, stems, leaves, seeds, fruits, and flowers.

IntechOpen

The role of explant is to initiate culture in a nutrient media, provided that they must be able to de-differentiate into totipotent cells.

The term 'Tissue Culture' or micropropagation is the technique to propagate plants under sterile conditions, often to produce clones of a plant. The cultured of explant on a solid medium produces mass of protoplasmic cells which later be induced to become a complete plant. However, in organ culture, for example, excised roots, the cultured explant (plant material) maintains its morphology identity, more or less, similar with the same physiology as in vitro of the parent plants. Plant organs are referred to part of plants that possess vascular tissues such as root, shoot, and leaves. Embryo is an independent structure and does not have vascular supply, thus, embryo is not supposed to be the plant organ.

2. What is totipotency?

Totipotency is the unique feature of plant cells where every single cells has biological potential to develop normal root, shoots and embryoids, which ultimately forming a plant. The term totipotency is defined as the capability of each living cell to carry out independent development to regenerate into a complete plant, provided with suitable conditions [1]. Many research has been done using tissue culture as a medium to study multicellular organism, because it is rather difficult when using the organism as biological units of study.

3. Problems in oil palm tissue culture

The oil palm clones are frequently generated via indirect somatic embryogenesis, in which the plantlets were produced from the growth of callus and differentiate into somatic embryos. The process is usually induced from seeds, hence, making seed propagation most common for oil palm somatic embryogenesis. However, seeds propagation often cause difficulties in terms of low germination of the seeds. Due to the hard woody and sturdy structure of oil palm, there is very few choices of explants that can be used to induce somatic embryogenesis. Apart from the seeds, immature leaves are another explant of oil palm that is often be opted.

Oil palm is one of recalcitrant plants, meaning that its explants are more difficult to be developed into plantlets as compared to other plant species. While somatic embryogenesis is well known for its low efficiencies in producing small number of plantlets, the other main setbacks of this method in oil palm breeding are the difficulties in initiating somatic embryogenesis out of embryo cell cultures as well as to fully convert those somatic embryos into whole plantlets. It was reported that immature leaves explants of oil palm has been able to generate callus up to 52%, while other explants such as zygotic embryos and immature inflorescence produced higher rate of callus induction [2]. Despite of these reported adequate efficiencies, the callus induction rate from an explants are highly dependent of its genotype. Many explants could only regain totipotency during the culture process at around 1–5% in order to initiate callus, thus, making somatic embryogenesis is very inefficient procedures [3]. The development of somatic embryogenesis into whole plantlets is also often problematic. This is due to the failure of shoot or root development or induction, whereby, results in low regeneration rate of oil palm clones [4, 5].

The combination of low efficiency of somatic embryogenesis induction, limited explants choices, and low regeneration rate into complete explants, making oil palm breeding via tissue culture is impossible without embryos proliferation process is carried out. However, by producing large number of embryos in one culture may produce risk of somaclonal variation. Another way to increase regeneration rate in oil palm culture is by using explant of prolific breed. Although the selection of this breed requires large screening process at molecular level to identify the best genotype, it could be developed as a tool that possibly provide long-term benefits [6, 7].

4. Regeneration of cultured plants

4.1 Organogenesis

There are several methods to regenerate cultured plants, one of it can be through organogenesis, either through direct organogenesis or indirect organogenesis. Organogenesis is the development of individual plant organ such as shoots or roots from the cells in culture (can be callus (indirect) or plant tissue (direct)) by the process of differentiation. Organogenesis in plant tissue culture involves two stages: Dedifferentiation and redifferentiation. Dedifferentiation starts shortly after explant initiated rapid cell division and consequently forms a mass of undifferentiated cells (callus). Redifferentiation, also referred as budding, is the process where the callus starts to differentiate to form plant organ (organ primordia). This primordia organ is then develop into small meristems (which contains of large nuclei).

In direct organogenesis, the formation of plant organs such as shoot and root is straight from organized tissue (explant) without undergoes callus formation. Progenies that are produced through this technique have identical genetic content as parent. However, this technique is depending on several factors such as explant type, age of explant and size of explant. If meristem tissues is used as explant, the genetic content of the progeny (offspring) are identical as parent. Conversely, if embryos are used as explant, the genetic content between progeny and parent are not identical as embryo formed through fertilization of gamete cells (male and female gametes) and some plants have dormancy period.

In indirect organogenesis, callus is produced first from explant. Callus is disorganized group of cells, rapid dividing, and undifferentiated into specialized tissues such as shoots and roots. Callus can be induced from callus culture, explant (tissues) or cell suspension culture of that callus. Subsequently, organ formation is induced from the callus where shoots and roots are stimulated by plant growth hormones. However, the disadvantage is the changes or variation in the genetic content of somatic cells of the progeny (somaclonal variation) due to repetitive subcultures. Direct organogenesis can be opted to avoid somaclonal variation. Direct organogenesis is often used to plants that are difficult to propagate and do not have abundance of meristem tissues. Therefore, these plants are propagated by using leaves, stems, and root tips as explants.

The rule of thumbs in organogenesis technique are the proportion of growth hormones combinations in the culture medium used to stimulate the respective organs. In direct organogenesis, high ratio of auxin to cytokinin will produce roots while high ratio cytokinin to auxin will induce shoots. In indirect organogenesis, balance or same ratio of auxin and cytokinin (1:1) will produce callus.

Organogenesis starts with induction process caused by the plant hormones in the medium, substances carried over from the explants and endogenous hormones produced by the explants. Organogenesis was first induced by Skoog in 1944 on the formation of root and inhibition of shoot after the addition of auxin. It was then proposed that the regulation of organogenesis is depending on the balance between auxin and cytokinin. The research team then later discovered that high ratio of auxin to cytokinin stimulated the root formation in tobacco callus, but a low ratio of auxin to cytokinin led to shoot formation (**Table 1**).

Auxin (mg/l)	Cytokinin (mg/l)	Organogenesis
0.0	0.2	No growth
0.03	1.0	Shoots
3.0	0.02	Roots
3.0	0.2	Callus

Table 1.
Standard concentrations of auxins and cytokinins to induce in vitro *organogenesis.*

4.2 Somatic embryogenesis

Other than organogenesis, somatic embryogenesis is another major regeneration technique in plant tissue culture. Embryo production is an important feature of the flowering plants. The process of embryo formation is called embryogenesis which starts from a single embryogenic cell and subsequently develops into either a zygote or undifferentiated callus cells. Embryo that develops from zygotes is called as zygotic embryos. Meanwhile, embryo that develops from somatic cells is called as somatic embryos where it is artificially induced in cultured plant tissues.

Somatic embryogenesis was first induced in cell suspension culture and callus culture of carrot. Other plants like *Coffea Arabica*, *Citrus cincensis*, *Nicotiana tabacum*, *Pinus ponderosa* and *Cocos nucifera* are among successful species in somatic embryogenesis. In plant tissue culture, somatic embryogenesis is the formation of somatic embryoids from somatic tissues of callus or cells of suspension culture, which can then develop into complete plants in a similar way to the zygotic embryos (sexual reproduction). Somatic embryoid (asexual embryo) is small and well-organized structure that is resemblance to zygotic embryo (sexual embryo), which is produced from embryogenic somatic cells. Somatic and zygotic embryoids share the same pattern of development where both undergo globular, heart, torpedo shaped and cotyledon stage for dicots and conifers. Embryo growth is bipolar which produces a shoot and a radicular pole at the other end. When encapsulated with suitable nutrient, somatic embryos become artificial or synthetic seeds and they as they can produce plantlets and planted directly into the field.

Somatic embryos can be produced through direct or indirect somatic embryogenesis. In direct somatic embryogenesis, the embryo is induced directly from cells or tissues without the formation of intervening callus. However, this technique is rare and uncommon compared to indirect somatic embryogenesis. In indirect somatic embryogenesis, callus if first formed from explant. Somatic embryos can be then induced from the callus or cell suspension culture of that callus. The embryoids are initiated from superficial callus aggregates where the cells contain large vacuole, dense cytoplasm, large starch granules and nucleus.

Two types of medium with different compositions are required to induce somatic embryoids. First medium contains auxin to initiate embryogenic cells. Second medium is lacked or reduced of auxin, is needed to support the development of the embryogenic cells into embryoids and plantlets. Similar to zygotic embryos, the embryogenic cells pass through 3 different stages i.e. globular, heart shaped, and torpedo shaped, to form embryoids. The embryoids can be separated from the non-embryoids callus cells by using glassbeads or filter paper.

The importance of somatic embryogenesis in agriculture, horticulture, and plant conservation is the zygotic and nucellar embryogenic can be obtained separately from the polyembrogenic plants such as citrus. Since somatic embryo has no food reserves, they can be preserved as encapsulated seeds (surrounded with

Zygotic embryo	Somatic embryo
Fertilized egg or zygote	Somatic cells
Contain seed coat	No seed coat
Produce seed	Produce embryo only
Plantlets are healthy	Plantlets are weaker
Not indentical to mother plant	Identical to mother plant
Propagation rate is low	Propagation rate is high

Table 2.
Comparison between zygotic and somatic embryo.

nutrients). This makes international exchange of germplasm possible. This artificial seeds provide an advantage for embryos of big and heavy fruits like coconut which can be preserved in a test tube for months and then cultured on medium. In addition, some plants that are crossed interspecific or intergeneric are failed to develop at maturity stage, therefore, before the embryos undergo maturity, they can be taken and cultured on artificial medium and grown into whole plants. As somatic embryogenesis produces many somatic embryos in cell culture, this technique is regarded as the ideal mass propagation system. The somatic embryo is a bipolar system which can develop directly into complete plant, hence, there is no need for separate rooting and shooting induction steps. Plants that derived from somatic embryo may be free of viral and pathogens. Therefore, it is another option in disease-free plants generation (**Table 2**).

4.3 Somaclonal variation

Somaclonal variation produces phenotypic variation in the somaclones either through genetic variation or epigenetic. In oil palm tissue culture where it is propagated through somatic embryogenesis, this technique is often lead to somaclonal variation [3]. Somaclonal variation refers to genetic variability generated during tissue culture and can be detected as genetic or phenotypic traits. Several features to identify somaclonal variation in somaclones are by examining the number and structure of chromosomes. Somaclones with altered chromosomes are usually exhibit changes in leaf shape and color, growth rate and sexual fertility. It is a heritable mutations and persist in next generations even after plantation into the field.

Somaclonal variation can be developed in tissue culture through genetic and epigenetic mechanisms. In genetic mechanism, variations are presence in somatic cells of explant which may be caused by DNA changes and mutations. In epigenetic mechanism, somaclonal variation is generated during tissue culture and results in temporary phenotypic changes. Somaclonal variation can also occurs due to physiological effect such as exposure to plant growth hormones and the culture conditions.

4.4 Screening work-flows for somaclonal variants

There are various methods to select somaclonal variants.

i. Analysis of morphological traits.

- Qualitative: Plant height, maturity date, flowering date, and leaf size
- Quantitative: Number of flowers, leaves, and seeds.

ii. Cytological studies

Feulgen staining can be used to stain nuclei of the somaclones to measure DNA contents by using cytophotometer. Number of chromosomes can also be measured using the same technique on meristematic tissues such as root tip and shoot tip.

iii. Gel electrophoresis

This technique can be used to detect variation among somaclones in terms of the concentration of protein, pigments and amino acids through the observation of amplification pattern.

iv. Disease resistance trait

Pathogen or toxin that is responsible for the disease resistance can be used to select disease resistance clones.

v. Herbicide resistance trait

Plantlets that are grown on culture medium containing the particular herbicide possess the herbicide resistance trait.

vi. Stress tolerance trait

Detection for stress tolerance trait has been done on tobacco cell lines for high salt tolerance, and drought tolerance in tomato.

4.5 Advantages of somaclonal variations

Somaclonal variations help in crop improvement. In India, a somaclonal variant *Citronella java*, a medicinal plant that has been named as "Bio-13", produce yield increment by 37% oil content and 39% more citronellon (type of terpenoid that can be used in skin products) than the control plant. In US, a new somaclonal variant was developed and called as "Supertomatoes" where these clones capable in reducing shipping and processing costs. Other advantages of this technique are as following:

- Allow the generation of genetic variations
- Increased and improved production of secondary metabolites
- Enable the generation of plants resistant to toxins, pesticides, herbicides and plants that tolerant to stress (unfavorable condition) such as drought and high salt concentration.
- Suitable for breeding of tree species

4.6 Disadvantages of somaclonal variations

The main setback that has made this breeding technique difficult to carry out is the difficulties to obtain a uniform clones as this feature is important in mass propagation of the plants especially in horticulture and forestry industries. In order to select uniform clones, extensive and extended field trials may be required. The genetic variations in somaclonal variant plants are also unstable, thus, the desired traits have tendency to disappear in the next generation. There is also risk where the clones may exhibit undesirable traits.

5. Types of explant and explant selection

Explant is a sterile excised fragment of plants from which cultures are initiated. Generally, all types of plant cells or tissues can be used as an explant, however, it is preferable to use young and immature tissues that is rapidly dividing (at early stage of development) as an explant such as shoot tip, root tip, and young leaves. There are several factors need to be considered during explant selection process:

i. Explant age – Physiologically, younger tissues are more responsive to the medium in order to induce cell division. Younger tissues are also more suitable as explant because the tissues surface are softer and that helps to ease sterilization process whereby, helps in preventing contamination.

ii. Explant size - The smaller the explant is the better to minimize possibilities of contamination from bacteria, fungus, and viruses. However, small sized explant has lower survival rate in the medium as compared to larger explant due to the lack of reserve nutrient available to sustain the culture.

iii. Season – The season of the year may have effects on explants survival rate in the medium and contaminations. For example, shoots or leaves that are taken during the Spring season are more responsive compared to other seasons. In Malaysia, which is a tropical country, this factor may not need to be considered.

iv. Plant quality – It is more suitable to select explant from plants that are healthy rather than plants that are under stress conditions such as water-stress or nutritional stress or plants that exhibit disease symptoms.

Apart from the above factors to select the best explant, another thing that is need to consider in selecting an explant is the goal of the experiment. The choice of explant tissues is depending on what type of response desired from the cell culture. For example, if the aim is to carry out clonal propagation, explant of shoot or root tip is suitable to achieve it. For callus induction, fragment of cotyledon, hypocotyl, stem, leaf, and embryo can be used as explant. For protoplast fusion, leaf tissue from aseptically germinated seeds are preferable.

6. The advantages and limitations in tissue culture technique

Plant tissue culture technique permits the growing plants in test tube or closed container in vitro under controlled environment. This technique is devoted to solve two problems: 1) To keep the plant cells free from microbes. 2) To grow the desired plants by providing suitable nutrient medium and other environmental conditions. Several advantages of plant tissue culture techniques are as following:

i. Uniform growth – As tissue culture plants (clones) possess the same genetic content, the progenies undergo the same growth pattern.

ii. True to type – Tissue culture plants are grown asexually from somatic cells of mother plant (explant). Therefore, the progenies possess the same genetic content as the explant.

iii. Increase availability of plants – Some plants are difficult to grow. Tissue culture provides the solution to multiply plants in large scales and in uniform growth. Industries that uses plants as raw materials such as in pharmaceutical or medicinal products require large quantity of plants, which can be produced by using tissue culture.

iv. Continuous supply of plants – Some diseases are fluctuate depending on changing of seasons, climate and crop diseases. Therefore, there is a need for continuous supplies of the medicinal plants for production of drugs, that can not be synthesized synthetically. Novel plant – Tissue culture facilitates in developing novel plants with desired traits that are generated through gene transformation.

v. Disease free and desired propagule – Plants can be grown in disease free environment in large scale, and the desired propagule such as buds and stems can be transported to other places without any damage.

vi. Biosynthetic pathways – Tissue culture can be used for detecting the production of secondary metabolites using labeled precursor in the medium. For example, the production of anthocyanin in apple callus culture.

vii. Immobilization – Tissue culture technique can be used for plant preservation. Through the production of artificial seeds, it allows the immobilization (entrapment) of tissue whereby minimize transportation handling and cost.

viii. Continuous production of medicinal drugs – Some medicinal plants are may be seasonal, and are difficult to obtain due to climate change, etc. Tissue culture acts as a tool to allow the production of the natural compounds independent of soil, climate change and seasons.

ix. Seedless propagation – Some plant seeds are difficult to germinate, tissue culture enables plant propagation without the need of seeds.

6.1 Limitations in plant tissue culture

Despite of various advantages of plant tissue culture, this technique has some limitations.

i. High level of expertise are required – Tissue culture technique require excellent level of handling skills as a small error can lead to damage of products or plants. Companies or institutions need to invest on staff training and that makes it costly, as it requires a long-term capital investment.

ii. Expensive – Other than high cost in expertise, tissue culture is also require high cost in chemicals which must contains in high purity.

iii. Low production of secondary metabolites – Amount of secondary metabolites that are harvested from tissue culture plants are often negligible. Therefore, it requires large scale production of plants in order to increase yield.

iv. Instability – Despite having an advantage of producing clones that are identical to the explant, there are times where genetic variation occurs among clones lines which cause changes on phenotype and genotype of the clones. This is called as somaclonal variation.

v. Prone to contamination – Aseptic technique are need to maintained throughout the *in vitro* plant growth. If contamination occurs, the plant growth is impaired and has low survival rate.

7. Shoot culture and micropropagation

In shoot culture, apical meristem (located at shoot apex and root apex) is cultured, and this culture is also known as meristem culture due to large size of the explant (5–10 mm). Shoot culture is widely applied in horticulture, agriculture and forestry. Murashige, from Morel research team (1960), has significantly established the technique for micropropagation and its further biotechnological application. Due to the small size of explant to be propagated and it occurs *in vitro* as compared to conventional propagation, this technique is also known as micropropagation. Stages in micropropagation are as following:

Stage I Selection and preparation of explant – Suitable explant is selected and inoculated into nutrient medium. This step is done is extreme aseptic condition.

Stage II Multiplication of cells – Growth of culture takes up to 2 months, followed by repeated subcultures.

Stage III Shooting and rooting induction – Culture is transferred to nutrient medium with suitable composition to induce multiple shoots. This step may take about 4 weeks. Subsequently, the culture is transferred on medium suitable for root induction and incubated for about 3 weeks.

Stage IV Transfer to soil – After about one month of culture, plantlets are aseptically removed from test tube environment to natural and harsh environment. At this stage, roots should be fully functional in potting soil mix. During this step of transplantation, plantlets have tendency to fail to survive due to desiccation i.e. from 100% humidity of test tubes to low humidity under ambient conditions), unfavorable environment, invasion of soil microorganisms, as well as fails to adapt changes from dependent (artificial medium) to independent environment.

8. Mass propagation

8.1 Acclimatization

In order to minimize the failure of transplantation, it is necessary to develop acclimatization capability in plantlets before transfer them into soil. Acclimatization is the process where an organism adapts to changes in its environment such as change in temperature, humidity, photoperiod, and pH, to allow it to survive in a range of environmental conditions. This can be done by:

i. Induction to develop some normal and functional leaves

ii. Induction of functional roots

iii. Exposing the *in vitro* cultures to harsh environment in two weeks before planting out.

8.2 Plantlets transplantation from in vitro culture to polybags

The following procedures are standard practice for transplantation:

i. Plantlet is removed from test tube and washed under running tap water.

ii. Soil mixture is prepared with ratio 3:2:1 of soil, peat soil, and fine soil.

iii. Polybags and small planting pots (8 cm x 8 cm) are prepared.

iv. Cocoa peat soil is filled into the base of polibags or planting pots.

v. The soil mixture is then added into the polybags and planting pots.

vi. Roots of plantlets are dipped into IBA rooting powder.

vii. The plantlets are planted into the soil mixture until the root area is covered.

viii. The plantlets are watered and then covered with transparent plastic, with holes on the surface to allow gas exchange.

ix. The plantlets are placed under up to 60% shading area, and strictly monitored in nursery for 2–3 months.

x. The plantlets are watered using mist sprayer and the plastic cover is removed after 1 month of planting.

xi. After 3 months, the plantlets are removed to bigger size of polybags or planting pots with the same soil mix composition.

xii. The plantlets are continuously watered and monitored until 4 months old before it is transferred to the orchard.

8.3 Transfer of plantlets to orchard (mass propagation)

After acclimatization, the plantlet is ready to be transferred to orchard in larger planting area. This step is called mass propagation.

i. After 3 months old of the plantlet, the shading is reduced to 20%.

ii. The plantlet is fed with NPK fertilizer 20:20:20 (Nitrogen:Phosphorous:Kalium). The fertilizer is diluted beforehand with water, to reach the concentration between 150 ppm to 200 ppm.

iii. After the plantlet reach 4 months old, it is then ready to be transferred to orchard.

iv. The planting holes are digged beforehand. The depth is depending on the plantlet size, to cover all the root system into the soil.

v. Mixture of organic fertilizers per plantlet can be added into the planting hole such as chicken or cows manure (100 g), triple superphosphate (100 g), and lime fertilizer (calcium carbonate) (100 g).

vi. Plantlet is removed from polybag and planted into the hole. The fertilizer mixture is added until the top of the soil and covers the area surrounds of plantlet.

vii. Dried leaves or grasses is placed around the base of the plantlet in order to maintain humidity.

viii. The plantlet is watered twice a day i.e. in early morning and late afternoon for 4 months.

9. Conclusion and future works

This reviews discussed the application of tissue culture techniques as an alternative mean of asexual propagation of important plants. The advantages of tissue culture techniques allow the propagation of recalcitrant plants including oil palms, endangered plants species as well as seasonal dependent plants [8]. This technique required small amounts of plant tissues to propagate large scales of plant clones, thus, making it a convenient method for plant breeding. Plantlets regeneration from cultured plant cells and tissues has been achieved in many species of high economic value. Many of studies are aimed to carry out large scale propagation of important trees yielding fuel, timber, pulp, oils and fruits [9–12]. Therefore, tissue culture techniques became an alternative for tree improvement.

Author details

Siti Khadijah A. Karim
Faculty of Applied Sciences, Universiti Teknologi MARA (UiTM) Jengka, Jengka, Pahang Darul Makmur, Malaysia

*Address all correspondence to: khadijahkarim@uitm.edu.my

References

[1] Thorpe, T. (2012). History of plant tissue culture. In Plant Cell Culture Protocols (pp. 9-27). Humana Press, Totowa, NJ.

[2] Corrêa, T. R., Motoike, S. Y., de Souza Andrade, A. P., Coser, S. M., Queiroz, V., Granja, M. M. C., ... & de Toledo Picoli, E. A. (2016). Accelerated in vitro propagation of elite oil palm genotypes (Elaeis guineensis Jacq.) by substituting cytokinin with putrescine. African Journal of Biotechnology, 15(50), 2767-2775.

[3] Gomes, H. T., Bartos, P. M. C., & Scherwinski-Pereira, J. E. (2017). Dynamics of morphological and anatomical changes in leaf tissues of an interspecific hybrid of oil palm during acquisition and development of somatic embryogenesis. Plant Cell, Tissue and Organ Culture (PCTOC), 131(2), 269-282.

[4] Weckx, S., Inzé, D., & Maene, L. (2019). Tissue culture of oil palm: Finding the balance between mass propagation and somaclonal variation. Frontiers in plant science, 10, 722.

[5] Gomes, H. T., Bartos, P. M. C., Balzon, T. A., and Scherwinski-Pereira, J.E. (2016). Regeneration of somatic embryos of oil palm (Elaeis guineensis) using temporary immersion bioreactors. Ind. Crops Prod. 89, 244-249. doi: 10.1016/j.indcrop.2016.05.021

[6] Ishak, Z., Hashim, A. T., Rosli, S. K., Bakar, D. A., Ooi, S. E., Mohd, N., & Ong-Abdullah, M. (2020). Oil Palm Tissue Culture: Fast Tracking Elite Commercial Lines. In The Oil Palm Genome (pp. 47-68). Springer, Cham.

[7] Siti, K., & Nik, M. S. (2018). Construction of molecule markers for prolific oil palm tissue culture. Sains Malaysiana, 47(8), 1701-1708.

[8] Oseni, O. M., Pande, V., & Nailwal, T. K. (2018). A review on plant tissue culture, a technique for propagation and conservation of endangered plant species. Int J Curr Microbiol App Sci, 7(7), 3778-3786.

[9] Aroonluk, S., Roytrakul, S., & Jantasuriyarat, C. (2020). Identification and Characterization of Phosphoproteins in Somatic Embryogenesis Acquisition during Oil Palm Tissue Culture. Plants, 9(1), 36.

[10] Bonga, J. M., & Durzan, D. J. (Eds.). (2012). Cell and Tissue Culture in Forestry: Volume 2 Specific Principles and Methods: Growth and Developments (Vol. 24). Springer Science & Business Media.

[11] Shekhawat, K., Gupta, P. C., Kumar, R., Kumar, A., Kumari, R., & Kumawat, S. (2019). Role of tissue culture techniques in social forestry. Recent Advances in Chemical Sciences and Biotechnology, 113.

[12] Suman, S. (2017). Plant tissue culture: A promising tool of quality material production with special reference to micropropagation of banana. Biochem. Cell. Arch, 17(1), 1-26.

Chapter 12

Chemical Composition of Parenchyma and Vascular Bundle from *Elaeis guineensis*

Sitti Fatimah Mhd. Ramle

Abstract

Elaeis guineensis is an alternative source of raw materials for renewable energy in Malaysia. Thus, to enhance the use of the abundant biomass generated by the oil palm industry in Malaysia, a study was conducted in view of exploring the chemical composition such as sugar potential of this industrial byproduct. In this context, oil palm trunks were separated into individual cell that are parenchyma and vascular bundle to investigate the fundamental research about oil palm trunk. The aim of this study was to examine the chemical composition of parenchyma and vascular bundle of oil palm trunk. The oil palm trunk was kept under shade at room temperature of 28–30°C for 0, 45, and 60 days. The chemical composition analysis was carried out according to TAPPI methods. Based on storage time and different part of oil palm trunk, the result has shown that the sugar content was higher in parenchyma compared to vascular bundle and increase at the storage time of 0, 45, and 60 days while amount of starch showed decrease at the same storage time. It shows that conversion or fermentation of starch to sugar occur in oil palm trunk during storage times of 0, 45, and 60 days, respectively.

Keywords: *Elaeis guineensis*, oil palm trunk, parenchyma, vascular bundle, storage

1. Introduction

Oil palm tree is one of the perennial oil crops that generate economic growth in Malaysia. It belongs in the species *Elaeis guineensis* under *Palmacea* family that comes from the tropical forests of West Africa. Palm oil production has almost doubled in the past decade. It is produced in 42 countries around the world at about 27 million hectares [1]. Oil palm cultivation has also become the first fruits of the world in terms of production for nearly 20 years.

Currently, Malaysia is the regional leader in biodiesel production with an output of 540 million liters per annum as of 2009 [2]. Meanwhile, Indonesia is second with the production of 400 million liters in 2010 [3]. During the process of replanting in Malaysia, large quantities of oil palm trunks (OPT) and oil palm fronds (OPF) waste are produced in oil palm plantations. The trunks are normally left in the field without any utilization thereafter. They are usually cut into pieces and burnt down to avoid insect and incidences [4].

Oil palm trunks have such special characteristics as high moisture content (1.5 to 2.5 times the weight of the dry matter), low cellulose and lignin content and high content of water soluble and NaOH soluble in comparison with rubberwood and bagasse. Physical properties of trunks showed heterogeneity and varied depending on both radial and vertical directions. Some difficulties in utilizing oil palm trunks also lie in extremely tough outer bark and high content of decay able parenchyma cells [5].

From the previous study [1, 6], found that the sugar content in the sap of felled palm trunks increased during storage after logging. This suggests that oil palm trunk can be a promising source of sugar as proper treatment after logging. Sap analysis can be an efficient raw material for bioethanol [6, 7]. In addition, oil palm trunk was considered as a useful material for pulp and paper properties which have been studied [5, 8, 9]. However, the physical, morphological properties and chemical compositions of individual cell of oil palm trunk such as parenchyma and vascular bundle have not been well studied [10].

The aim of this study therefore was to examine the chemical composition of parenchyma and vascular bundle of oil palm trunk that were separated based on storage time [11]. Fibers useful for materials occur in vascular bundle, while living cells containing sugars and starch useful for energy and livestock foods mainly exists in parenchyma. The outcome of this study forms the basis in realization of the full potential of the chemical compositions and it can guide us for particular applications and uses.

2. Overview of *Elaeis guineensis*

In Malaysia, the production of palm oil has tremendously increased since the 1970s. Government policy on crop diversity has led to an increase in hectares of oil palm trees as shown in **Figure 1**. In 1996, 2.6 million hectares were used for the cultivation of oil palm and this figure is staggering. In 2005 and 2010 the percentage changes in the areas of oil palm tree is about 20% for each of the next five years. The year 2011 has reached 5 million hectares but only three percent different from 2010. The rapid growth in oil palm cultivation has seen in five years of 1965–1970, 1970–1975, and also in 1975–1980. This is due to the crop diversification program [12].

Figure 1.
Oil palm plantation in Pahang, Malaysia.

2.1 Botanical description of *Elaeis guineensis*

The classification of the botanical description of oil palms has presented some problems; the rules of plant taxonomy demand that the name first correctly applied to a species must be used (though the generic name can be changed if research indicates that a species has been placed in the wrong genus). The African oil palm was named *Elaeis guineensis* by [13] and this name is not questioned. According to [14] the first valid name for the American species is Corozo oleifera, but [15] considered that the American and African species are so closely related that separation in different genera is not justified. This is supported by the production of viable seeds in controlled pollination trials between the two species [16]. The correct name for the American oil palm is therefore *Elaeis oleifera* (H.B.K.) Cortes. In the hybrids, *Elaeis oleifera* shows dominance over *Elaeis guineensis* for several characters, suggesting that *Elaeis oleifera* is more primitive and *Elaeis guineensis* represents a derived

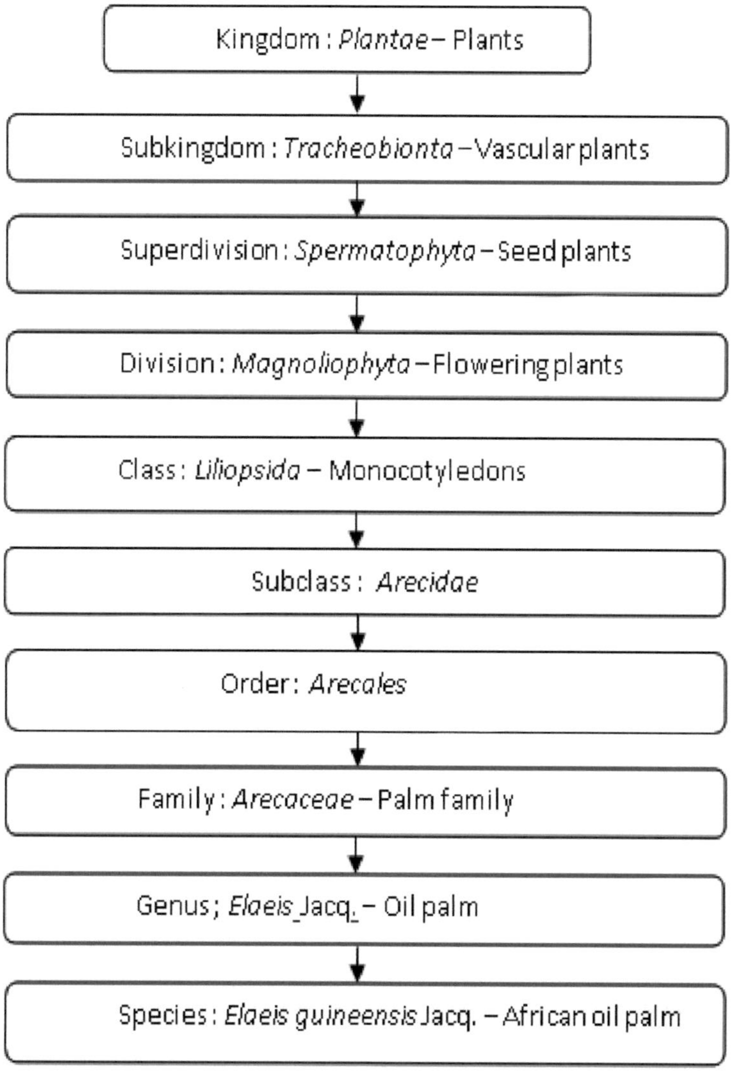

Figure 2.
Botanical description of oil palm.

species [17]. **Figure 2** present the oil palm profile according to the Plant, United State Department of Agriculture USDA [18].

2.2 Anatomy structure of *Elaeis guineensis*

Nonwood fibers are derived from selected tissues of various mono- or dicotyledonous plants [19] and are categorized botanically as grass, bast, leaf, or fruit fibers as can be seen in **Figure 3**. Nonwood fibers are classified as fibers such as sugar cane bagasse, wheat straw and corn stalks are byproducts. Other non-wood fibers are grouped as fiber plants which are plants with high cellulose content that are cultivated primarily for the sake of their fibers such as jute, kenaf, flax, cotton and ramie [20].

There are several types of nonwood produce useful byproducts, for example, oil of kenaf and flax seed. From nonwood fibers can also be used to make paper, although the quality varies as it depends on the fiber source [21]. The combination of wood and nonwood fiber can reduce the amount of chemicals needed for pulp. It also shortens the time the pulp, thus saving energy. High cellulose content of cotton linters (85–90%) compared to wood (35–49% cellulose), and low lignin content of hemp (3%) make valuable nonwood fibers for paper making [5].

In this study oil palm trunk was used and categorized in non-wood tree. Oil palm tree does not have cambium, secondary growth, growth ring, ray cell, sapwood, and heartwood or branches and knots. The anatomical structure of oil palm consists of parenchyma and vascular bundle, in contrast to hardwood and softwood which the cells consist of mostly fibers, trachieds, vessel parenchyma and

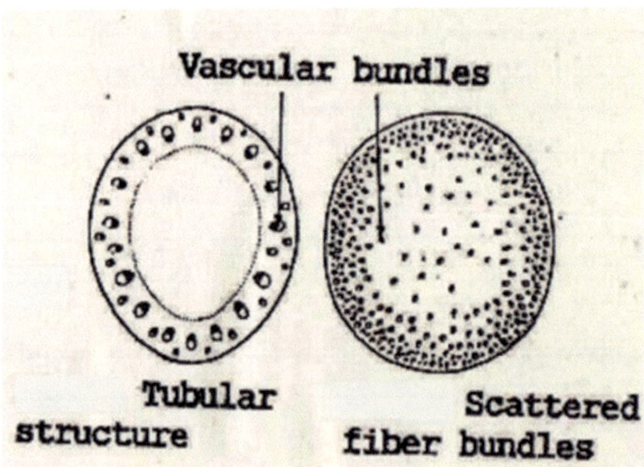

Figure 3.
Structure of non-wood.

Figure 4.
Separation of oil palm trunk into parenchyma and vascular bundle.

ray parenchyma cells. In this study focused on chemical compositions of separated sample of oil palm trunk which are parenchyma and vascular bundle as shown in **Figure 4**.

2.2.1 Parenchyma and vascular bundle from Elaeis guineensis

Parenchyma cells are the oldest type of eukaryotic cells that have been the first cell a grow. There is only a thin primary wall and lack of a secondary wall in the cell. Vascular bundle cells are called as a transportation tissue and grouped together with vascular tissue. Vascular tissue is made up from different types of plant cells and also called as a transportation system. The function of this tissue is to transfer the water, organic and inorganic molecules from synthesized or absorbed in the plant, to be used or stored by plant. The arrangement of the vascular tissue in a plant stem differs in dicotyledonous and monocotyledonous plants [22]. In this study we only focus on monocotyledon plants, which are oil palm trunk that was used in this study is one of the monocotyledon plant.

2.3 Utilization of biomass *Elaeis guineensis* to renewable energy

Biomass waste from palm oil is one of the solutions to the renewable energy because is believed to have availability and continuity. In the present situation, the oil palm biomass is one of the problems that have yet to be exploited. There are many considerations such as the economy, energy balance, environmental technology and the best solutions to meet the oil palm biomass utilization. All economic activity begins with physical materials and energy carriers (fuel and electricity). In the present era of transformation, we need a reliable source for sufficient sustainable energy requirements [23]. Otherwise have the materials, food, shelter technology will lead to no energy, no work and no economic activity [24].

The income and population growth have increased demand for energy. Energy is needed in almost all aspects of life, including agriculture, health care, drinking water, lighting, telecommunications and industrial activities. At this time, the demand for energy is fulfilled by fossil fuels (ie, coal, petroleum and natural gas). The world's current production rate of liquid fossil fuels (petroleum and natural gas) decrease in 2012 [25].

According to [26] due to high energy consumption and greenhouse gas emissions also cause a huge impact on global climate change. Based on existing scientific evaluation, observation records show that from the industrial era to the present day global average temperature has increased by between 0.3 and 0.6°C since the late 19th century, whereas sea levels have risen between 10 and 25 cm over the same period.

Biomass contributes about 12% on a worldwide scale. This causes the main energy supply increased to 40% and 50% of the country's most developed nations. In Malaysia, there are some examples of energy derived from biomass, including crop production resulting in starch, such as sorghum, or cane sugar as artichoke. Cellulose such as poplar, eucalyptus trees or other wood treelike form and sunflower oil is included in the energy-producing plants. There is a plentiful supply of palm oil waste and gives the reasons to choose the first biomass as a renewable and will be developed for large-scale applications, mainly in the palm oil industry [27]. In addition, palm oil waste can be considered as energy crops. This is because the palm oil industry has more than 40 years of experience in biomass power generation system operations and has working experience in use of palm oil waste to heat and power generation in the country [28].

3. Chemical composition of *Elaeis guineensis*

Chemical composition varies from species to species and within different parts of the same wood species. Chemical composition also varies within woods from different geographic locations, ages, climates and soil conditions [29].

Based on previous study, oil palm is a lignocelluloses material that contains a high level of carbohydrates [30]. These carbohydrates are mainly in the form of sugar containing cellulose, starch, hemicelluloses, and lignin [30] also found that the chemical compositions of oil palm biomass consists of high holocellulose, lignin, starch, and sugar contents that have been found to aid in the production of binderless panels.

According to [30] starch content is high in parenchyma cells. It is due to the starch of the oil palm trunk is stored inside the parenchyma cells of the coarse vascular bundles, which contain a high percentage of lignin. However, the data about the individual cell of parenchyma and vascular bundle are still limited and need future investigating. In this study, the chemical compositions are focused on starch and sugar of oil palm trunk during storage time.

3.1 Sugar content of parenchyma and vascular bundle at different part of oil palm trunk by HLPC

The **Figure 5** shows the results of the sugar content for separated samples (parenchyma and vascular bundle) and a non-separated sample of oil palm trunk at different part during storage time. From the figure showed that the parenchyma gave the most concentration of sugar compared to the vascular bundle. Non-separated sample at the bottom part of oil palm trunk showed a slightly highest of sugar content compared to the non-separated samples on the top part. Different parts such as inner, middle and outer part showed that the parenchyma in the middle contains the highest sugar content compared to others.

Based on this study, the highest sugar content was found in the parenchyma and bottom part of the oil palm trunk for the non-separated sample at the different part

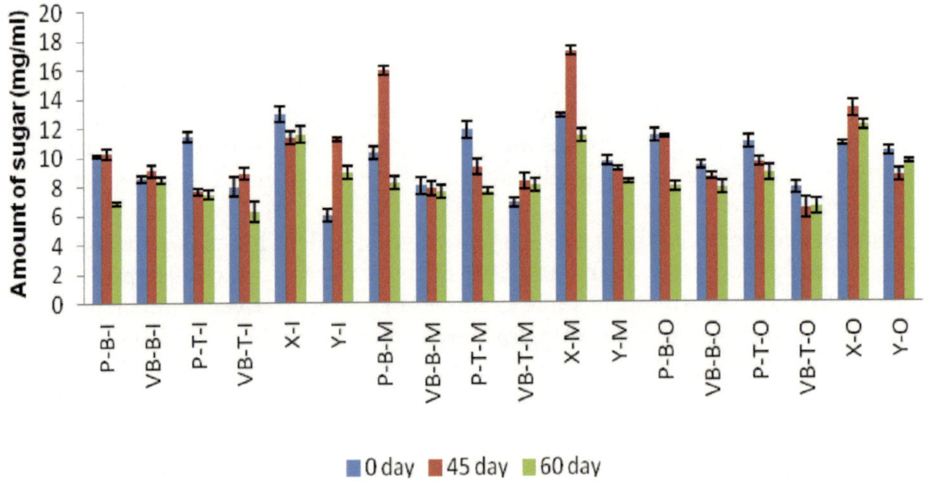

Figure 5.
Amount of sugar content of bottom and top part oil palm trunk during storage time. P-B, parenchyma-bottom; VB-B, vascular bundle-bottom; P–T, parenchyma-top; VB-T, vascular bundle-top; X, bottom non-separated; Y, top non-separated; I, inner; M, middle; O, outer.

during storage time which contains more parenchyma compared to the top part of the non-separated samples. The difference between these is probably due to the function of parenchyma as a storage organ and contain abundant amount of sap and nutrients which is rich in oligosacchides compare to vascular bundle that function as a mechanical support of the oil palm trunk [31].

The studies of 0, 45 and 60 day had been chosen to observe the potential sugar increase to optimum yield of storage time that was obtained. This storage time was chosen based on previous studies which 0, 45 and 60 days. From the **Figure 5** also showed the sugar content with the average at 0 days was 8–10 mg/ml and slightly increases at 45 days with the sugar content average at 10–17 mg/ml. At the 60 day, the sugar content decrease to 8 mg/ml and this pattern is similar to the separated sample and non-separated sample of oil palm trunk. The figure also showed that the bottom part of the middle part of the oil palm trunk of individual parenchyma and non-separated has the highest sugar content at 45 days compared to others. Amount of sugar in the individual parenchyma at the middle part is 16.0 mg/ml at the bottom part whereas non-separated sample of the bottom part of oil palm trunk at the middle part is 17.3 mg/ml. The lowest sugar content at 45 days are shown in the individual vascular bundle at the top part of the outer part of the oil palm trunk with the amount of sugar content is 6.5 mg/ml whereas for the non-separated sample show that the top part of the outer part of the oil palm trunk contain 8.73 mg/ml of sugar content.

Based on the results it was observed that the pattern results of sugar content for all samples increased as the duration of storage increased. Based on the findings of this study, the sugar content at day 0 initially increased and decreased rapidly around day 45 to 60. This changing pattern may happened because of the starch in the oil palm trunk probably was being converted to glucose and other fermentable sugar by enzyme activities. These activities probably involved degrading enzyme and sucrose metabolism enzymes [7]. The error bars in the **Figure 5** represent the differences were statistically significant for parenchyma and vascular bundle of oil palm trunk based on storage time.

3.2 Starch analysis of parenchyma and vascular bundle of oil palm trunk

Figure 6 showed the percentage of starch content for separated sample (parenchyma and vascular bundle) and non-separated (mixture parenchyma and vascular bundle) sample from inner to outer of oil palm disks. The duration of the storage time was determined between 0, 45 and 60 days were used similar as the determination of sugar content of oil palm trunk.

Based on this study, the middle part of the oil palm trunk had the highest starch content compared to the outer and the inner part. The amount of the starch content in the middle part showed with a range 3–5%, whereas in the inner part in a range 2–4% and in the outer part around 3–4% in the 0 days. Parenchyma at the bottom part in the middle part showed the highest starch content compared to others which is 4.54% at 0 days whereas parenchyma at the top part in the inner part showed the highest starch content which is 3.5% at 45 days. Referring to previous studies on the starch content for a 0 day, as indicated by Tomimura (1992), the starch content for separated sample of vascular bundle and parenchyma was 2.4% while in parenchyma was 55.5% respectively.

The pattern results of starch content for all samples were reduced as the duration of storage increased. Based on the findings of this study, the starch content at day 0 initially to increase and decrease rapidly around day 45 to 60 and became almost negligible after 60 days. This changing pattern may happened because of the starch probably been converted to glucose and other fermentable sugar by enzyme

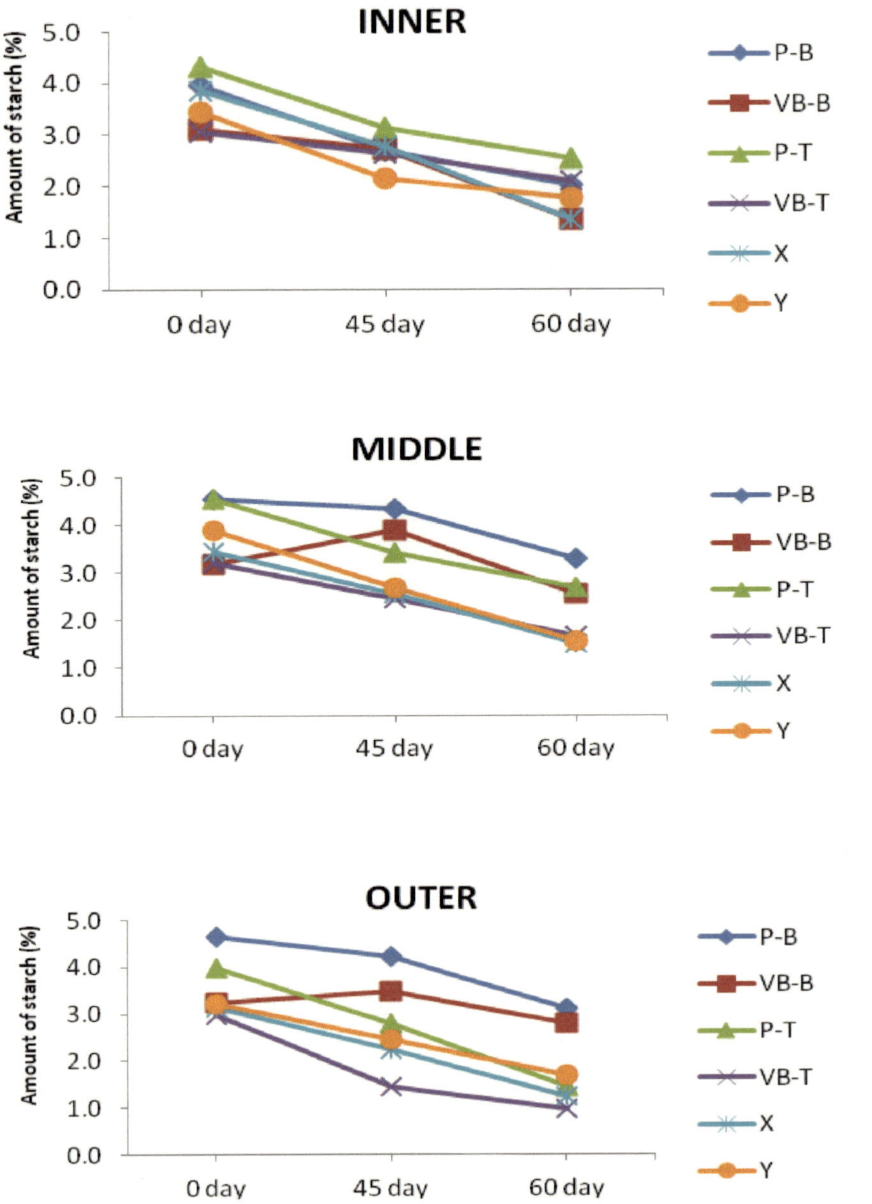

Figure 6.
Amount of starch content at different part of the oil palm trunk during storage time. P-B, parenchyma-bottom; VB-B, vascular bundle-bottom; P–T, parenchyma-top; VB-T, vascular bundle-top; X, bottom non-separated; Y, top non-separated.

activities. These activities probably involve degrading enzyme and sucrose metabolism enzymes [10]. In this study we only focused on 0, 45 and 60 day because it is the optimal day according to the previous study.

This research showed that the higher starch content had been found in the parenchyma comparatively to vascular bundle. Abundant amount of parenchymatous tissue which is rich in starch content accumulates in these parts [32]. According to [32] study, the carbohydrate content in oil palm trunks reported that peripheral cortex contains high starch level. This study showed that the middle part contained the highest starch content which is being different from the previous study. This may be due to the different types of cultivar has been used.

According [32] found that high level of starch content was located dominantly in the top part of oil palm trunk. This phenomenon relates to an older cells distribute in basal part of the oil palm trunk compared to top part that contain young cells. As young cells, it needs a lot of carbohydrates which is consisting of the abundant amount of starch content, reserved for the growing process [33]. This statement was supported by [31] in his report that starch stored in the upper part of the tree is purposely used for flowering process.

4. Conclusions

The chemical compositions that were determined in this study consisted of starch, and sugar of parenchyma and vascular bundle of oil palm trunk. Regarding in this study, parenchyma showed higher extractive content compared to the vascular bundle during the storage time while, parenchyma on the top parts showed slightly higher instructive compared to the parenchyma bottom. Vascular bundle on the bottom and top part of the trunk have the highest percentage of holocellulose which are range 80–85% compared to the parenchyma on the bottom and top part that around 65–75%.

Acknowledgements

The authors would like to express their gratitude for the USM-RU-PRGS grant (1001/PTEKIND/844063) for sponsoring this research. The authors gratefully appreciate the Japan International Research Center for Agricultural Sciences for partially sponsoring this research (304/PTEKIND/650481/J122). Last but not least, to the Faculty of Bioengineering and Technology, Universiti Malaysia Kelantan, to support my research indirectly from the beginning until the end.

Conflict of interest

I declare that this chapter entitled "Chemical composition of parenchyma and vascular bundle from *Elaeis guineensis*" is the results of my own research except as cited in the references.

Notes/thanks/other declarations

Special thanks to my husband, my children, my mom, my grandfather, and my in-laws; it is impossible to acknowledge inappropriate language the sacrifices you had made to support and encourage me to keep me devoted to my research.

Author details

Sitti Fatimah Mhd. Ramle
Faculty of Bioengineering and Technology, Universiti Malaysia Kelantan,
Jeli, Kelantan, Malaysia

*Address all correspondence to: fatimah.m@umk.edu.my

References

[1] Ramle, S. F. M., Sulaiman, O., Hashim, R., Hamid, Z. A. A., Arai, T., Kosugi, A., Murata, Y. Chemical characterization from parenchyma and vascular bundle at different parts of oil palm trunk. AIP Proceeding. 2019: 020039. DOI:10.1063/1.5089338

[2] Teoh, Cheng Hai. 2010. "Key Sustainability Issues in the Palm Oil Sector," Discussion Paper for Multi-stakeholder Consultations commissioned by the World Bank Group.

[3] Slette, J., and Wiyono, I. E. (2011). Indonesia Biofuels Annual: 2011. Global Agricultural Information Network (GAIN), US Department of Agriculture.

[4] Husin, M. (1985). Potentials of oil palm by-products as raw materials for agro-based industries, proceed. Nat. Symp. On oil Palm by-products for agro-based industries, Kuala Lumpur. Palm oil res. Inst. Malaysia Bull. 11, 7-11.

[5] Tomimura, Y. (1992). Chemical characteristics and utilization of oil palm trunks. JARQ. 25, 283-288.

[6] Yamada, H., Tanaka, R., Sulaiman, O., Hashim, R., Hamid, Z. A. A., Yahya, M. K. A., Kosugi, A., Arai, T., Murata, Y., Mirasawa, S., Yamamoto, K., Ohara, S., Mohd Nor Mohd Yusof, Ibrahim, W. A., and Mori, Y. (2010). Old oil palm trunk: A promising source of sugars for bioethanol production. Biomass and Bioenergy. 34, 1608-1613.

[7] Kosugi, A., Tanaka, R., Magara, K., Murata, Y., Arai, T., Sulaiman, O., Hashim, R., Hamid, Z. A. A., Yahya, M. K. A., Yusof, M. N. M., Ibrahim, W. A., and Mori, Y. (2010). Ethanol and lactic acid production using sap squeezed from old oil palm trunks felled for replanting. Journal Bioscience and Bioengineering. 4, 4C.

[8] Abdul Khalil, H. P. S., Siti Alwani, M., Ridzuan, R., Kamarudin, H., and Khairul, A. (2008) Chemical composition, morphological characteristics, and cell wall structure of malaysian oil palm fibers. Polym-Plast Techno. 1(47), 273-280.

[9] Wan Rosli, W. D., and Law, K. N. (2011). Oil palm fibre as papermaking materials: Potential and challenges. BioResources 6(1), 901-917.

[10] Abe, H., Murata, Y., Kubo, S., Watanabe, K., Tanaka, R., Sulaiman, O., Hashim, R., Sitti Fatimah, M. R., Zhang, C., Noshiro, S. and Mori, Y. (2013). Estimation of the ratio of vascular bundle to parenchyma tissue in oil palm trunks using NIR spectroscopy. BioResources 8(2), 1578-1581.

[11] Baker, E. S., Sahri, M. H., and H'ng, P. S. (2008). "Anatomical characteristics and utilization of oil palm wood," In: The Formation of Wood in Tropical Forest Trees – A Challenge from the Perspective of Functional Wood Anatomy. T. Nobuchi and M. H. Sahari (eds.). Penerbit Universiti Putra Malaysia, Serdang, 161-178.

[12] MPOB. Malaysia Palm Oil Board, Retrieved May 06, 2021, from (http://www.mpob.gov.my)

[13] Jacquin (1763). *Elaeis Guineensis* Select. Stirp. Amer. Hist. 280, t. 172

[14] Bailey, L. H. (1940). The great carossier. Gentes Herbarium, 4, 262-265.

[15] Wessels-Boer, J. G. (1965). Flora of Suriname 5(1). Palmae. E. J. Brill, Leiden.

[16] Hardon, J. J., Corley, R. H. V., and Lee, C. H. (1987). Breeding and selection for vegetative propagation in oil palm. Improvement of Vegetatively Propagated Crops (Abbott, A. J., and

Atkin, R. K., eds.). Academic Press. p. 64-81.

[17] Hardon, J. J., Williams, C. N., and Watson, I. (1969). Leaf area and yield in the oil palm in Malaya. Experimental Agriculture 5:25-32.

[18] Wunderlin, R. P., and B. F. Hansen (2002). Atlas of Florida Vascular Plants. University of South Florida, Tampa.

[19] Parham, R. A., and Kausftinen, H. M. (1974). Papermaking Materials. An Atlas of Electron Micrographs. The Institute of Paper Chemistry, Appleton, WI.

[20] Palma, M. T., Bocchio, L., Capretti, G., Gastaldi, G., Orlandini, S., Focher, B. (1996). Fourth European Workshop on Lignocellulosics and Pulps: Advances in Characterization and Processing of Wood, Non-Woody and Secondary Fibers. Stresa, Italy. September 8-11, pp. 273-278 (abstracts).

[21] Rowell, R. M. (1975). Chemical Modification of Wood: Specialty Treatments. 19, 5-10.

[22] Smith, A. G. (2001). Embryonic stem cells. In: Stem Cells, ed. D. Marshak, R. L. Ardner, D. Got-Tlieb, Pp. 205-30. New York: Cold Spring Harbor Lab. Press.

[23] Cheng, C. K., Hani, H. H., and Ismail, K. S. K. (2007). Bioethanol Glucose Biomass Fermentation: Palm Oil Biomass Energy Alcohol as Fuel. Oil Palm 1st Conference on Sustainable Materials (IcoSM) 9th 12th June, 2007. Universiti Malaysia Perlis: 69.

[24] Alam, M. Z., Kabbashi, A. A., Nahdatul, S., and Hussin, I. S. (2009). Production of bio-ethanol by direct bioconversion of oil-palm industrial effluent in a stirred-tank bioreactor. Journal of Industrial Microbiology and Biotechnology. 36, 801-808.

[25] Dunagani, R., Jawaid, M., Abdul Khalil, H. P. S., Jasni., Sri Aprila., Hakeem, K. R., Hartati, S., and Islam. A Review on Quality Enhancement of Oil Plam Trunk Waste by Resin Impregnation: Future Materials. Bioresources.com, Peer Review Article. 2013.

[26] Esther, O. U. (1997). Anaerobic digestion of palm oil mill effluent and its utilization as fertilizer for environmental protection. Renewable Energy 10(213): 291-294.

[27] H'ng, P. S., Wong, L. J., Chin, K. L., Tor, E. S., Tan, S. E., Tey, B. T., and Maminski, M. (2011). Oil palm (*Elaeis guineensis*) trunk as a resource of starch and other sugars. Journal of Applied Polymer Science 11, 3053-3057.

[28] Shuit et al., 2009.

[29] Cai, L., Han, S., and Deng, W. (1992). Chemical and physical changes of steam treated wood chip. Journal of Northeast Agricultural University 4(1), 48-52.

[30] Sulaiman, O., Salim, N., Nordin, N. A., Hashim, R., Ibrahim, M., and Sato (2012). The potential of oil palm trunk biomass as an alternative source for compressed wood. Bioresources 7(2), 2688-2706.

[31] Corley, R. H. V. (1996). What is the upper limit to oil extraction ratio? Proc. of the International Conference on Oil and Kernel Production in Oil Palm ñ a Global Perspective (N. Rajanaidu; Henson, I. E. and Jalani, B. Seds.). ISOPB and PORIM. p. 256-269.

[32] Henson, I. E. (1999). Notes on oil palm productivity. IV. Carbon dioxide gradients and fluxes and evapotranspiration, above and below the canopy. J. Oil Palm Research Vol. 11 No. 1: 33-40.

[33] Husin, M. (2000). Utilization of oil palm biomass for various wood-based and other products, edited by Basiron, Y., Jalani, B. S., and Chan, K. W. In: Advance in oil Palm research. Malaysian Palm Oil Board. 2(32): 1346-1354.

Chapter 13

Processing and Properties of Oil Palm Fronds Composite Boards from *Elaeis guineensis*

Razak Wahab, Mohd Sukhairi Mat Rasat, Norashikin Mohd Fauzi, Mohamad Saiful Sulaiman, Hashim W. Samsi, Nasihah Mokhtar, Ros Syazmini Mohd Ghani and Muhammad Haziq Razak

Abstract

Oil palm fronds are one of the biomass residues originating from oil palm plantations. It has great potential to be used as an alternative material for the composite boards industry to reduce dependency on wood-based raw materials. The fronds are obtainable all the year round and in big quantity. The oil palm fronds had been processed as compressed oil palm fronds to form such a potential composite board in this topic. A composite board from compressed oil palm fronds was produced by removing the fronds' leaflets and epidermis. The sample was sliced longitudinally into thin layers and compressed into an identical thickness at about 2 to 3 mm. Pieces of the sample were dry using the air-dried method. They were then mixed with phenol and urea-formaldehyde of resins in the range of 12-15% and compressed again with another layer forming a composite board. Standard outlined by the International Organization for Standardization (ISO) tested for their physical and strength properties of composite board. Found that the physical and strength aspects' properties show that the composite board possessed characteristics at par or equivalent. The composite board from compressed oil palm fronds has good prospects to be used as an alternative to wood. Thus, this characteristics can overcome the shortage in materials supply in the wood-based industry.

Keywords: Oil Plam Fronds, Anatomy Characteristics, Chemical Compsotions, Physical & Mechanical Properties, Composite Panel

1. Introduction

Oil palm seems like another promising alternative to reduce the dependency on timber since there are plenty of oil palms in Malaysia's plantation and have high potential to be future composite boards through their residues. Agricultural fibre, especially oil palm fronds used, can be easily crushed and may be used as substitutes for wood-based raw materials. Explore the use of local natural fibre for composite boards and has an excellent potential to compete with other commercial products. The oil palm fronds are usually left rotting between the rows of palm trees. This method served mainly for soil conservation, erosion control, and ultimately the long-term benefit of nutrient recycling [1, 2].

Many studies on the oil palm fronds showed the potential of utilizing this agricultural residue for several types of value-added products. The value-added products are the manufacturing of pulp and paper, animal feed, and fibreboard in the wood-based industry [3–5]. Intensive research work is ongoing using various technologies to convert oil palm fronds for the manufacture of commercially viable composite board products [6–9]. They were chosen because they are residues and abundant in oil palm plantations. It will optimize the uses of biomass by-products, especially from oil palm fronds. Moreover, it gives better understanding and knowledge of non-wood material for future sustainability.

Currently, there is a limited supply of timber in the wood industry to cater to permanent structural use. This issue is due to poor and inconsistent quality, high and fluctuating costs associated with the supply shortage [10]. Due to the depletion of forest resources, there is a shortage of wood suppliers required by the industry. It is not surprising that alternative bio-materials are getting popular in reducing the over-dependence on the local timber industry. Furthermore, previous researchers worldwide seem to focus more on woody plants, and they paid less attention to non-woody plants. For some reason, there is a need to expand the knowledge to non-wood properties.

Additionally, the use of these oil palm fronds still lacks its contribution to commercializing [11]. Simultaneously, by maximizing the knowledge, especially on oil palm crops, the world will know that Malaysia is taking seriously in minimizing the effect of agricultural residues on the environment and reducing the consumption of natural forest. It should be done as Malaysia complained about foreign countries because of its mass production of palm oil, and Malaysia is known as the largest producer in the palm oil industry. Novel technologies with improved efficiencies and reduced environmental impacts need to be established timely to utilize a large amount of waste [12].

2. Oil palm fronds

Oil palm fronds are considered to be one of the most abundant agricultural by-products in Malaysia. Oil palm fronds are currently considered waste from oil palm plantations, and their biomass is not used entirely. In Malaysia, the total production of these felled and pruned oil palm fronds is estimated at 24.4 million mt dry matter per year [13], and this was almost doubled within a decade to about 40 million mt in 2004. Recycling is needed to produce something that can be used and avoid the pollution of the environment. Pruning activities of these oil palm fronds will be made from time to time depending on the individuals who manage them and the available quantity of oil palm fronds. The available amount is depending on the age of the oil palm tree. It is estimated that about 10,400 kg/ha of oil palm fronds can be produced yearly [14]. Meanwhile, about 14,500 kg/ha of oil palm fronds are produced by replanting activity annually [15].

The average economic life span of the oil palm is 25 years [16] and generally was replaced after 25 to 30 years [6, 17]. Oil palm mass cultivation began in 1960 [18]. The years 1990 and beyond marked a peak in the replanting of the oil palm trees. This information presents an excellent opportunity to harness the lignocellulosic biomass or by-products of the oil palm, including the fronds. Oil palm fronds are available throughout the year when the oil palms are pruned during the harvesting of fresh fruit bunches for palm oil production. Besides, many fronds are produced by replanting each year, making these fronds show an up-and-coming source for composite panels and ensuring their abundance and availability.

The oil palm trees are typically planted about 145 oil palm trees per ha. People can harvest about 25 pieces of fronds from a single palm tree roughly. Each frond weighs about 8 kg [19]. This frond weighs resulted in about 200 kg of fronds that can be obtained in a year. Malaysia can produce about 30 tons of frond biomass in 1 ha in a year. Other reviews show that under standard practice, around two palms are pruned per palm per month. Under the current plantation system in Malaysia, the oil palm density per ha is about 136 palm trees per ha. Therefore, this would yield at least 3200 pruned fronds per ha per year, yielding at least 18 mt of fronds biomass per year [20].

Oil palm fronds have great potential for use as a roughage source or as a component in compound feed for ruminants, either fresh or processed as silage or pellet [1]. This is because oil palm fronds contain the right level of nutrients, which is 70%, and the rest is carbohydrate content [21]. According to the Husin et al. [22], for soil conservation, increase fertility, improve the amount of water retention in the ground, erosion control and provide a source of nutrient to the growing oil palm trees, the fronds were left to rot in between the row of oil palm trees in the plantation. Oil palms by-products are available in a large quantity sufficient for industrial raw materials in agro-based industries. The endless and consistent supply of lignocellulosic materials from the oil palm industry, especially oil palm fronds, should be considered as new bio-resources. New product development from oil palm is now at their stage of research to be developed later on.

3. Properties of oil palm fronds

The fronds are found around the trunk in two spirals which are left-handed or right-handed. Individual mature frond has rachis, leaflets, and thorns [23]. Oil palm fronds, the aerial part of the oil palm tree comprises two central portions; they are the petiole and the leaflet. The petiole, which is the woody part of the frond, represents more than 70% of the whole frond, whereas the weight of the leaflet is less than 30%. Therefore, the proportion of petiole and leaflet portions, which is determined by the age of the frond at the time of harvest, would be the dominant factor determining the fibre composition of maturing oil palm fronds [24].

The moisture content of oil palm fronds is very high, up to 60% on a wet basis. The leaves are found at the top of the plant arranged like a crown containing 40 or more fronds. Each palm frond has 20 to over 150 pairs of roughly 2.5 cm wide leaflets arranged in two rows along each side of the petiole [25]. The frond lengths decrease from the bottom to the top level of the crown, reaching the length of about 4 m. In a cross-section, the frond shows a triangle shape with the width decreasing from the base to the end of the petiole and from the bottom to the top fronds [23]. A fruiting branch that contains thousands of fruits is held in the axils of the leaves and arranged in a rosette pattern around the crown [26, 27].

3.1 Anatomical characteristics of oil palm fronds

Anatomical characteristics of oil palm fronds are different from other woody structures in that they have four essential cell elements: parenchyma, vascular bundles, sclerenchyma, and epidermis [17]. Oil palm fronds do not possess cambium, sapwood, heartwood, and growth ring; hence their 'wood' is the primary tissue itself [28]. The fronds are primarily composed of parenchymatous tissues with numerous fibrous strands and vascular bundles. The oil palm fronds consist of a mass of discrete vascular bundles embedded in parenchymatous tissues [29]. The growth and increase of the fronds result from the overall cell division and cell enlargement in the parenchymatous tissues, together with the enlargement of the fibre of the vascular bundles [30].

Like other monocotyledon plants, oil palm has inner and outer vascular bundles, and the same goes for its fronds. The outer region of living tissues is differentiated into a narrow cortex from the wide central cylinder. **Figure 1** shows the anatomical characteristics of the oil palm fronds. In the cylinder, vascular bundles were found concentrated at the outer and scattered at the inner division. The distribution of the vascular bundles throughout the fronds has become an essential factor influencing the oil palm fronds' anatomical features and physical properties [30]. **Figure 2** shows the oil palm fronds in the transverse section. The morphological of oil palm frond structures presented in **Figures 3–5** were observed under high-performance microscopy with different magnification.

The parenchymatous tissues comprise a short chain of polysaccharides, starch, and also a soft structure. However, the fibrous strands are principally dense cellulose which is challenging to degrade. A study found the weight ratio of the parenchymatous tissues on fibre strands in the range 24-29% to 71-76%. The riches of starch discovered in the parenchyma structure are about 55% while 2.4% in the fibre. Then, the lignin content recorded a slightly comparable value, 20% in fibre and 15.7% in the parenchyma [31]. Furthermore, regarding the orientation of

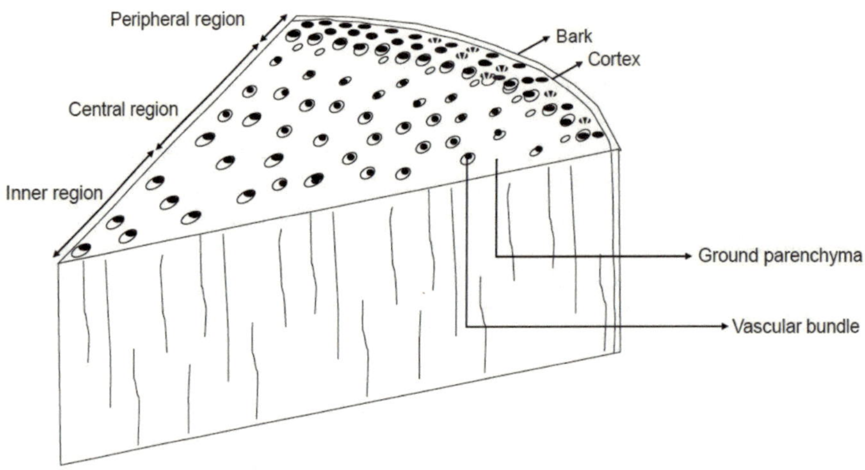

Figure 1.
Structure of anatomical characteristics of oil palm fronds.

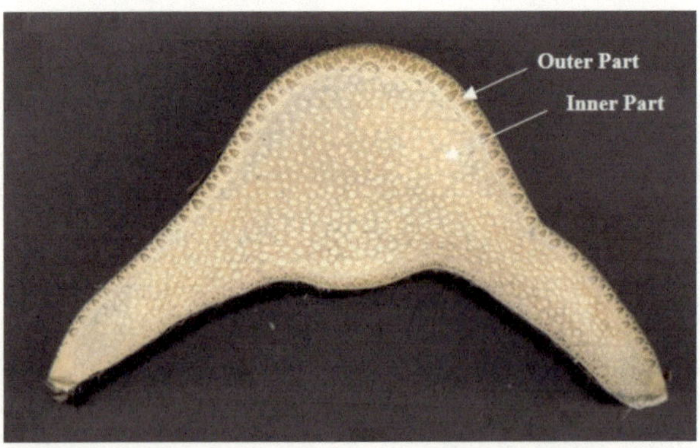

Figure 2.
Oil palm fronds at the transverse sectional view.

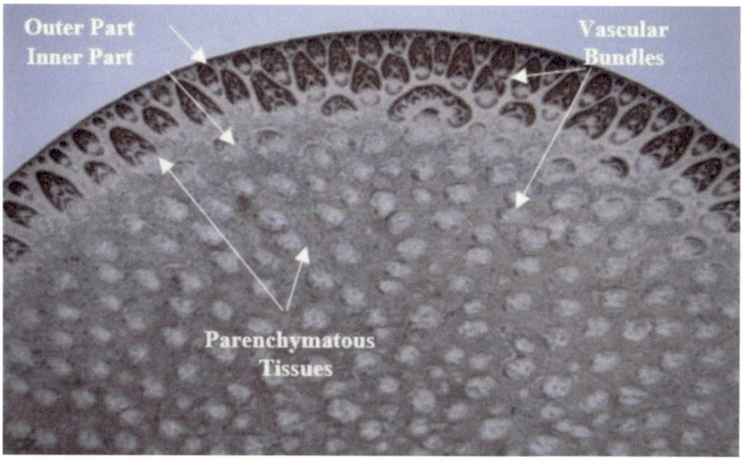

Figure 3.
High-performance micrograph of oil palm fronds at the transverse sectional view (0.75× magnification).

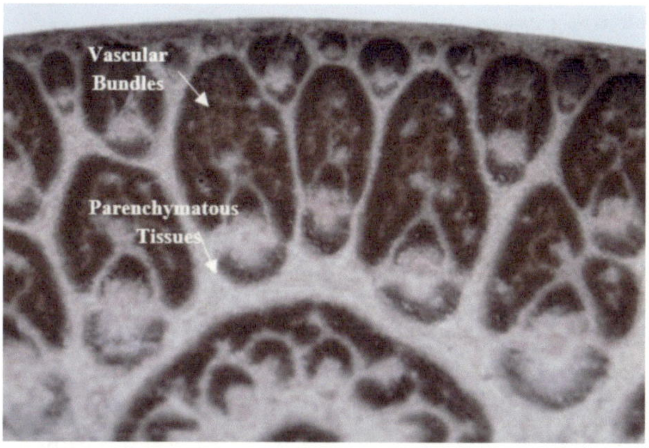

Figure 4.
High-performance micrograph of outer part on oil palm fronds at the transverse sectional view (4× magnification).

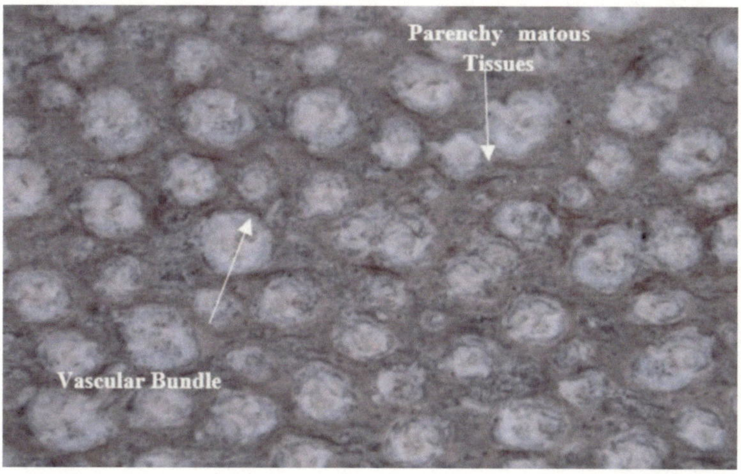

Figure 5.
High-performance micrograph of inner part on oil palm fronds at transverse sectional view (2× magnification).

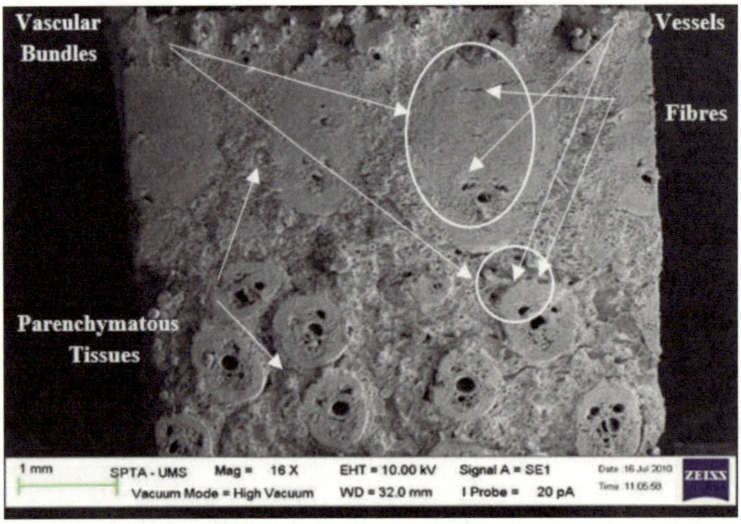

Figure 6.
SEM of oil palm fronds at the transverse sectional view (16× magnification).

vascular bundles over the transverse section, most of the vascular bundles were oriented randomly. **Figure 6** clearly shows the difference in sizes and population of vascular bundles towards the outer to the central part of the oil palm fronds using scanning electron microscopy (SEM).

3.1.1 Vascular bundles

Vascular bundles serve as the supporting structure and the transport system of the oil palm fronds. Typically, the vascular bundles are composed of fibre, vessel or metaxylem, protoxylem, protophloem or sieve tubes, axial parenchyma, stigmata, and companion cells [30, 32]. Most of the vascular bundles are composed of one or two-vessel cells. Though uncommon, vascular bundles with more numerous than three vessels designed tangentially or in batches can also be observed scattered, especially in the core section. Widespread protoxylem lessened vascular tissue, and little bundles with tiny fibrous tissues are also regularly seen scattered between the broader bundles in the core section. The arrangement of fibrous strands depends on the amount of bundles present [33]. Phloem can be found located between the vessel and fibre sheath. Protoxylems are present at the outer part adjacent to the vessel. The number of vascular bundles for the frond is the same and permanent. The growth is happening on vascular bundles diameter only and not its quantity [34].

Based on the visual investigation under SEM, the vascular bundle consists of one or two large vessels and surrounded by fibres cells (**Figure** 7). These vascular bundles are embedded around parenchymatous tissue. The vascular bundles are surrounded by parenchymatous ground tissue. Therefore the wood material from this species is not comparable to the woods produced from both dicotyledons and gymnosperms species which developed from the secondary xylem [35]. The fibres that compose the vascular bundles are arranged in a crescent-shaped sheath surrounding the phloem. The fibres are irregular in length and wall thickness. The number of fibres associated with vascular bundles and the number of secondary walls deposited in them on the location of the bundles within fronds [36].

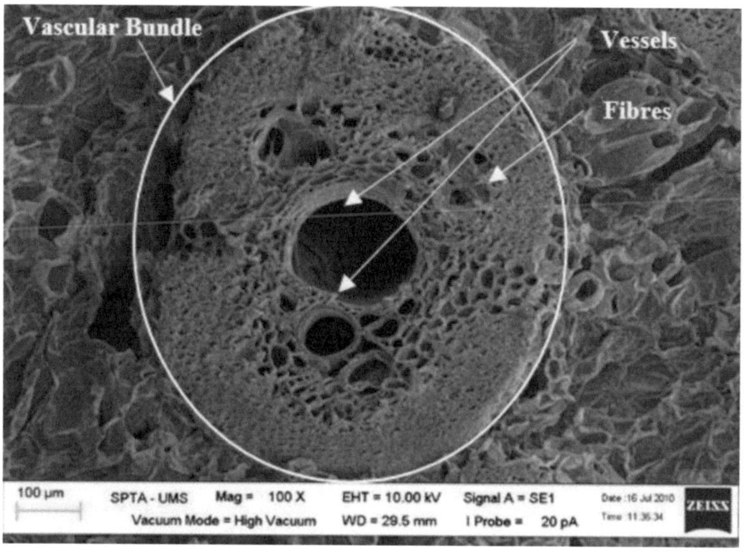

Figure 7.
SEM of vascular bundle in oil palm fronds at the transverse sectional view (100× magnification).

3.1.2 Vessel structure

According to the vascular bundle structure presented in **Figure 7**, the presence of large vessels varies from one to three vessels. The term of large vessel here is the tracheary elements of a vascular bundle like the oil palm trunk [35]. The extensive surveys of tracheary elements in palms are conducted by Tomlinson [30], and Bierhorst and Zamora [37], and these studies can be applied to oil palm fronds. Further, Parthasarathy and Klotz [28] found the palm vascular bundle clusters tracheary elements display a gradation in morphology from protoxylem within beginning to final metaxylem. The end walls of the tracheary part manifest a rising degree of evolutionary specialization, i.e., they enhance decreasingly tracheid-like. The protoxylem tracheary parts and any of the narrow early metaxylem elements approximately evermore appear to be tracheids. The remaining thin metaxylem elements and the broad late metaxylem elements present varying degrees of specialization, depending on the organ and the species.

Large vessels with very thick vessel-wall were predicted as the main component responsible for transporting the nutrient. This statement was in agreement with the result from Lim and Khoo [33] that has been studied on the oil palm trunk. **Figures 8** and **9** show the vessel on a close view in the vascular bundle from two different sectional views at transverse sectional and at the longitudinal views.

3.1.3 Fibre structure

The oil palm fronds consist of primary vascular bundles embedded in parenchyma ground tissues. Oil palm fronds fibres spread beyond the vascular bundles and loaded by the parenchyma cells. Fibres have a tight end, mainly influenced. The composition of fibres was essentially comparable to common woods, including softwood and hardwood, which comprise pits, cell walls, and lumen. Fibres in the oil palm fronds played a vital role in the strengthening mechanism of the composite when stress was transferred between the matrix and fibres. The SEM of fibres contains in vascular bundles is displayed in **Figures 10** and **11.**

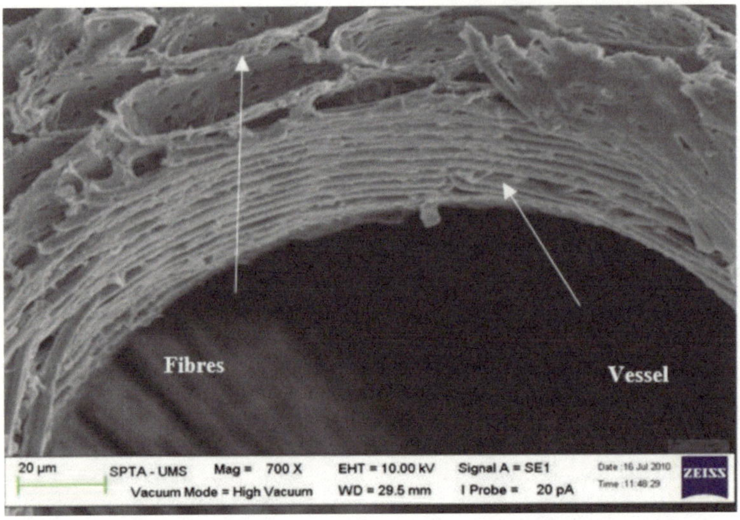

Figure 8.
SEM of the vessel in oil palm fronds at the transverse sectional view (700× magnification).

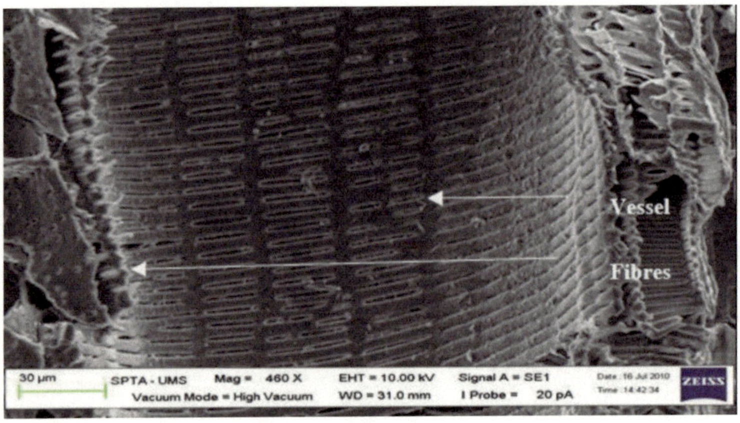

Figure 9.
SEM of the vessel in oil palm fronds at the longitudinal sectional view (460× magnification).

The transverse sectional view of oil palm fronds in **Figure 10** shows that the fibres attached to the others in very compact formations like fibres in oil palm trunk. **Figure 11** presented the structure of the fibre in oil palm fronds closely. Various sizes and shapes were distinguished, e.g. spherical, triangular, and rectangular. The presence of pits also identified at the fibre wall and companion cells, like shown in **Figure 11**. The walls might be thick or thin, and the small or large lumina. The primary function of fibres was predicted to provide strength support to the living oil palm fronds.

3.1.4 Parenchymatous tissues

Parenchymatous tissues are not only found in the trunk but also found abundantly in the fronds. They are food storage elements and must, therefore, remain alive for an extended period. Similar to vascular bundle shapes, parenchyma tissues show two different forms, which are isodiametric-shape and elongated-shape. The isodiametric cells have thin walls, while the elongated cells have thicker

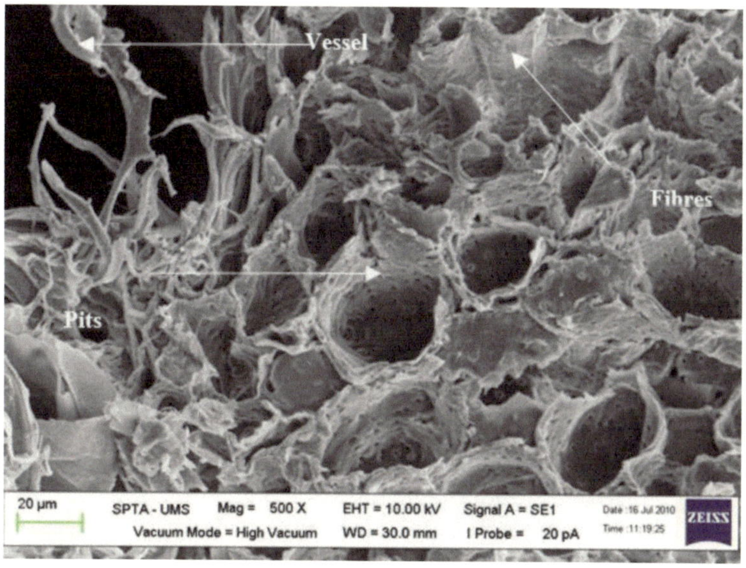

Figure 10.
SEM of fibres in oil palm fronds at the transverse sectional view (500× magnification).

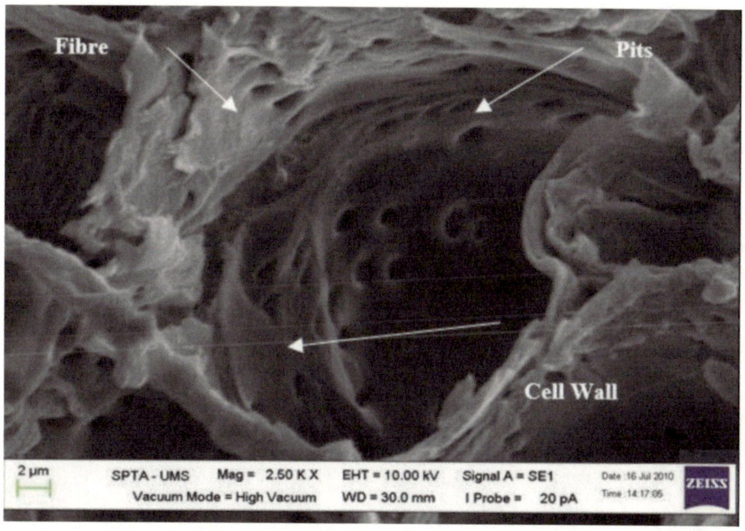

Figure 11.
SEM of fibres in oil palm fronds at the transverse sectional view (2500× magnification).

poly-laminate walls [38]. Parenchyma tissues are also found associated with the vascular bundles in the axial position, an elongated shape. They are found at the inner tip of the vascular bundles related to the vessels, protophloem, and protoxylems. They may have a special function in water transport, like the vessel-associated cells in the dicotyledonous trees [33]. These parenchymatous tissues consist mainly of ground parenchyma, parenchyma strands, sieve elements, and companion cells [36].

The parenchymatous ground cells consist mainly of thin-walled spherical cells, except in the vascular bundles. The walls are progressively thicker and darker from the inner to the outer region. Parenchyma tissues contain much sugar and starch as food storage elements, which are soluble in water and NaOH [39]. Parenchyma cells of oil palm fronds functioned as the ground tissues that make up the bulk of oil palm fronds structures and are used as storage for food. Parenchyma cells of oil palm fronds were mainly in the form of spherical cells with thin-walled and brick-like formations, but in the narrow space or area between vascular bundles, they were familiar as elongated cells and oval-cells, and this is similar to oil palm trunk. Physically, this tissue was spongy and moist in green condition and very lightweight and easy to separate one cell from the others. Parenchyma cells also contain some amount of chemical composition [40]. **Figure 12** shows detail of the structure of parenchyma cells under SEM.

Figure 12 shows that many pits were observed on the primary cell wall, which functioned for water or nutrient transport purposes. Based on this fact, it is logically accepted why ground parenchymatous tissue was very hygroscopic. It was easy to evaporate when the temperature is rising and also easy to absorb the moisture in high humidity conditions. A study by Tomimura [41] and Sun et al. [42] found that parenchyma contained a high amount of starch and lignin compared to vascular bundles. Besides, parenchyma ground tissue which cements the vascular bundles together is undesirable for the manufacture of wood-based products like pulp and composite boards. In pulp manufacture, these tissues consume chemicals and produce fines that may be lost during screening. The rounded parenchyma also reduces the paper strength properties. In particleboard manufacturing, it may interfere with the bonding between particles and will reduce strength properties.

3.2 Chemical composition of oil palm fronds

The chemical composition of biomass significantly varies due to their diverse origins and types. Biomass is generally composed of cellulose, hemicelluloses, lignin, and inert ash. Chemically, fronds strands are rich in holocelluloses (83.5%) and also high in α-cellulose (49.8%) [43]. The extractive oil palm oil fronds are about 4.5%, and the lignin content is about 20.5% [43]. This lignin content shows that it is lower than generally found in commonly hardwood, for eucalyptus of 22%. **Table 1** shows the chemical composition of oil palm fronds.

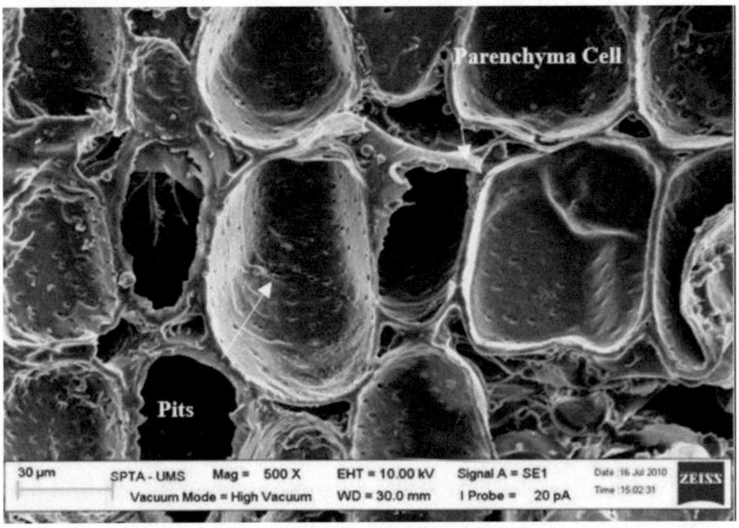

Figure 12.
SEM of parenchyma cell in oil palm fronds at the transverse sectional view (500× magnification).

Composition	Chemical composition (%)				
	Extractive	Holocellulose	Cellulose	Lignin	Ash
Oil Palm Fronds	4.5	83.5	49.8	20.5	2.4

Source: Abdul Khalil et al. [43].

Table 1.
Chemical composition of oil palm fronds.

This is not surprising since oil palm trees are non-woody, and the requirement for structural support is lower compared to woody trees. The functional significance of lignin has long been associated with strength support for plant organs that enables increased growth in height [44, 45]. Its lacking will no longer allow plants to be upright [46]. It should also be noted that the fronds strands, like other non-wood fibres, contain comparatively high ash content. This characteristic might contribute to abnormal strength wear of processing equipment. The monomer composition of polysaccharides shows only glucose, xylose, and other monosaccharides representing less than 6%. Thus, it is in broad similarity with that of hardwoods [47, 48].

3.3 Physical properties of oil palm fronds

Physical properties of oil palm fronds are essential in contributing its characteristics as a raw material of producing a product that required the properties to stand against load. Many different wood properties have some influence on the processing of wood products. Since it is not possible to evaluate all these properties for all species concerned, it is necessary first to identify the essential characteristic to determine the quality of the products that are finally used [49]. The main physical properties are the moisture content, density as well as basic density.

3.3.1 Moisture content

The moisture content of oil palm fronds is similar to oil palm trunks. The oil palm moisture content recorded ranges between 100% and 500%. A progressive rise in moisture content is shown onward with the trunk height and towards the central section, while the lower and outer positions were should far lower values than the other two sections [33]. An increase in the number of vascular bundles causes a decrease in the percentage of parenchyma cells, which have a high capacity in water absorption [50].

3.3.2 Density

Due to its monocotyledonous nature, there is a tremendous variety of density values at different oil palm fronds like the oil palm trunk. The density values range for oil palm trunk is from 200 to 600 kg/m^3 with an average density is 370 kg/m^3. The density of the oil palm trunk decreases linearly with the trunk height and towards the centre of the trunk [33]. These fluctuations are due to various factors. Beyond the trunk, the density is mainly influenced by the number of vascular bundles per square unit, reducing the centre. Nonetheless, vascular bundles existing younger at the top end influenced changes in density onward trunk height. Although the bundle is smaller in size and cell walls are thinner, proportionally, a higher per square centimetre is higher.

3.4 Oil palm fronds strength properties

The oil palm trunk strength properties are proportional to the density variation recognized in both vertical and radial directions. The highest bending strength values from the peripheral lower portion, and the top portion of the central core highlighted the most insufficient strength. Variation of the compression strength parallel to grain also follows the same trend as the bending strength [51]. But, for the oil palm fronds, there is no strength testing done in raw material because compared to the oil palm trunk, there are no uses of oil palm fronds in raw material directly like the trunk.

Some studies have been done in manufacturing composite boards from oil palm fronds compared to their strength properties. Laemsak and Okuma [7] mentioned that there seems to be a good possibility for producing binderless boards using steam-exploded fibres of oil palm fronds considering the chemical components of oil palm fronds which are rich in hemicelluloses. The strength properties such as modulus of rupture (MOR) and modulus of elasticity (MOE) of the boards increased linearly with increasing board density as the standard hardboard. They reported that the boards made from fibres treated under a steam explosion condition of 25 N/cm^2 (steam pressure) and five minutes (digestion period) exhibited the maximum strength.

The compatibility of oil palm frond cement mixtures was tested in the hydration test, with magnesium chloride ($MgCl_2$) as an accelerator at different water and cement ratios [52]. He reported that the optimum weight ratio of cement-wood increased with decreasing wood powder size based on the equal specific surface area ratio of cement/wood in the hydration test and board manufacturing. The addition of magnesium chloride improved the compatibility of oil palm fronds with cement, enhancing the cement hydration and ultimate board strength properties. The study dealt with the effects of a curing method that uses gaseous and supercritical CO_2 to see its impact on the properties of oil palm fronds cement-bonded board manufactured by the conventional cold-press setting method were carried out by Hermawan et al. [53]. The study showed that high-performance cement-bonded boards made from oil palm fronds were successfully manufactured using the CO_2 curing method.

4. Processing of compressed oil palm fronds

The oil palm fronds were obtained from the plantation and selected based on decay-free and no defect. The oil palm fronds were taken from the plantation were divided into three groups according to their maturity. They were the old maturity fronds that have been taken from the below of the fronds crown. The intermediate maturity fronds were obtained from the middle of the frond's crown. The third group was the young maturity fronds that have been taken from the above of the frond's crown. The difference between maturity groups is shown in **Figure 13**.

4.1 Compressed oil palm fronds preparation

Figure 14 shows the fresh fronds was obtained from the oil palm tree. Leaflets were removed from the fronds, and then each maturity group was divided into three portions: the bottom, middle and top portion shown in **Figure 15**. A disc about 10 cm in the middle was cut from every portion for the physical properties study for the raw of oil palm fronds, and the rest were peeled of their skin and sliced in the longitudinal direction as shown in **Figure 16**. These sliced fronds were then compressed using rollers compressed machine to increase their density before undergoing sun-drying which has been shown in **Figure 17**.

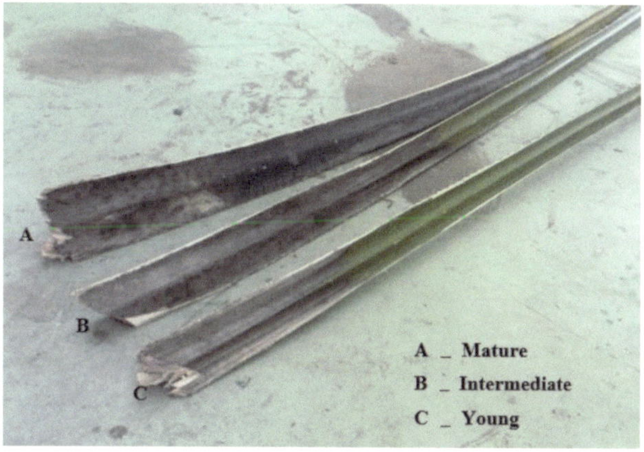

Figure 13.
Differences between maturities of oil palm fronds.

Figure 14.
Fresh oil palm fronds obtain from the oil palm tree.

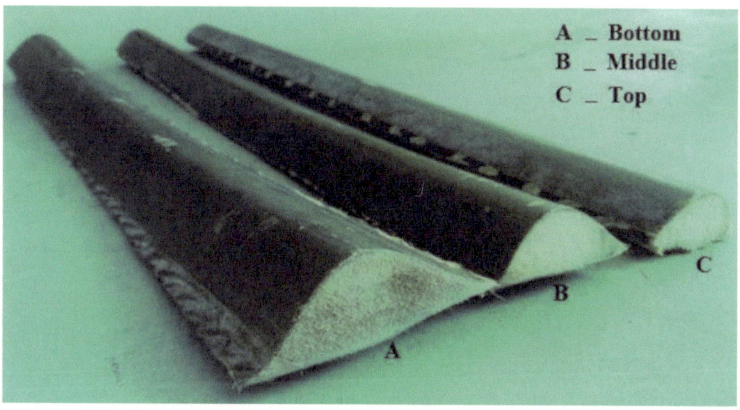

Figure 15.
Portion division from the oil palm fronds.

4.2 Air drying process

All the compressed oil palm fronds then have been dried in sun-drying for 12 hours until almost the moisture is removed from the fibre and their moisture

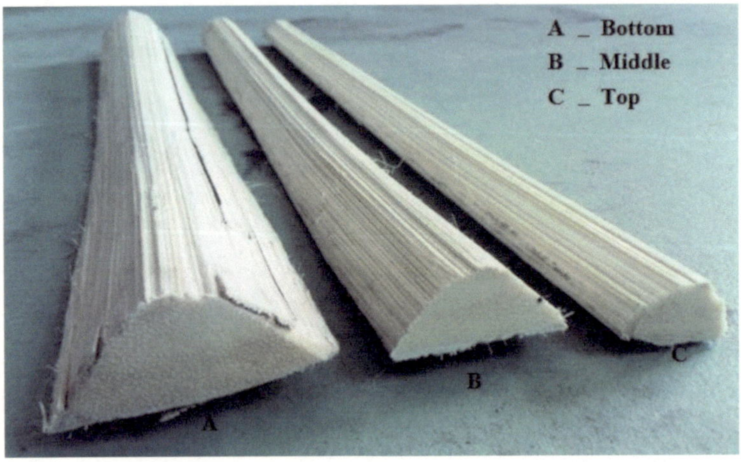

Figure 16.
The skin peeled off oil palm fronds.

Figure 17.
The compressed oil palm fronds.

Figure 18.
Compressed oil palm fronds under sun-drying.

content was about 12% measured by MC meter as shown in **Figure 18**. This air-drying mainly enhance their durability against fungi and insect attacks.

5. Manufacturing of composite board from oil palm fronds

The oil palm frond composite boards used in this study were made on a laboratory scale but follow the ISO standard. The oil palm fronds had been processed as compressed oil palm fronds to form such a potential composite board in this topic. The leaflets and epidermis are first manually removed from the fronds by a scraper. They were then sliced longitudinally into thin layers and compressed into 2-3 mm in thickness. The material was then air-dried under the shed before added with 12-15% of phenol and urea-formaldehyde resins and hot-pressed into composite boards.

Figure 19.
Layers of palm oil fronds being compressed using a hotpress machine.

Figure 20.
Composite board from compressed oil palm fronds were trimmed.

Hardner NH_4Cl at 1% later added forming layers which were compressed into 350 mm (length) × 350 mm (width) x 20 mm (thickness) boards. This was done by transferring the material to a single-opening hydraulic hot-pressed machine with a platen temperature of 125±5°C for phenol-formaldehyde resin and 100±5°C for urea-formaldehyde resin and pressed into the desired shape for subsequent testing.

The composite board from compressed oil palm fronds was pressed using a three-step-down method of pressing among 40 secs/mm for phenol-formaldehyde resin, meanwhile 30 secs/mm for urea-formaldehyde resin. Thickness spacing bars of 20 mm thick were inserted between the hot platens during hot pressing to ensure the desired board thickness is shown in **Figure 19**. All these composite boards were trimmed as shown in **Figure 20** and cut into various size test specimens and then conditioned at 20±3°C and 65±3% relative humidity (RH) for 72 hours prior for testing to produce equilibrium moisture content of about 12±1%. More samples of the composite board from compressed oil palm fronds were shown in **Figures 21–24**.

Figure 21.
Front view of the compressed oil palm fronds board.

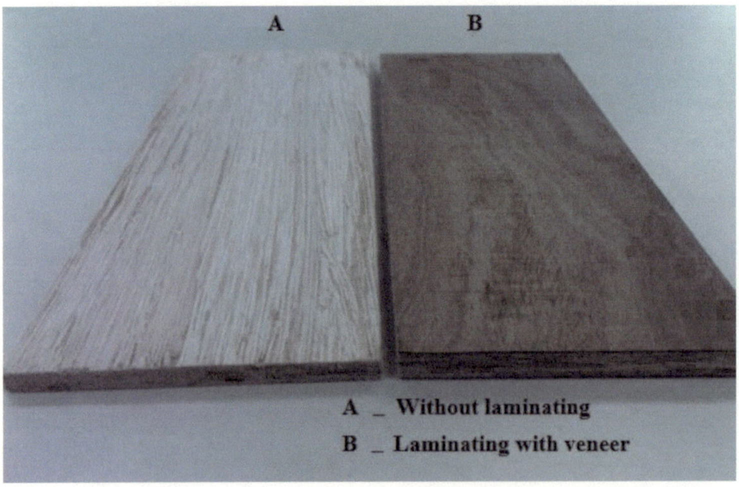

Figure 22.
The compressed oil palm fronds boards can be laminating with veneers.

Figure 23.
The compressed oil palm fronds boards at differences thickness.

Figure 24.
High-performance micrograph of the phenol-formaldehyde composite board from compressed oil palm fronds at the longitudinal sectional view (2 × magnification).

6. Physical properties of composite board from compressed oil palm fronds

To understand the composite board behaviours and performances, it is necessary to consider first some of the basic physical properties which are affecting its strength properties furthermore. Several physical properties of the composite board from compressed oil palm fronds were investigated, which are density and basic density-based based on its maturity groups, portions, and resin types of the composite board that had been produced.

6.1 Density of composite board from compressed oil palm fronds

Density is an excellent indicator of the amount of substance in a piece of wood [40]. The density of the composite board from compressed oil palm fronds will depend on the maturity groups, portions, and types of the resin that have been used for bonding this composite board. **Table 2** shows the mean value results for the density of the composite board for each maturity group, portion, and resin type. **Table 2**, showed that the highest density for this composite board came was from

Maturity Groups	Density (g/cm³) at different Portions		
	Bottom	Middle	Top
Matured			
Phenol Formaldehyde Composite Board	0.45	0.44	0.42
Urea Formaldehyde Composite Board	0.46	0.43	0.42
Intermediate			
Phenol Formaldehyde Composite Board	0.43	0.42	0.40
Urea Formaldehyde Composite Board	0.44	0.42	0.41
Young			
Phenol Formaldehyde Composite Board	0.42	0.41	0.40
Urea Formaldehyde Composite Board	0.42	0.41	0.40
Source: Rasat et al. [54].			

Table 2.
Value for density of the composite board.

the bottom portion for each maturity group followed by the middle and then top portions. The matured maturity group possessed the highest density values for every portion of the intermediate and young maturity groups.

According to the obtained results in **Table 2**, the decreased summarised composite board from compressed oil palm fronds density from the bottom to top portions for each maturity group and from the old to young maturity groups for each portion was influenced by internal structure in the oil palm fronds by its abundance of vascular bundles and parenchymatous tissues. The analysis of variance (ANOVA) in **Table 7** indicates the significant difference between density with maturity groups and portions. Nevertheless, no significant difference was indicated between the utilization of resin to produce composite boards. Hence, the density value of the composite board did not affect the types of resin.

Nonetheless, the compactness of the compressed oil palm fronds higher than the raw oil palm fronds density by the effect of resin penetration used in producing this composite board. It was found that the presence of both resins could increase the density of the composite board from compressed oil palm fronds that cause the increase in a material substance per unit volume in this composite board.

6.2 Basic density of composite board from compressed oil palm fronds

Table 3 indicates the mean values of basic density for the composite panel from compressed oil palm fronds for each maturity group, portion, and resin types. Decreases in values of the basic density from the bottom to the top portion for each maturity group and from the old to young maturity groups for each portion were caused by the high concentration of fibrous vascular bundles of the oil palm composite boards [55, 56]. The basic density differs according to their cell size, cell wall thickness, and relative amount of solid cell wall material. Mature and thickly cells occurred at the bottom part of the wood resulting in higher basic density values than other parts. The decreases in the basic density value from the fronds' bottom portion to the top is due to the growing differences in the anatomical cell maturity development [57]. Both density and basic density are the main factors affecting the strength properties of wood.

The analysis of variance (ANOVA) in **Table 7** indicates the significant difference between density with maturity groups and portions. Nevertheless, no significant

Maturity Groups	Basic Density (g/cm^3) at different Portions		
	Bottom	Middle	Top
Matured			
Phenol Formaldehyde Composite Board	0.38	0.36	0.33
Urea Formaldehyde Composite Board	0.39	0.35	0.32
Intermediate			
Phenol Formaldehyde Composite Board	0.36	0.35	0.32
Urea Formaldehyde Composite Board	0.37	0.34	0.31
Young			
Phenol Formaldehyde Composite Board	0.34	0.33	0.30
Urea Formaldehyde Composite Board	0.34	0.32	0.30

Source: Rasat et al. [54].

Table 3.
The basic density of the composite board.

difference was indicated between the utilization of resin to produce composite boards. Hence, the density value of the composite boards did not affect the types of resin. According to Paridah and Anis [58], the parenchyma acts similar to a sponge and easily absorbed moisture. On that account, this composite could effortlessly absorb phenol and urea-formaldehyde resin during the production process and increasing the basic density of the composite board of compressed oil palm fronds.

7. Strength properties of composite board from compressed oil palm fronds

The strength characteristics of wood are stratagems of its resistance to exterior forces, which direct deform its mass [35]. Such forces depend on their measurement and loading method (bending, compression, shear, tension, etc.). Tsoumis [59] declared that wood manifests various strength properties in different growth paths; therefore, it is strongly anisotropic. According to Bowyer et al. [60], In structural applications, strength properties are usually the most important aspects to define the products used. There are appointed as one of the primary criteria for selecting the material. Strength of wood structural practiced for wall sheathing, floor joint and rafters, and also subflooring application. [35].

Several strength properties were tested in this study, including static bending strength (modulus of elasticity (MOE) and modulus of rupture (MOR)) and compression strength. The testing was carried out based on International Organization for Standardization (ISO) standard for the strength properties evaluation. The analysis of strength properties of the composite board from compressed oil palm fronds has specifically looked into the effect of types of resin, maturity groups, and portions. Phenol and urea-formaldehyde were the resins used in producing this composite board.

7.1 Static bending strength of composite board from compressed oil palm fronds

According to Erwinsyah [35], the static bending strength refers to the tests performed. Bending stress is applied to the specimen to determine the stiffness or

MOE of the samples and the amount of force required to cause the sample to fail expressed as the MOR. Erwinsyah also postulated that the bending strength of wood is commonly expressed in MOR and is the most vital parameters that is occasionally used for engineering purposes [61].

The summary result in the static bending, which included the MOE and MOR strength, can be referred to **Tables 4** and **5**. Composite boards made from the bottom portion of the fronds possess the highest value for MOE and MOR strengths. The intermediate and young maturity groups follow this. The strength values of the boards both from phenol and urea-formaldehyde resin decreases from the bottom to top portions for every maturity group and from the old to young maturity groups for each portion, respectively. The MOE values of the maturity group from the bottom, middle, and top portions for the phenol-formaldehyde composite board were at 999.61, 952.29, and 844.18 N/mm^{2}, respectively. The MOE for the urea-formaldehyde composite board at 980.31, 949.40, and 840.40 N/mm^2. The MOE

Maturity Groups	**Static Bending MOE (N/mm^2) of Portions**		
	Bottom	**Middle**	**Top**
Matured			
Phenol-Formaldehyde Composite Board	999.61	952.29	844.18
Urea-Formaldehyde Composite Board	980.31	949.40	840.40
Intermediate			
Phenol-Formaldehyde Composite Board	979.15	942.44	817.29
Urea-Formaldehyde Composite Board	953.93	928.34	776.04
Young			
Phenol-Formaldehyde Composite Board	935.36	837.24	761.14
Urea-Formaldehyde Composite Board	936.24	836.67	666.30

Table 4.
MOE static bending strength of the composite board.

Maturity Groups	**Static Bending MOR (N/mm^2) of Portions**		
	Bottom	**Middle**	**Top**
Matured			
Phenol-Formaldehyde Composite Board	16.66	12.55	11.72
Urea-Formaldehyde Composite Board	15.40	12.38	11.63
Intermediate			
Phenol-Formaldehyde Composite Board	14.38	12.37	10.87
Urea-Formaldehyde Composite Board	12.62	12.07	10.51
Young			
Phenol-Formaldehyde Composite Board	12.61	11.62	10.27
Urea-Formaldehyde Composite Board	12.25	11.19	9.10

Source: Rasat et al. [54].

Table 5.
Mean value for MOR static bending strength of the composite board.

strength decreases from bottom to top portion for the maturity group either for phenol or urea-formaldehyde composite board, and the same situation was done for the other two maturity groups, the intermediate and young maturity groups.

Based on the results obtained from the study of the effect of resin types in static bending, it was found that the composite board from phenol-formaldehyde resin possessed a high value of both MOE and MOR test than urea-formaldehyde resin. The latter contained a high amount of solid content compared to phenol-formaldehyde resin.

The MOE strength of boards made from the bottom portion from the matured, intermediate, and young maturity groups of the phenol-formaldehyde composite board was 999.61, 979.15, and 935.36 N/mm^{2}, respectively. The MOE strength of the urea-formaldehyde composite board was at 980.31, 953.93, and 936.24 N/mm^{2} from the matured, intermediate, and young maturity groups. The MOE strength decreases from the matured to young maturity groups for the bottom portion for both resin types of the composite boards. The MOE strength when the specimen reached the breaking point and then could not recover its shape, where the load achieves its maximum value, is called MOR. This strength property is one of the significant parameters which usually used for engineering purposes. Relating to the resulting test of MOR of the composite board from compressed oil palm fronds at the different maturity groups, portions, and resin types, the summarised data of mean values is presented in **Table 5**.

The MOR of the compressed oil palm fronds composite boards increases in strength from the top to the bottom portions for the maturity group and from young to matured fronds for every portion. The MOR strength for the maturity group was at 16.66, 12.55, and 11.72 N/mm^{2}, respectively for the top, middle, and bottom portions for phenol-formaldehyde resin, while the MOR for the urea-formaldehyde were at 11.63, 12.38, and 15.40 N/mm^{2}. Similar trends were observed in the intermediate and young maturity groups from the bottom towards the top portions. The results in **Table 5** showed that the bottom portion for each maturity group (matures, intermediate, and young) and the portions grouping from the phenol-formaldehyde composite boards were at 16.66, 14.38, and 12.16 N/mm^{2} and, the MOR for urea-formaldehyde composite boards at 15.40, 12.62 and 12.25 N/mm^{2} respectively. The strengths decrease from the matured towards the bottom portion for both resin types used in the maturity groups. Similar trends were noted in the MOE values. The MOR decreases from bottom to top portions for each maturity group and from old towards young maturity groups for every portion.

The values of both the MOE and MOR for the oil palm frond compressed composite boards increases from the top to the bottom portions. Similar observations were noted in the frond maturity groups from young, intermediate, and matured groups. These occurred to both of the composite boards made from phenol and urea-formaldehyde resin. The decreases can explain the trend of variations in the MOE and MOR values and the tree height in the maturity of wood and fibre length from top to the bottom of the tree [62]. This is logically accepted due to vascular bundles that decrease from the bottom to top portions along with the oil palm fronds and the old to young maturity groups. A large amount of the vascular bundle in the oil palm fronds containing a higher quantity of fibre cells gives higher density and basic density values in both the composites. According to Haygreen and Bowyer [57], the woody materials with higher values density and basic density will directly possess higher strength. The bottom portion has a higher value for both MOE and MOR strengths compared to the middle and top portions for the maturity group in every portion [57]. Based on the results obtained from the study of the effect of resin types in static bending, it was found that the composite panel from phenol-formaldehyde resin possessed a high value of both MOE and MOR test than urea-formaldehyde resin. By the latter contained a high amount of solid content compared to phenol-formaldehyde resin [63].

The static bending was significantly affected by the density and basic density value [63]. This thus gives effect to the MOE and MOR strengths of the composite boards from top to bottom portions. The ANOVA in **Table 7** supports this statement. The result also showed that the composite boards from phenol-formaldehyde resin possessed a higher value of both MOE and MOR test than urea urea-formaldehyde resin, which contained a higher amount of solid content compared to phenol-formaldehyde resin.

Furthermore, the distribution of phenol-formaldehyde resin is located irregularly in the composite boards' structures [40]. When the stress was applied, the stress could not be transferred consistently between the fibre and matrix. Besides, the penetration of high viscosity of urea-formaldehyde resin probably breaks the cell wall of the composite board from compressed oil palm fronds [40].

7.2 Compression strength of composite board from compressed oil palm fronds

Compression strength is defined as the maximum stress sustained by compression of a specimen with the specimen having a ratio of length to smallest dimension [64]. In contrast, Ronald and Gjinoli [65] reported that the characteristic of the compression load-deformation curve was similar to those for static bending strength. The compression strength of the composite is strongly dependent on the effectiveness of the matrix in supporting the fibre against buckling [66].

The obtained data was examined using statistical analysis to define the effect of three parameters like static bending strength test, which were based on maturity groups, portions including types of resin to the compression strength of the composite board from compressed oil palm fronds. According to the testing result, the data is presented in **Table 6** which was based on mean value compression strength. From the display result in **Table 6**, the compression strength value of the old maturity group from bottom to top portions was 473.17, 395.93, and 260.22 N/mm^2 for the phenol-formaldehyde composite board, while for the urea-formaldehyde composite board, the result was 459.52, 344.60, and 260.00 N/mm^2 respectively. It can be observed that the compression strength was decreased from the bottom

Maturity Groups	**Compression (N/mm^2) of Portions**		
	Bottom	**Middle**	**Top**
Matured			
Phenol Formaldehyde Composite Board	473.17	395.93	260.22
Urea Formaldehyde Composite Board	459.52	344.60	260.00
Intermediate			
Phenol Formaldehyde Composite Board	453.67	318.88	196.71
Urea Formaldehyde Composite Board	431.88	274.90	190.70
Young			
Phenol Formaldehyde Composite Board	301.46	235.60	183.48
Urea Formaldehyde Composite Board	312.94	198.79	181.06

Source: Rasat et al. [54].

Table 6.
The value for compression strength of the composite board.

portion towards to middle and top portions for the old maturity group. Similar decrement distribution data have been done for intermediate and young maturity groups towards the bottom, middle, and top portions.

Table 6 showed that the trend for each oil palm fronds composite made from matured, intermediate, and young in maturity groups. The result at matured, intermediate, and young maturity groups was at 473.17, 453.67, and 301.46 N/mm^2 for the phenol-formaldehyde composite boards. The result for the urea-formaldehyde composite was at 459.52, 431.88, and 312.94 N/mm^{2,} respectively. It clearly shows the decrement from mature to intermediate and young groups for the bottom portion. The same trends were also observed in the others two portions (the middle and top).

This was due to the differences in vascular bundle abundance and oil palm fronds, affecting the value of density and basic density. The differences between the latter promote the distribution result of compression strength for the maturity groups and portions. The bottom portion scored higher results in compression strength as compared to others portions. **Table 7** tabulated the ANOVA results that indicate the significant difference between compression strength with maturity groups and portions. According to Oyagade AND Fasulu [67] and Nordahlia [68], some wood properties, including compression strength failure, typically occur in the low density of the wood.

The results obtained showed that the phenol-formaldehyde composite boards possessed higher values than the strength of the urea-formaldehyde composite. This can be attributed to the fact that properly cured phenol-formaldehyde composite resin is usually tougher than the bonded [69]. However, the differences in the compression strengths are not significant as shown in **Table 7**. Therefore, we can conclude that either resin's use does not matter as long as it is economically feasible to produce the mass quantity of the oil palm frond composite boards.

Source of Variance	Dependent Variable	Sum of Square	df	Mean Square	F-Ratio
Maturity	Density	0.01	2	0.01	7.94**
	Basic Density	0.01	2	0.02	28.75**
	MOE Bending	155675.00	2	77837.50	57.05**
	MOR Bending	79.02	2	39.51	40.39**
	Compression	255794.00	2	127897.00	63.81**
Portion	Density	0.01	2	0.01	8.26**
	Basic Density	0.04	2	0.01	28.75**
	MOE Bending	507856.00	2	253928.00	186.12**
	MOR Bending	157.72	2	78.86	80.62**
	Compression	565023.00	2	282512.00	140.95**
Resin type	Density	0.00	1	0.00	0.20 ns
	Basic Density	0.00	1	0.00	1.28 ns
	MOE Bending	11232.80	1	11232.80	8.23 ns
	MOR Bending	8.23	1	8.23	8.41 ns
	Compression	7538.01	1	7538.01	3.76 ns

Source: Rasat et al. [54].
***indicates citation of the researcher who conducted and produced the data/results.*

Table 7.
ANOVA on physical and strength properties of composite board.

8. ANOVA on physical and strength properties of composite board oil palm fronds

The analysis of variance (ANOVA) on the physical and the strength properties of the composite oil palm fronds is shown in **Table 7**. The ANOVA was used to determine the significant level between the physical and the strength properties with the dependent variables such as the maturity groups, portions, and types of resin used in the boards made. Based on the ANOVA (**Table 7**), there were significant differences between the physical and strength properties of the boards. Significant differences exist between the maturity groups and the portions. It shows that the physical and strength properties of the boards were affected and influenced by these factors—substantial differences at p-value $\leq$ 0.01exists between them. No significant difference exists between physical properties and strength properties with the application in the types of resin used. The phenol or urea-formaldehyde resin used to produce these boards give quite similar values in the test results.

9. Microstructural of composite board from compressed oil palm fronds

The microstructural observation of the composite board from compressed oil palm fronds was carried out with the help of high-performance microscopy, and a more detailed structure has been observed by scanning electron microscopy (SEM). More attention has been done to the microscopic resin penetration of the composite board from compressed oil palm fronds. The macroscopic structural characteristics of the composite board are the features that are using high-performance microscopy to magnify from 0.75 to 8.0 times. **Figures 24** and **25** show the roughly structural composite board from compressed oil palm fronds at the longitudinal sectional view for phenol and urea-formaldehyde composite board that had been observed under high-performance microscopy.

Figure 24 showed that the reddish colour of the composite board had been affected by phenol-formaldehyde resin colour naturally, and the same goes for the urea-formaldehyde composite board that shown whitish in the colour of its

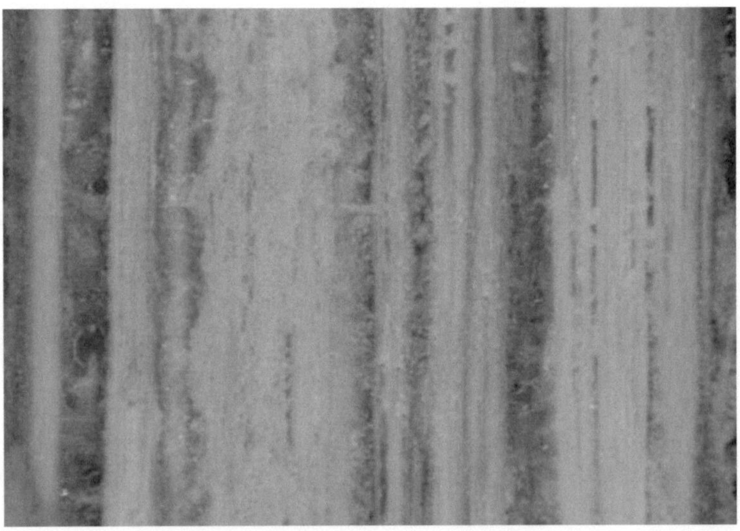

Figure 25.
High-performance micrograph of the urea-formaldehyde composite board from compressed oil palm fronds at the longitudinal sectional view (2× magnification).

appearances according to **Figure 25**. Appearances comparison among both of them resulted in the composite board from urea-formaldehyde resin, which gave better appearance quality than the composite board from phenol-formaldehyde resin, which is one reason why urea-formaldehyde resin has been used for internal usage of mostly wood-based composites. This composite board is formed layer by layer like laminated veneer lumber (LVL) because of its parallel arrangement among the compressed oil palm fronds. A more complex arrangement of this composite board has been presented in **Figures 26** and **27** under SEM observation at different magnification levels.

9.1 Resin penetration on composite board from compressed oil palm fronds

The high porous morphology of compressed oil palm fronds helps the resin be located and filled within the space, improving the characteristics of the composite board from compressed oil palm fronds. According to the obtained results, the density of the composite board was generally increased than oil palm fronds, and it

Figure 26.
SEM of the composite board from compressed oil palm fronds at longitudinal sectional view (15× magnification).

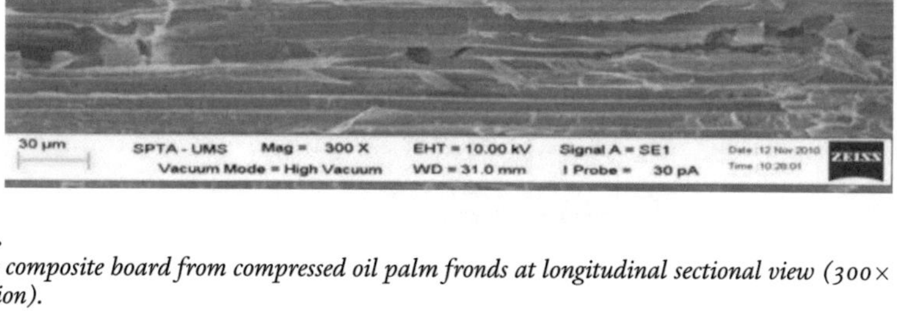

Figure 27.
SEM of the composite board from compressed oil palm fronds at longitudinal sectional view (300× magnification).

can be observed that an increase of resin penetration increased wood density and basic density result. This is because the resin reinforcement was applied to improve the wood features of the composite board and logically accepted the fact that the resin penetrated through the intercellular cavities of the composite board from compressed oil palm fronds, as shown in **Figure 28**.

The microscopic image of resin penetration on the composite boards from compressed oil palm fronds is presented in detail are shown in **Figures 29** and **30**. The strength increases with the enhancement in the density and basic density of the composite board. Most of the tested specimens increase their strength properties, including MOE of static bending strength and MOR of static bending and compression strength that have been tested in this study. It is attributed that the applied resin to the compressed oil palm fronds to form them into composite board was affecting positively to improve the strength properties of the composite board from compressed oil palm fronds. However, the different types of resin used in

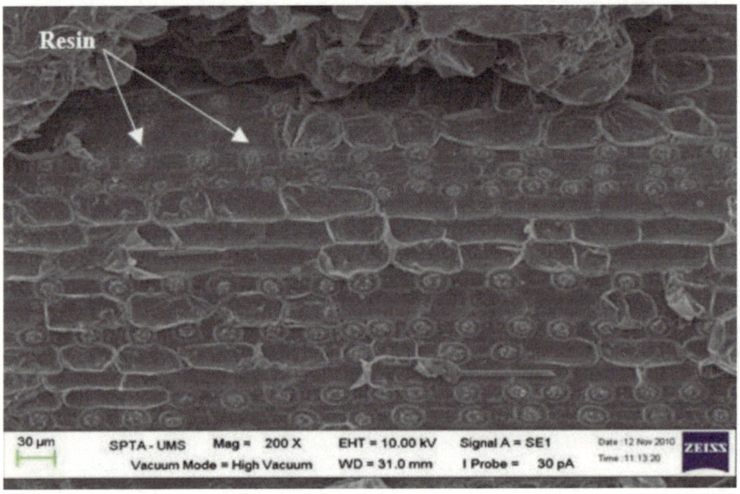

Figure 28.
SEM of resin penetration on composite board from compressed oil palm fronds (200× magnification).

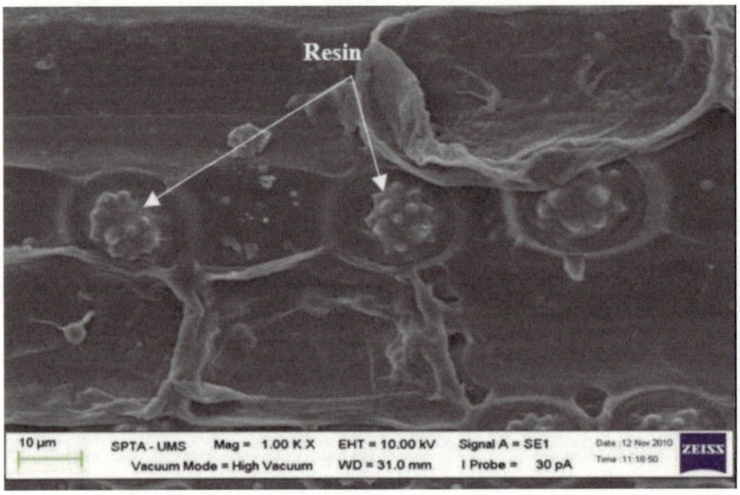

Figure 29.
SEM of resin penetration on composite board from compressed oil palm fronds (1000× magnification).

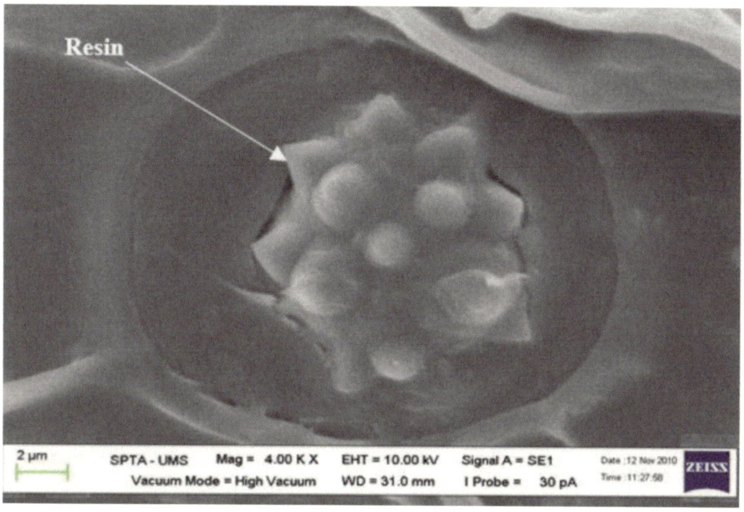

Figure 30.
SEM of resin penetration on composite board from compressed oil palm fronds (4000× magnification).

producing this composite board did not significantly differ for density and basic density values and strength properties discussed above.

10. Conclusion

The results represent a correlation between the maturity of the oil palm frond and the portion of the oil palm fronds composite boards experimented in expressions of the physical properties and strength of the composite boards. The highest density using an air-dried method for the boards was highlighted at the matured bottom portion, followed by intermediate middle and young top portions. Besides, the density under the oven-dried method represented the values were reduced from the bottom to the top portion for each maturity group and highlighted that the mature group possessed the highest density values for every portion compared to others, followed by the intermediate and young maturity groups. Nonetheless, the MOE strength showed reduced values from the bottom to the top portion for mature level with either phenol or urea-formaldehyde resins. The MOE strength also decreased with a proportional decrease of maturity level. In addition, the MOR represented decreased value from the bottom to the top portion for each maturity level and matured to the young maturity levels for every section. The results include both types of resin used for composite board manufacturing.

Furthermore, the compression strength decreases from the bottom portion towards the middle and top section for the maturity and bottom, middle and top portions. Highlighted also the decreasing trend from mature towards intermediate and young maturity groups for the bottom portion, and this polar was also observed for the middle and top portion. Statistical analysis found significant differences between physical properties and mechanical properties across varying maturity levels and a different board section.

Finally, we can conclude that the oil palm fronds composite boards is suitable for furniture manufacturing based on physical and strength properties. The oil palm frond can produce quality composites board that possesses the physical and mechanical strength values that are better than the composite boards made from oil palm stem and less than rubberwood.

Author details

Razak Wahab[1,2*], Mohd Sukhairi Mat Rasat[3], Norashikin Mohd Fauzi[3], Mohamad Saiful Sulaiman[1,2], Hashim W. Samsi[4], Nasihah Mokhtar[1,2], Ros Syazmini Mohd Ghani[1,2] and Muhammad Haziq Razak[5]

1 University College of Technology Sarawak (UCTS), Sibu, Sarawak, Malaysia

2 Centre of Excellence in Wood Engineered Products, UCTS, Sibu, Sarawak, Malaysia

3 Universiti Malaysia Kelantan (UMK), Jeli, Kelantan, Malaysia

4 Forest Research Institute Malaysia (FRIM), Kuala Lumpur, Malaysia

5 Universiti Sains Malaysia (USM), Penang, Malaysia

*Address all correspondence to: drrazakw5181@ucts.edu.my

References

[1] Hassan, O. A., Ishida, M., & Mohd Sukri, I. (1995). Oil palm fronds (OPF) technology transfer and acceptance, a sustainable in-situ utilization for animal feeding. In *Proceedings of the 17th Malaysian Society of Animal Production (MSAP) Annual Conference, 28-30 May 1995, Penang* (pp. 134-135).

[2] Pamin, K., & Guritno, P. (1997). The utilisation of oil palm empty fruit bunch fiber for oil palm seedling pot in pre-nursery. *Jurnal Penelitian Kelapa Sawit (Indonesia)*.

[3] Sreekala, M. S., George, J., Kumaran, M. G., & Thomas, S. (2002). The strength performance of hybrid phenol-formaldehyde-based composites reinforced with glass and oil palm fibres. *Composites science and technology*, *62*(3), 339-353.

[4] Abdul Khalil, H. P. S., Ismail, H., Ahmad, M. N., Ariffin, A., & Hassan, K. (2001). The effect of various anhydride modifications on strength properties and water absorption of oil palm empty fruit bunches reinforced polyester composites. *Polymer International*, *50* (4), 395-402.

[5] Henson, I. E. (1999). *Comparative ecophysiology of oil palm and tropical rain forest* (No. L-0537). MOPGC.

[6] Sulaiman, O., Hashim, R., Wahab, R., Samsi, H. W., & Mohamed, A. H. (2008). Evaluation on some finishing properties of oil palm plywood. *Holz als Roh-und werkstoff*, *66*(1), 5-10.

[7] Laemsak, N., & Okuma, M. (2000). Development of boards made from oil palm frond II: properties of binderless boards from steam-exploded fibers of oil palm frond. *Journal of wood science*, *46*(4), 322-326.

[8] Chew, L. T. (1987). Particleboard manufacture from oil palm stems a pilot-scale study. *Particleboard manufacture from oil palm stems: a pilot scale study.*, (4/87).

[9] Ho, K. S., Choo, K. T., & Hong, L. T. (1987). Processing, seasoning and protection of oil palm lumber. In *National Symposium on Oil Palm By-products for Agro-based Industries, Kuala Lumpur, 5-6 Nov 1985*. IPMKSM.

[10] Bledzki, A. K., & Gassan, J. (1999). Composites reinforced with cellulose based fibres. *Progress in polymer science*, *24*(2), 221-274.

[11] Gurmit, S. (1999). The Malaysian oil palm industry: progress towards environmentally sound and sustainable crop production. *Industry and Environment*, *22*(2/3), 45-48.

[12] Yang, H., Yan, R., Chen, H., Lee, D. H., Liang, D. T., & Zheng, C. (2006). Pyrolysis of palm oil wastes for enhanced production of hydrogen rich gases. *Fuel processing technology*, *87*(10), 935-942.

[13] Dahlan, I. (1996). Oil palm by-products: its utilisation and contribution for livestock Industry. In *Proceedings of the Agricultural Conference on International Palm Oil Congress* (pp. 23-28).

[14] Zin, Z. Z., & Tarmizi, A. M. (1983). Penggunaan bahan sampingan industri kelapa sawit. *Proc. Bengkel Pemindahan Teknologi Perushaan Kelapa Sawit. Masaalah Kehendak dan Keutamaan. PORIM*, 121-131.

[15] Chan, K. W., Jalani, B. S. & Basiron, Y. (2000). *Advances in Oil Palm Research VII.* Kuala Lumpur: MPOB.

[16] Ismail, A. R., Hoi, W. K., & Puad, E. (1990). Economics and processing of oil palm trunks as ruminant feed. In *Proceedings of the 13th Malaysian Society*

of Animal Production Annual Conference (pp. 6-8).

[17] Killmann, W., & Koh, M. P. (1988). Oil palm stem densification using ammonia treatment: a preliminary study. *Journal of Tropical Forest Science*, 1-10.

[18] Kamaruddin, H., Abdul Halim, H., & Riazuan, R. (1991). Mohd. Zain J. Improvement of Oil Palm Fronds CTM-Pulp Brightness-A Preliminary Study. In *Proceeding of Seminar on Oil Palm Trunks and Other Palmwood Utilization (Oil Palm Tree Utilization Committee of Malaysia), Kuala Lumpur, Malaysia*.

[19] Misha, S., Mat, S., Ruslan, M. H., Salleh, E., & Sopian, K. (2016). Performance of a solar-assisted solid desiccant dryer for oil palm fronds drying. *Solar Energy*, *132*, 415-429.

[20] Hong, L. S., Ibrahim, D., & Omar, I. C. (2012). Oil palm frond for the production of bioethanol. *International Journal of Biochemistry and Biotechnology*, *1*(1), 007-011.

[21] Hamdan, A. M. (1997). Machinery system for in-field collection and handling of oil palm fronds (OPF). *Proceedings of Oil Palm Residue: Progress Towards Commercialisation*, April 29 – May 1, 1997, Oil Palm Tree Utilisation Committee (OPTUC), Kuala Lumpur, Malaysia. 54-62.

[22] Husin, M., Ramli, R., Mokhtar, A., Wan Hassan, W. H., Hassan K., Mamat, R. & Abdul Aziz, A. (2005). *Malaysian Palm Oil Board Bulletin: Research and Development of Oil Palm Biomass in Wood Based Industries*. Kuala Lumpur: MPOB.

[23] Hishamudin, M. J. (1987). The Oil Palm Industry in Malaysia. A Guide Book.

[24] Ng, S. K., Helmunt von Uexkull & Rolf, H. (2003). Botanical Aspects of the Oil Palm Relevant to Crop Management. *Fairhurst, T. and Hardter, R. Oil Palm. Management for Large and Sustainable Yields*, 13-26.

[25] Cobley, L. S. (1976). *An introduction to the botany of tropical crops* (No. 2. ed.). Longman.

[26] Opeke, L. K. (1982). *Oil Palm: Tropical Tree Crops*. New York: John Wiley and Sons.

[27] Rehm, S. & Espig, G. (1991). *Oil Palm: The Cultivated Plants of the Tropics and Subtropics*. West Germany: Verlag.

[28] Parthsarathy, M. V., & Klotz, L. H. (1976). Palm "wood" I. Anatomical aspects. *Wood Science and Technology*, *10* (3), 215-229.

[29] Corley, R. H. V., & Tinker, P. B. (2008). *The oil palm*. John Wiley & Sons.

[30] Tomlinson, P. B. (1961). Anatomy of the Monocotyledons. II. *Palmae. Clarendon Press. Oxford. England*.

[31] Ibrahim, W. A., & Kadir, W. R. (2007). Turning oil palm residues into products. *Research Pamphlet-Forest Research Institute Malaysia*, (127).

[32] Lim, S. C. & Fujii, T. (1997). A note on the structure of oil palm trunk by scanning electron microscopy. *Journal of Tropical Forest Product*. *3*(1): 105-109.

[33] Lim, S. C. & Khoo, K. (1986). Characteristics of oil palm trunk and its potential utilisation. *Journal of the Malaysian Forester*. *47*(1): 3-22.

[34] Tan, K. S. (1987). *The Botany of Oil Palm*. Kuala Lumpur: The Incorporated Society of Planters.

[35] Erwinsyah. (2008). *Improvement of oil palm wood properties using bioresin* (Doctoral dissertation, Verlag nicht ermittelbar).

[36] Jusoh, M. Z., & Imamura, Y. (1989). Anatomical characterisation of vascular bundles in oil palms. In *Proc. The 2nd Pacific Regional Wood Anatomy Conference. Laguna, Philipinnes* (pp. 430).

[37] Bierhorst, D. W., & Zamora, P. M. (1965). Primary xylem elements and element associations of angiosperms. *American Journal of Botany*, *52*(7), 657-710.

[38] Hong, L. T., & Killmann, W. (1989). Some aspects of parenchymatous tissues in palm stems. In *Proc. The 2nd Pacific Regional Wood Anatomy Conference. Laguna, Philipinnes* (pp. 449-455).

[39] Rahayu, I. S. (2001). *Basic properties of vascular bundles and parenchyma tissues of the oil palm trunk*. (Master Dissertation, Bogor Agriculture University, Indonesia).

[40] Abdullah, C. K. (2010). *Impregnation of oil palm trunk lumber (OPTL) using thermoset resins for structural applications* (Master Thesis, Universiti Sains Malaysia).

[41] Tomimura, Y. (1992). Chemical characteristics of the oil palm trunk. *Bulletin of the Forestry and Forest Products Research Institute, Ibaraki*, (362), 133-142.

[42] Sun, R., Mott, L., & Bolton, J. (1998). Isolation and fractional characterisation of ball-milled and enzyme lignins from oil palm trunk. *Journal of agricultural and food chemistry*, *46*(2), 718-723.

[43] Abdul Khalil, H. P. S., Siti Alwani, M. & Mohd. Omar, A. K. (2006). Cell walls of tropical fibers. *Journal of Biology Resources*. *1*(2): 220-232.

[44] Boudet, A. M. (2000). Lignins and lignifications: Selected issues. *Journal of Plant Physiology Biochemistry*. *38*: 81-96.

[45] Douglas, C. J. (1996). Phenylpropanoid metabolism and lignin biosynthesis from weeds to trees. *Journal of Trends Plant Science*. *1*(6): 171-178.

[46] Zhong, R., Taylor, J. J. & Ye, Z. H. (1997). Disruption of interfascicular fibre differentiation in an *Arabidopsis* mutant. *Journal of Plant cell*. *9*: 2159-2170.

[47] Timmel, T. E. (1967). Recent progress in the chemistry of wood hemicelluloses. *Journal of Wood Science and Technology*. *36*(1): 45.

[48] Wallis, A. F. A., Wearne, R. H. & Wright, P. J. (1996). Chemical analysis of polysaccharides in plantation eucalypt woods and pulps. *Journal of Appita*. *49*(4): 258-262.

[49] Laurila, R. (1995). Wood properties and utilisation potential of eight fast-growing tropical plantation tree species. *Journal of Tropical Forest Products*. *1*(2): 209-221.

[50] Prayitno, T. A. (1995). Trunk Shape and Physical Properties of Oil Palm Trunk. *Buletin Fakulti Kehutanan UGM*. (29): 43-59. Yogyakarta, Indonesia, Universitas Gajah Mada.

[51] Killmann, W. & Choon, L. S. (1985). Anatomy and properties of oil palm stem. *Bulletin PORIM*. (11): 18-42.

[52] Hermawan, D., Subiyanto, B. & Kawai, S. (2001). Manufacture and properties of oil palm frond cement-bonded board. *Journal of Wood Science*. *47*: 208-213.

[53] Hermawan, D., Hata, T., Kawai, S., Nagadomi, W. & Kuroki, Y. (2002). Manufacturing oil palm fronds cement-bonded board cured by gaseous or supercritical carbon dioxide. *Journal of Wood Science*. *48*: 20-24.

[54] Rasat, M. S. M., Wahab, R., Kari, Z. A., Yunus, A. A. M., Moktar, J., &

Ramle, S. F. M. (2013b). Strength properties of bio-composite lumber from lignocelluloses of oil palm fronds agricultural residues. *International Journal on Advanced Science, Engineering and Information Technology*, *3*(3), 9-19.

[55] Mohamad, H., Zin, Z. Z. & Abdul Halim, H. (1985). Potentials of oil palm by-products as raw materials for agro-based industries. *Proceedings of National Symposium on Oil Palm by-product for Agro-based Industries*, Kuala Lumpur, Malaysia. 34-42.

[56] Rowell, R. M. (1994), J.L. O'Dell, and T.G. Rials (1994). Chemical Modification of Agro-Fiber for Thermoplasticization. USDA Forest Service, Forest Products Laboratory, One Gifford Pinchot Drive,Madison, WI 53705, andSouthern Forest Experiment Station, 2500 Shreveport Highway, Pineville. LA,71360.

[57] Haygreen, J. G. & Bowyer, J. L. (1930). *Introduction to Forest Product and Wood Science*. Subtitled by Suhaimi Muhammed and Sheikh Abdul Karim Yamani Zakaria. Kuala Lumpur: Ampang Press Sdn. Bhd.

[58] Paridah, M. T. & Anis, M. (2008). Process optimisation in the manufacturing of plywood from oil palm trunk. *Proceedings of 7th National Seminar on the Utilisation of Oil Palm Tree,* Oil Palm Tree Utilization Committee, Kuala Lumpur, Malaysia. 12-24.

[59] Tsoumis, G. (1991). *Science and Technology of Wood – Structure, Properties and Utilisation*. New York: Van Nostrand Reinhold.

[60] Bowyer, J. L., Shmulsky, R. & Haygreen, J. G. (2004). *Forest Product and Wood Sciences – An Introduction* (4th Edition). London: Blackwell Publishing Company.

[61] Walker, J. C. F., Butterfield, B. G., Langrish, T. A. G., Harris, J. M. & Uprichard, J. M. (1993). *Primary Wood Processing* (1st Edition). London: Chapman and Hall.

[62] Rulliarty, S. & America, W. A. (1995). Natural Variation in Wood Quality Indicators of Indonesian Big Leaf Mahogany (*Swietenia macrophylla*. King). *Proceedings of XX IUFRO World Congress*, Tampere. 76-83.

[63] Desch, H. E. (1968). Timber, Its Structure, properties and utilisation. New York: London and Basingstoke Associated companies.

[64] Thanate, R., Tanong, C. & Sittipon, K. (2006). An investigation on the strength properties of trunks of palm oil trees for the furniture industry. *Journal of Oil Palm Research*. 114-121.

[65] Ronald, W. W. & Gjinoli, A. (1997). The Use of Recycled Wood and Paper in Building Applications. *Proceedings of Forest Product Society No. 7286*. 8491.

[66] John, R. L. & Reid, R. J. (1969). Compressive Strength of Boron Composites. *Journal of Composite Materials*. *3*: 48-8.

[67] Oyagade, A. O. & Fasulu, S.A. (2005). Physical and Strength Properties of *Trilepisium madagascariense* and *Funtumia elastica* Wood. *Journal of Tropical Forest Science*. *17*(2): 258-264.

[68] Nordahlia, A. S. (2008). *Wood Quality of 10-Year-Old Sentang (Azadirachta excelsa) Grown from Seedlings and Rooted Cuttings*. Master Thesis. University Putra Malaysia.

[69] Baldwin, R. F. (1995). *Adhesives and Bonding Techniques. In: Plywood and Veneer- based Products Manufacturing Practices*. California: Miller Freeman Inc.

Chapter 14

Environmental Impacts of the Oil Palm Cultivation in Cameroon

Tchindjang Mesmin, Ludovic Miaro III, Fideline Mboringong, Gilles Etoga, Eric Voundi and Emmanuel Pierre Jonathan Ngom

Abstract

Since 1990, oil palm cultivation, because nibbling large zones in dense forest areas of Cameroon, becomes the main driver of deforestation. It leads to the loss of plant and animal biodiversity as well as engaging soils and water pollution, which raises questions about its sustainability. Nowadays, palm plantations occupy almost 400 000 ha shared between agro-industries, elites and small farmers while annual palm oil production increased from 150, 000 tons in 2000 to 413,000 tons in 2018 against a demand that peaked at 1.179 million tons in 2018. This would assess the impacts of the oil palm exploitation in Cameroon. The objective of this article is to analyze the four dimensions of impacts closely linked to sustainability dimensions (ecological, sociocultural, economical and institutional) dimensions of sustainability of the oil palm sector in Cameroon. The approach is based on field surveys carried out in various production basins, particularly in the South-West, Littoral and Central regions. They also take into account the resolutions of various workshops bringing together stakeholders on the matter of sustainability in the oil palm sector in Cameroon. Satellite images were also used to map the spatial evolution of oil palm in the production basins. The result is a boom and a considerable expansion of the oil palm to which we can note a lack of adequate policy due to the constraints and hesitations of the Cameroonian administrations. Such a situation requires a better articulation of the tensions between development and environmental issues in Cameroon.

Keywords: Cameroon, deforestation, environmental impacts, oil palm, sustainability

1. Introduction

In recent years, many developing countries worldwide have been tapping renewable resources for food security. Such a tendency has been spurred by high demand in some commodities and also and increasing concerns in agriculture feedstock. Agriculture is one of the main causes of the degradation of natural ecosystems [1, 2]. It accounts for 24% of global greenhouse gas emissions [3]. The resulting climate changes affect the whole humanity [4]. Agriculture is also the primary anthropogenic cause of deforestation and desertification. It greatly participates in the degradation of water resources with the increased use of chemical inputs [5].

IntechOpen

These negative impacts are mainly attributable to industrial agriculture, practised over large areas and without taking into account the basic principles of sustainability. Artisanal agriculture also presents unsustainable practices such as shifting slash-and-burn agriculture [1, 6]. Among the most incriminated is oil palm cultivation [2, 7]. This plant, which is native to the Gulf of Guinea, has experienced strong expansion around the World [8, 9]. It is planted for its oil, which is currently the first vegetable oil seconded by soybean oil [10]. Southeast Asia (Malaysia and Indonesia) accounts for 80% of world production of palm oil [11]. Effort to render palm oil production sustainable (Round Table on Sustainable Palm Oil and Belgian Alliance for Sustainable Palm Oil) has led to 19% of its worldwide production certified as organic [12]. In recent years, Africa has consolidated its position as the Third production pole. There is increasing rush for its production by national and international investors, attracted by the availability of land [13]. Such a rush raises concerns captured by this paper such as reconciling increasing production of palm oil and preserving the forest and biodiversity for the sustainability of the oil palm sector in Cameroon.

WWF report estimates that palm oil supplies 35% of the world's vegetable oil on just 10% of the land. In Cameroon, oil palm exploitation is taking a global scale. Since economic crisis of the year 1990s, Cocoa and coffee have already lost ground while the rubber tree is floundering. The palm oil is reassuring, because of its many uses.

People are getting more involved in the activity because of its economic importance (money like cash crop at any time, cultural uses and benefit, etc.). Estimations on oil palm plantations reach 375 000 ha shared between agro-industries, elites and small farmers. Annual palm oil production increased from 270,000 tons in 2013 to 413,000 tons in 2018 against a demand that peaked at 1.179 million tons in 2018 [14]. Oil palm expansion in Cameroon has been driven by rising global demand for vegetable oils for consumption and cosmetics. While making a significant contribution to national economies, the expansion of oil palm plantations is a cause for environmental concerns.

Most plantations, as well as CPO productive basins, are located in the rainiest area of the country, being South West, Littoral and some part of the Centre and South regions. Between 2000 and 2020, more than 10,000 ha of oil palm plantations were established annually. Gradually, new lands available were allocated for industrial plantations while concerns on deforestation were raised up.

This paper assesses and analyzes the environmental impacts and risks associated with this activity. The main assumption is that oil palm cultivation generates ecological and socio economic impacts which put its sustainability to question. The approach is based on field surveys carried out in various production basins, particularly in the South-West, Littoral and Centre Regions. The study reviewed resolutions of various workshops bringing together stakeholders on the matter of sustainability in the oil palm sector in Cameroon. It emerges that impacts are assessed in four domains: ecological, economic, social and institutional. The latter implies a better articulation of the tensions between development and environmental issues.

2. Methods

2.1 Research sites

Within the oil palm production basins of Cameroon, three main sites were chosen to drive this study namely Sanaga Maritime Division in 2013–2014, Ngwéi (2016–2018) and Ekondo-Titi (2016) Subdivisions (**Figure 1**). The diversity of the biophysical environment favours the cultivation of a wide variety of food (cassava, maize, millet, macabo, rice, etc.), and cash (sugar cane, cotton, palm oil, rubber, cocoa, etc.) crops (Tchindjang et al. 2015–2016). Regarding oil palm, it develops preferentially in the coastal area qualified as the "elaeisfarming" belt of Cameroon (**Figure 1**). Administratively, oil palm plantations and concessions are set up in the maritime facades of the southern, coastal and southwestern regions [10, 15, 16].

As shown by Tchindjang et al. [17] and Ndjogui [18], oil palm belt offers suitable conditions for the development of oil palm: low altitude (less than 500 m); sufficient rainfall (more than 1800 mm/year); favorable temperature between 22 and 30°C; low thermal amplitude; rich and deep soils; etc. Agro-industries are also common in this area and constitute the major producers of palm oil (**Table 1**). At the edge of this "elaeisfarming" belt, there are a few small marginal farms both on the vast southern Cameroonian plateau and in the Western Highlands where oil palm could be grown with limited success.

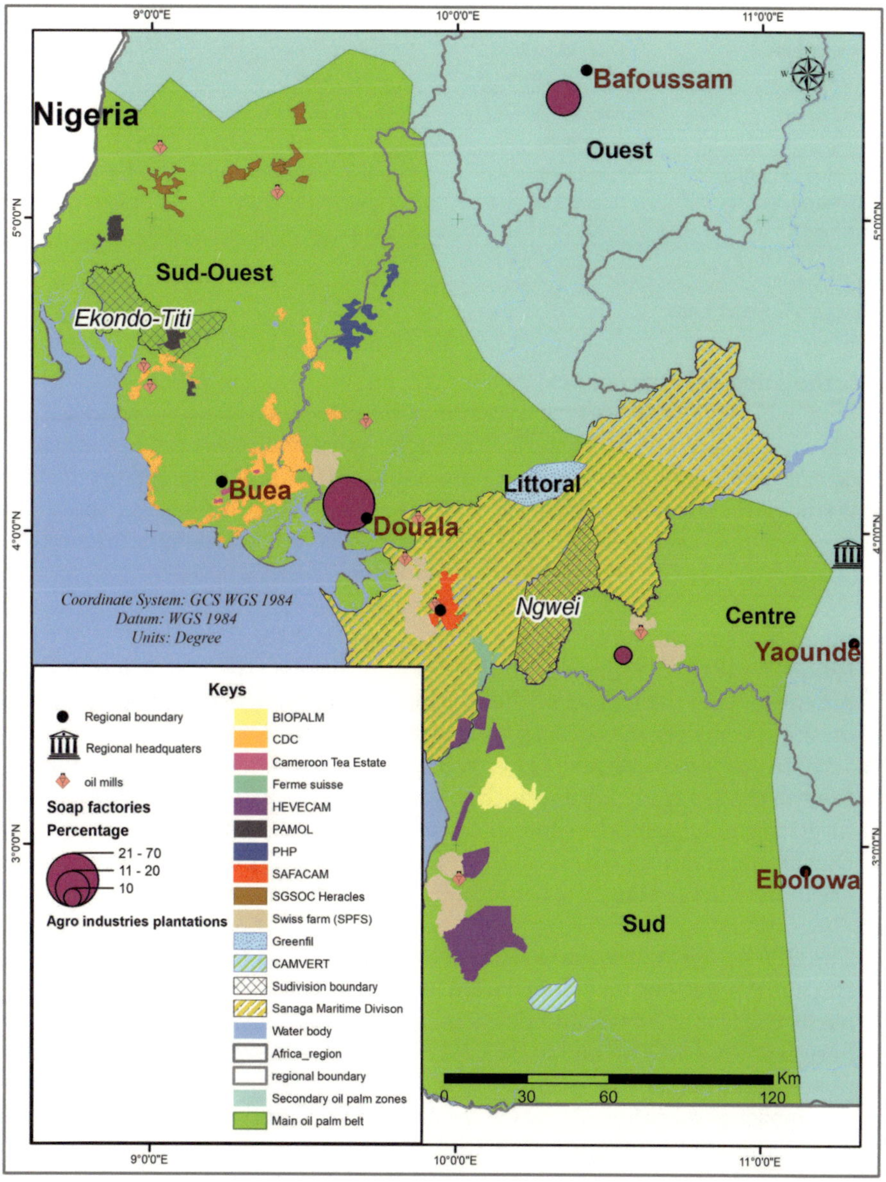

Figure 1.
"Elaeiscultivation" areas, oil pam farms and industries of Cameroon.

Parameters	Sanaga Maritime	Ngwéi	Ekondo-Titi
Area in sq. km	9311	848	652
Number of Protected areas	3 (179,661 ha)	0	0
Number of agro industries	3	1	2
Name of agro industries	SAFACAM, SPFS, SOCAPALM	SOCAPALM	PAMOL and CDC
Planted areas (ha)	30,000	5,000	25,000
Labor regime	Workers and farmers	Workers and farmers	Workers and farmers
Number of Elite oil palm owners	5	1	2
Number of Small holders	3000	398	550
Type of Soils	Oxisoils, yellow ferralitic soils	Oxisoils, yellow ferralitic soils	Volcanic soils and peat lands
Land cover	Primary and secondary forest	Primary and secondary forest	Primary, secondary forests and mangroves
Number of Landsat satellite images processed	9 Landsat, 5 Spot, 3 Ikonos, 1 Geoeye	3 Landsat, 1 Google Earth 1 map box 1 Geoye	3 Landsat, 1 Google Earth, 1 map Box

Table 1.
Summary of the research sites.

2.2 Methodological approaches

This work constitutes an analytical and conclusive synthesis of the research carried out, participative field investigations and works undertaken in production basins since 2013 (from 2013 to 2018) under the framework of the palm and forest project of Cameroon (PALMFORCAM-IRD) as well as Oil palm Adaptative Landscape (OPAL). The main missions included:

- Investigations carried out within the framework of the PALMFORCAM project funded by IRD, took place in 2013–2014 [5]. The objective was to monitor by remote sensing the impact of the development of village oil palm plantations on the forest cover in the Sanaga Maritime Division. The methodological activities set up concerned: (1) Carrying out an inventory of the palm plantations in the Sanaga Maritime with collection of GPS waypoints, without forgetting to identify the types of owners and the sizes of the plantations; (2) Diachronic mapping (Landsat, SPOT, IKONOS and GEOEYE images) of land use in the Sanaga Maritime Division in 1975, 2000 and 2013.

- The OPAL project drive us to study the environmental impact of village/elitist palm plantations on deforestation in Sanaga Maritime and in the Ndian basin: the case of Ngwéi and Ekondo-Titi subdivisions. This study, carried out between October 2016 and March 2017 by an interdisciplinary team (geographers, botanists, environmentalists and geomatician specialists), highlighted the impact of oil palm cultivation on the landscape and the range of tools used during this study are resumed in **Table 2**.

The quadrate method is advantageous because it helps in studying the dynamics of the fauna and flora in a quantitative and qualitative approaches. As flora recording is concerned for the analysis of the impact of oil palm farming on

Methods	Operational work
Satellite image processing	LANDSAT image processing (MSS, TM, ETM+ & 8 de 1975 to 2015), MAP Box images (1.5 m resolution)
Botanical Survey on quatrates and transects	Two quadrates and two transects in each palm plantation visited on the field (village, elitist et industrial); a quadrates in the dense forest.
Environmental impacts assessment	Interaction matrix and impact sheet per receiving environment
Questionnaire Survey	330 questionnaires in Sanaga Maritime (2013) 290 and 260 copies of questionnaires administered in Ngwéi and Ekondo-Titi Sub-divisions respectively (2016)
Landscape methods	Application of the SEPL exercises in these areas.

Table 2.
Methods used in assessing the impact of oil palm plantation in the environment.

plant biodiversity, experimental plots were applied on four types of vegetation's namely: a village or smallholders oil palm plantations, an elitist oil palm plantations, an industrial palm plantation and a forest area. To this effect, the team realised an experimental plot where recorded plant species were immediately identified.

Criteria such as the nature, probability of occurrence, scope or extent, magnitude or intensity, reversibility and duration of the identified impacts were combined for an appropriate impact assessment. The rating grid of Leduc and Raymond [19] was used for impact assessment. Ratings from 1 to 5 were assigned to the indicators (**Table 3**) depending on the degree of impact. The absolute importance represents the average of the impact ratings over the total number of rated indicators. The nature of an impact can be positive (+) ▲ or negative (−) ▼ on the environment concerned.

The absolute importance or significance of the impacts is determined by calculation by taking the product of all the ratings assigned to each indicator over the total number of indicators. This is illustrated by the following equation:

Value	Occurrence of the impacts	Territorial scope (extent) of the impacts	Duration of the impacts	Intensity of the impacts	Reversibility of the impacts	Final rafting
1	Very unlikely	Very reduced space (10%)	Very short	Very weak	Immediately reversible	1–2 non significant or negligible
2	Unlikely	Reduced space 15–20%	Temporary	Low	Quickly reversible	
3	Likely	Fairly extensive 25–40%	Long enough	intermediary	Reversible	2.1–2.9 insignificant
4	Certain	Extended 50%	Long	High	Little reversible	3–4 significant
5	Very certain	Very extensive 60–100%	Very long	Very high	Irreversible	4.1–5 very significant

Table 3.
Impact assessment indicators and rating of their impacts.

$$\text{Absolute importance} = \frac{\sum Ratings\left(\text{intensity x reversibility x extent x duration xoccurence}\right)}{5} \quad (1)$$

After rating the impacts were qualified according to the results obtained.

- The rating between [1–2] represents the insignificant or negligible impacts;
- The rating between [2.1–2.9] represents the insignificant impacts;
- The rating between [3–4] represents the significant impacts;
- The rating between [4.1–5] represents very significant impacts.

The critical impact threshold is established when the rating value is greater than or equal to the average of the grid: 3.

To complete methodology and tools used, it is worth mentioning that oil palm is one of the most studied agricultural speculations in Cameroon today. Its cultivation and oil production are of interest to economists, sociologists and anthropologists because of its income-generating character and the by-products are used in traditional pharmacopoeia. With further investigations carried out in the Sanaga-Maritime oil production basin (2018–2020), under the coordination of WWF and EPFL Switzerland, three work packages have been developed since 2018 till 2020 on the intercropping of oil palm:

- Analysis of the conversion of forests into "elaeisfarming" agrosystems, dynamics of the oil palm in this new environment.
- Analysis of the impacts of palm plantation management on soil fertility.
- Socio-economic impacts of the management of palm plantation in the main "elaeisfarming" production basins of Cameroon.

This study also take into account various meetings held with the actors of the Africa Palm Oil Initiative (APOI), whose objective is to seek and support the transition to a sustainable palm oil not linked to deforestation in Central and West Africa.

3. Results and discussion: environmental impacts of the oil palm in Cameroon

Paragraphs below allow to address the main impacts identified into the oil palm sector being ecological, economic, social or institutional. Fieldwork observations show that most of the components of the biophysical environment are affected by this activity. Globally, it is noticed the clearing of the forest for the establishment of new plots, the erosion of the land during exploitation, the pollution of air and water during processing as well as relative poverty and inequality among peasants. One can include grievances related to the distribution of benefits and the development of neighboring.

3.1 Ecological impacts of oil palm

There are so many ecological impacts of oil palm identified namely deforestation, loss of biodiversity, erosion as well as soils and water pollution or contamination.

3.1.1 Oil palm cultivation and deforestation

Ecology of the oil palm shows that the forest area has the suitable conditions (soil, rainfall, temperature, relief and insolation) for its development. It is in this vein that all old and new palm plantations (from all actors being agro industries, small farmers and elites) are located in forest areas. This is because forest milieu guarantees them a sustained production over a long period while others areas like fallows and abandoned farms do not bring the expected results (**Figure 2**).

It is worth notice that agro industrial palm plantations/concessions and properties are 100% created on forests. The areas of industrial oil palm producers increased from 46,850 ha in 2009 to 63,200 ha in 2014 and more than 176,600 ha in 2019. It means 35% increase in about 5 years and 73% in 10 years. Similarly, remote sensed data in Ngwéi and Ekondo-Titi Sub-divisions show that more than 83% of smallholders palm plantations were created in both primary and secondary forests. Increase in smallholders palm plantations is estimated at more than 50% in 10 years between 2000 and 2010. The latter do no longer used fallow land and other abandoned fields. Because there are so many smallholders actors more than 90% for only 15% of area, the accelerated deforestation process by atomizing the forest. Atomization of the forest by unsustainable oil palm worsen practices (construction or building, clearing, palm plantations and wasteland) contributes enormously to the decline of the forest which is suffocated. Any parcel of forest located between two or more of these plots is doomed to disappearance.

The threat of palm plantations on the original forest is all the more serious as certain industrial concessions granted in recent years are adjacent to protected areas. This is the case of Sithe Global Sustainable Oils Cameroon (SGSOC); a subsidiary of the American multinational Herakles Farms. The concession acquired by the latter is located near protected areas (Korup and Mounts Barossa national parks, Rumpi hill reserve and Banyang-Mbo fauna sanctuary) recognized as High conservation value forests (HCV) and also endemic for its biodiversity. The same situation was observed in the Greenfil case whose palm plantations are located very

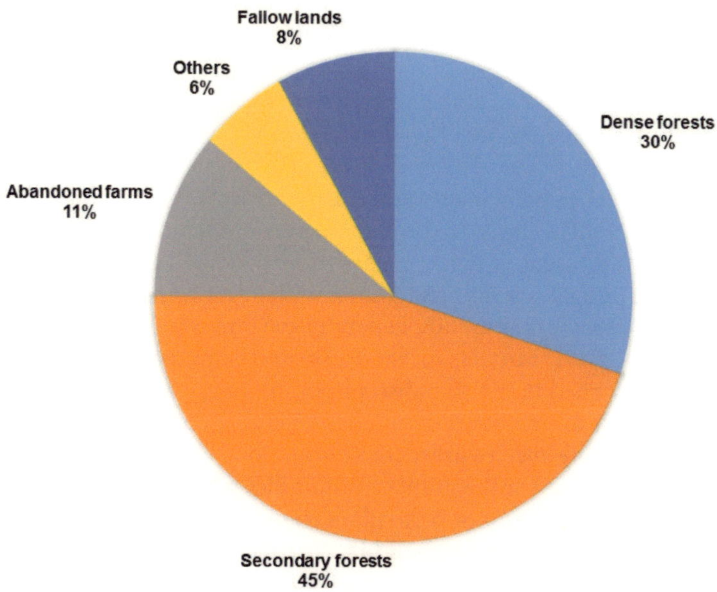

Figure 2.
Type of land use chosen for the creation of oil palm plantations in the main production basins (Source: field survey 2013–2020).

close to the Ebo forest which plays host to a wide variety of wild animals, especially the western gorilla, the Nigeria-Cameroon cross boarder chimpanzees, drills and several other primates as well as many endemic plant species. Another case is the 2019 de-gazettement of Forest Management Unit (FMU) number 09–025 near the famous Campo Ma'an National park for its transformation into oil palm (60,000 ha) plantation by CAMVERT. This National Park is recognized as Model forest and Biosphere Reserve. The proposed declassified area encompasses two blocks covering 40,000 ha to the north and 20,000 ha to the south bordering Dipikar Island (Campo Ma'an National Park) where there is a gorilla habituation project ongoing.

Deforestation caused through palm planting also fragments the habitat of endangered species and disturbs wildlife corridors usually used by forest elephants. Such a situation bring confusion and more and more anger, disappointment and land conflicts due to ambiguous governance of the forest. This issue is discussed in the following paragraph.

As the ecological impacts are concerns, there is a high degree of confidence that the expansion of palm oil cultivation has resulted in deforestation. Numerous authors and reports have emphasizes on oil palm as a driver of deforestation and land-use change in tropical countries [20]. Globally, oil palm crop development is responsible for less than 0.5% of deforestation, but in parts of the tropics this figure can reach 50% [21]. For Indonesia, the proportion of direct and indirect deforestation linked to the expansion of elaeis farming is estimated to be between 11% (2000 to 2010) and 16% (1990 to 2005). At the same time, in 2016, in the same country, 45% of oil palm plantations were on land which, in 1989, was forests [22, 23]. This phenomenon is very marked in Malaysia and Indonesia because they are the two largest palm oil producers in the world. With an average forest loss of 350,000 hectares annually, deforestation is particularly dramatic on the island of Borneo, where about half of the deforestation between 2005 and 2015 was directly linked to industrial oil palm plantations [21, 24]. As shown by [25], 2/3 of deforestation in the South-West region of Cameroon during the same period was caused by the expansion of oil palm farms and the installation of new actors. The peculiarity of oil palm basins areas compared to the rest of Cameroon is that they host most of the agro-industrial activities (oil palm, rubber, plantain and sugar cane). To this should be added the share of deforestation generated by small producers, which is still unclear. In addition, oil palm plantations are responsible for significant greenhouse gas emissions.

3.1.2 Elaeiscultivation and the loss of biodiversity

The establishment of palm plantations generally involves the almost total clearing of the forest. This leads to the loss of species, including those that are endogenous in certain areas of high conservation value. Oil palm is generally grown in pure culture. It doesn't tolerate association with other commodities. The other species found in palm plantations are generally the ombrophilous grasses. **Table 4** built with quadrates and transects methods, shows the numbers of species, families, and individuals recorded in the oil palm cultivation basins of Ngwéi and Ekondo-Titi compare to those of protected areas bordering these basins. One can observed that less than half of the species present in humid forests are present in industrial plantations. It is the same for the number of families which decreases between the mangroves and different forms of palm plantations. From the table below, smallholders palm plantations conserve their biodiversity better than industrial one. This is explained by the solicitation of other ecosystem services such as traditional pharmacopoeia and the harvesting of non-timber forest products by populations.

From the **Table 4** above, out of the 18 families in moist and dense forest and 20 families in the mangrove, only 5 families are found in all palm plantations (village,

Type of area	Number of species	Number of families	Number of individuals
Ekondo-Titi – 2016			
Dense and humid forest	48	18	162
Mangrove	37	20	178
Smallholders' plantations	45	24	76
Elites plantations	38	25	159
Industrial plantation	21	15	78
Ngwéi – 2016			
Dense and humid forest	50	26	142
Smallholders' plantations	31	19	39
Elites plantations	60	32	88
Kribi, Campo, Douala areas – 2020			
Campo Ma'an National Park	108	36	98
Kribi Marine Park	72	26	65
FMU 09–025	69	27	60
Douala Edéa National Park	59	31	54

Fieldworks quadrates and transects Survey 2016 & 2020.

Table 4.
Floristic diversity between palm plantation and forest of Ngwéi and Ekondo Titi subdivisions and some protected areas around Kribi and campo.

elitist and industrial). These include *Annonaceae, Apocynaceae, Euphorbiaceae, Fabaceae* and *Loganiaceae* as well as *Moraceae* found in wet and dense forests and in the elitist and villager oil palm plantations and *Phyllanthaceae* found in the mangroves and in the elitist and smallholders plantations. This constitutes either a real quantitative and qualitative decrease in biodiversity outside the *Fabaceae* families. The number of families, however, increased in the elitist (25) and smallholders (24) oil palm plantations of Ekondo-Titi. In protected areas and FMU, families range from 26 to 36 meaning that oil palm is a driver of deforestation. Therefore, the least diversified plots are the smallholders' farms, followed by industrial and finally elitist palm plantations. This can be explained by the regular maintenance of industrial palm plantations, the mixed food crops grown in some village palm plantations and the irregularity of the maintenance in the elitist oil palm plantations.

Conclusively, the clearing of hundreds or thousands of hectares of land for oil palm cultivation is one of the most important factors in the destruction of vegetation cover and consequently of biological diversity. Deforestation and degradation are the root cause of a considerable loss of flora species, fragmentation and disturbance of the natural habitat in these areas. The original evergreen natural forest has disappeared in favour of the mono-specific oil palm plantations, which occupy three-tenth of the territory, but catalyse deforestation. Also, aggressions on forests and fallow lands for oil palm establishment create enormous pressure on traditional and modern land reserves and protected areas.

3.1.3 Floristic diversity index

The measurements from quadrates and transects allow to calculated many indices. The Simpson index which measures rare species is roughly equal across the four project sites. The equitability of Pielou, which provides information on the distribution of species, is approximately equal in the different sites sampled. The Shannon index, which takes into account floristic diversity, is higher in PNCM (4.01) and similar in PMK, PNDE and UFA (**Table 5**). The Shannon index (**Table 5**) shows significant biological diversity for dense forests and for mangroves (0.28). The Shannon index is also high for the industrial palm plantation of Ekondo-Titi, relatively less for the village palm and elitist palm plantations. The Simpson index is 0.08–0.09 in mangrove, moist and dense forest compared with 0.07 in the industrial and elitist palm plantations of Ekondo-Titi against 0.01 in the smallholders and elitist palm plantations of Ngwéi. Simpson index shows the degree of land use in the two districts. This is due to the fact that ecosystems are profoundly affected by agricultural practices and especially by oil palm cultivation (and even cocoa farming with exotic species), which reduces density and specific diversity locally.

3.1.4 Dynamics of plant biological and fauna diversities

Before concluding this section, it would be important to highlight the variation in floristic and wildlife biodiversity from natural environments to oil palm plantations. This would give and idea of the real impact of oil palm plantations on the biodiversity decrease within the landscapes studied. Taking the floristic level, **Tables 4** and **5** show that biodiversity in terms of species and families is so important in protected areas and dense forest than anywhere else. This is quite conspicuous in Campo Ma'an National Park. Hence the advantage of avoiding installing oil palm plantations next to protected areas or in dense forests because they considerably reduce biodiversity.

As fauna is concerned, the survey show that forest degradation is one of the major infringements to the loss and decline of wildlife for more than 50% of the surveyed population. It happens through the clearing of hectares of forest land which drive to

Types of land use	Shannon index	Equitability of Pielou	Simpson index
Ekondo-Titi			
Wet and dense forest	0.27	0.01	0.08
Mangrove	0.28	0.01	0.09
Smallholders palm plantations	0.20	0.00	0.02
Agro industries palm plantations	0.27	0.01	0.07
Elitist palm plantations	0.25	0.01	0.06
Ngwéi (Makondo)			
Wet and dense forest	0.28	0.01	0.09
Smallholders oil palm plantations	0.21	0.01	0.01
Elitist oil palm plantations	0.18	0.00	0.01
Protected areas			
Kribi Marien Park	3.56	0.83	0.95
Campo Ma'an National Park	4	0.85	0.96
Douala-Edéa National Park	3.56	0.87	0.96
FMU 09–025	3.57	0.84	0.95

Table 5.
Biological diversity index.

the destruction of wildlife habitats and the disappearance of species. There are almost any game (porcupines, monkeys, antelopes, etc.) and some species have already completely disappeared from the area (such as the elephant that disappeared from Njock-Loumbe, *Njock* meaning the elephant). In addition, the oil plantation guards confiscate the small mammals such as rodents that are caught in these single-crop farming.

Biodiversity impacts is the most documented facet of environmental oil palm effects. Land clearance for oil plantations removes, fragments and damages important wildlife habitats, leading to a high loss of species. The species these forests support are highly adapted to rainforest habitats and are often unique. Clearing tropical forests for oil palm results in strong local and regional biodiversity declines [21]. It is link to the fact that oil palm is commonly produced in monocultures which affect the habitat of great mammals and their biodiversity declines from 47 to 90% [26] or 65–90% [23]. Also, the mammal diversity in oil palm strongly depends on the proximity of natural forests [21, 26]. In Cameroon, great APES are endangered by the spread of oil palm plantations around protected areas like Campo Ma'an, Ebo, Korup etc. And in other areas, due to oil palm expansion, elephant have disappear like in Njockloumbe village at Ngwéi. The IUCN Red List of Threatened Species documents 321 species for which oil palm is a reported threat [21]. Meijaard et al. [26] added that species those threatened are made up 3.5% of the taxa threatened by annual and perennial non-timber crops (9,088 species) and 1.2% of all globally threatened taxa (27,159 species) in 2019.

As we saw on **Table 4**, the highest diversity of animal species in oil palm areas, however, is generally found in the wider landscape that includes remnant patches of native vegetation. Factors that are likely to positively influence biodiversity values in both industrial-scale and smallholder plantations include higher landscape heterogeneity, the presence of large forest patches and connectivity among these and the plant diversity and structure of undergrowth vegetation.

It is clear that oil palm becomes the source of deforestation and land degradation. The statistics compute from image processing help calculating the deforestation rate in the main studying sites. In Sanaga Maritime, from 1986 to 2013, deforestation rate is estimated at 23.61%. Ngwéi deforestation is estimated at 45.94% in 38 or 40 years, with an overall rate of 697.22 ha/year between 1975 and 2013. Deforestation in Ekondo-Titi is accessed at a rate of 22.74% in 37 years, i.e. 0.61% per year and especially 150.34 ha/year of Atlantic forests against 67.07 ha/year for mangroves. Let's take the detail case of both Ekondo-Titi and Ngwéi subdivisions to illustrate the results of images processing (**Figures 3** and **4**).

Table 6 emphasizes the synthesis of deforestation linked to the expansion of oil palm. The estimate of total deforestation varies according to administrative units and the dynamics of elaeis cultivation, and in this sense, the Ngwéi landscape appears to be more threatened.

Table 7 summarizes the perception of the populations in terms of ecological impacts. This shows the illusion that the people of Sanaga Maritime and Ngwéi have of thinking that the situation is not changing and of underestimating the deforestation linked to palm oil. However, they have clearly seen the decrease in wildlife. On the other hand, Ekondo-Titi recognizes the impact of oil palm cultivation both on the forest and wildlife.

Table 8 below shows negligible positive impact (3.5%) on fauna and NTFPs with overwhelming negative impact (55) (96.5%), meaning that oil palm cultivation largely undermines the resilience of the natural environment. The impact is more on surface water, flora and fauna (biodiversity), soils, natural habitats and non-timber forest products (NTFPs).

Based on remote sensing techniques, one can deduce that in the various literature, the case of Indonesia and Malaysia have been well identified compared to African

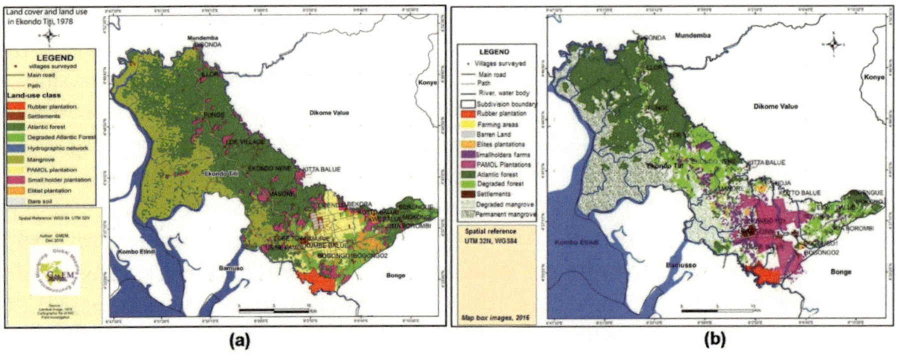

Figure 3.
Land cover and land use in Ekondo-Titi between 1978 and 2016 thanks to Landsat (1978), Map Box images (1.5 m resolution) and Google Earth (2016). The original forest and mangrove has disappeared everywhere apart from the North western part of the map. Smallholders' farms are spreading north western wards.

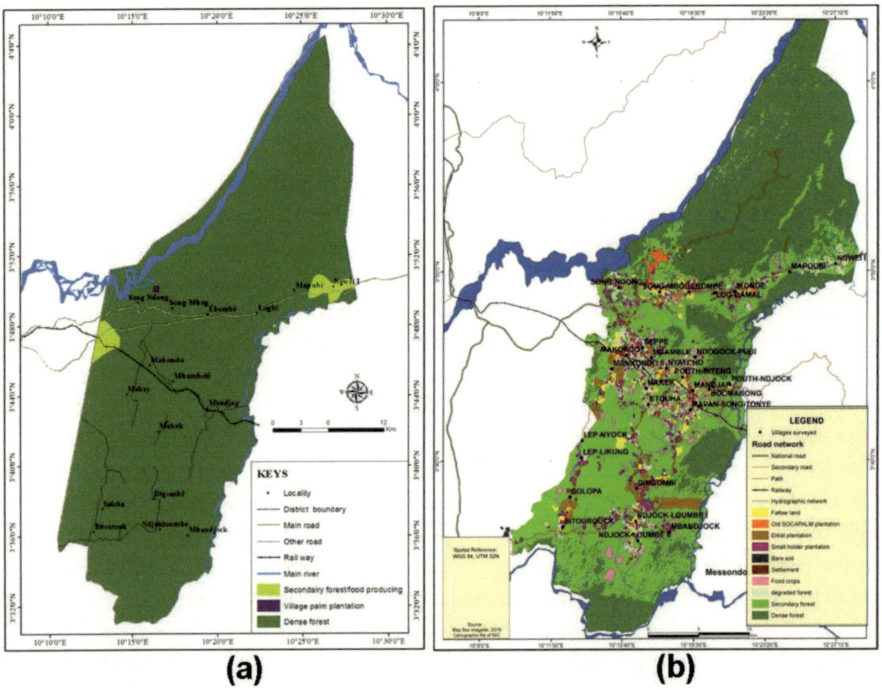

Figure 4.
Land cover and land use in Ngwéi District between 1975 and 2016 from Landsat (1975) and Google Earth image. A small portion southwards and a great areas northwards of the images remains intact and need to be preserved. Numerous smallholders' farms oil palm widespread and scattered in the central part of the image show that the Ngwéi District is the hold almost more than 30% of the areas producing red oil within the Sanaga Maritime Division.

countries producers like Nigeria, Ghana, Ivory Coast and Cameroon. Remote sensing studies of a subset of plantations in 20 countries suggests that around 45% of oil palm plantations in Southeast Asia came from areas that were forests in 1989. The estimates vary from one region to another being at 31% in South America, 7% in Africa and 2% in Central America. For Indonesia and Malaysia, the estimates were 54% and 40% respectively [22, 26, 27]. Another estimation gave during the last 40 years, 47% and 16% of total deforestation by oil palm in Malaysia and Indonesia, respectively [21, 28]. Those statistics could be compared to what we observed in Ngwéi (45,94%)

Parameters	Sanaga Maritime	Ngwéi	Ekondo-Titi
Area in sq. km	9311	848	652
Total deforestation rate	28% in 40 years	45,94% in 40 years	22,74% in 37 years
Mean annual rate	0,7% per year	1.15% per year	0,61% per year
Net deforestation (ha)	121,043 ha	11,872	7,882
Deforestation due to oil palm (ha)	65,177	7,632	3,977
Time span projected for the disappearance of the forest	50–70 years	37–50 years	125–189 years
% Oil palm expansion occurring at the expense of the forest	70%	90%	60%

Table 6.
Deforestation and oil palm expansion in the three sites.

Area	Sanaga Maritime (n = 335)			Ngwéi (n = 290)			Ekondo-Titi (n = 260)		
Impact	N	No change	P	N	No change	P	N	No change	P
Threat on forest	44.5	33.4	22.1	33.2	43.2	23.6	20.39	28.85	50.77
Economic	15	7	78	5	5	90	7	3	90
Animal BD	65.51	23.24	11.25	47.52	33,28	19.2	58.53	25	16.47

Table 7.
Perception of livelihood ecological impacts through questionnaire and landscape methods.

and Ekondo Titi (22,74%) as well as Sanaga maritime (23,61%) productive basins. According to Ordway et al. [25], oil palm expansion dynamics in sub-Saharan Africa have been overlooked. They proved that 67% of oil palm expansion from 2000 to 2015 occurred at the expense of forest in the Southwest region of Cameroon.

Coincidentally, these are priority areas often safeguarded by the policies of the World Bank and the African Development Bank (ADB), because they are elements of the natural heritage of a country.

3.1.5 Soils and water pollution

Water samples were collected upstream, at the spillway and downstream of the palm oil extraction sites in Ngwéi and Ekondo Titi. Chemical and bacteriological analyses were carried out on 09 water samples. The results revealed contamination and pollution, including even groundwater. Overall the surface waters analyzed are basic with pH values all above the recommendations of the WHO standard for drinking water. Chemical oxygen demands are high and reflect pollution. Suspended matters are present in all samples. In addition, two Ekondo-Titi samples have concentrations of ammonium ($NH4^{+}$) ions relatively higher than the recommended value (≤0.50). At the microbiological level, six (06) samples showed concentrations of fecal coliforms not complying with guideline values (0UFC/100 ml). As a result, these waters would be under the influence of a major source of pollution, making them unfit for human consumption without prior treatment. Well water and groundwater are also contaminated.

Component of the affected environment	Activities sources of impacts	Impacts	Characterization parameters and rafting						Final assessment
			Nature	Occurrence	Intensity	Spatial extent	Duration	Reversibility	Importance
Air	Land clearing/ deforestation Storage & preparation of nut Oil extraction	Degradation of air quality	▼	3	2	1	1	2	1,8
Surface water	Land clearing/ deforestation Oil extraction Waste management	Degradation of water quality / contamination, pollution	▼	3	3	4	4	4	3,6
Underground water	Land clearing/ deforestation Oil extraction Storage & preparation of nut	Contamination, water table attack, pollution	▼	3	2	1	3	3	2,4
Soil	Land clearing/ deforestation Staking, hole punching Planting Storage & preparation of nut Oil extraction	Degradation of soil quality Contamination, pollution	▼	4	4	2	3	4	3,8
Naturel habitat	Land clearing/ deforestation Staking, hole punching	Fragmentation, destruction of natural habitats	▼	3	3	4	5	5	4

DOI: http://dx.doi.org/10.5772/intechopen.97862

Component of the affected environment	Activities sources of impacts	Impacts	Characterization parameters and rafting						Final assessment
			Nature	Occurrence	Intensity	Spatial extent	Duration	Reversibility	Importance
Flora	Land clearing/ deforestation Plant maintenance Harvesting bunch Felling old palm plants	Deforestation, fragmentation	▼	5	5	4	5	4	4,6
Fauna	Land clearing/ deforestation Staking, hole punching Plant maintenance	Fauna habitat disturbance Migration and loss of fauna species	▼	4	3	3	5	5	4
Non-timber forest products (NTFPs)	Felling old palm plants	Increase /decrease in NTFPs Loss of medicinal species	▲▼	3	2	2	5	5	3,4

Source: fieldwork, 2016–2020.

Table 8.
Absolute importance of the impacts of the oil palm cultivation on the biophysical milieu.

Also, field surveys show that waste oils emanating from SOCAPALM and PAMOL mills flow into rivers and streams close to village dwellings. These rivers and streams remain the most fishing and living places estimated by local populations (consumption, bathing, etc.). As a result of these liquid waste, local populations are not only deprived of much of their fishing resources, but they are also exposed to health risks. Another negative aspect is the environmental impact of artisanal mill units whose process is polluting because the discharges are not treated. Finally, deforestation exposes the soil surface and accentuates its leaching. The oil palm plantation establishment modifies the soil texture as well as its biological characteristics, which is often partly responsible for the degradation of plant diversity in oil palm plantations. This degradation of soil quality is at the origin of the loss/fragmentation of the wildlife natural habitat as well as the destruction of the soil micro-fauna.

Unfortunately, the issue in water and soil pollution, is the mostly poor assessed aspects in many studies. However, greenhouse gas emissions occur from mill and plantation activities, and especially from Palm Oil Mill Effluent (POME), a liquid waste from the initial processing of fresh fruit bunches. Little is known about the pollution of waterways by fertilizers, pesticides and other chemicals used in oil palm plantations, as well as their impact on human health, aquatic species and fisheries [21]. Qaim et al., [29] found that forest conversion to oil palm plantations also affects ecosystem functions. Among others, the functions affected include carbon storage, nutrient cycles, soil regeneration, and air and water purification. Releasing POME into waterways harms aquatic ecosystems by creating highly acidic environments or causing eutrophication and this is in line with our results.

Landscapes dynamics have been assessed through the populations rating. The synthesis is shown by **Table 9** interpreted the lines after.

The high score (34.62) for the "slow increase" trend in Ekondo Titi reflects the illusion of the population's margin for maneuvering the resources of their territory (sea, Atlantic forest, dense forest and mangrove) in Ekondo-Titi. It is worth mentioning that the fallacy of the people of Ngwéi and Sanaga Maritime, of whom approximately 2/5 believe that the landscape has not changed (43.20 & 40.75). However, they objectively acknowledge (23.4 & 22.57%) that negative changes (landscape degradation, resource depletion, poverty) are more significant than positive changes (15.4 & 18.15%). This means forest depletion is a reality even though if people succeed in getting a cash benefit from oil palm activities.

3.2 Economic impacts of oil palm

In Sanaga Maritime, 51% of the population admitted that oil palm enable them to validly meet their existential needs [5]. For the elite, this is a sector where people

Area	↑ steep upward	↗ slow increase	→ no change	↘ slow decrease	↓steep downward	Total
% synthesis Ekondo Titi	*16.15*	*34.62*	*28.85*	*13.85*	*6.54*	*100*
% synthesis Ngwéi	*8.2*	*15.4*	*43.2*	*23.4*	*9.8*	*100*
% synthesis Sanaga Maritime	9.25	18.15	40.75	22.57	9.28	100
Mean Total	**11.2**	**22.72**	**37.6**	**19.94**	**8.54**	**100**

Table 9.
Landscapes trend arrows and scores in Ekondo-Titi, Ngwéi and Sanaga Maritime.

invest to earn extra income or prepare for retirement. In the elaeisfarming areas of Cameroon, an abundance of direct or indirect activities linked to this sector makes it possible to more or less effectively rule out the specter of unemployment and poverty. In terms of employment and the local economy, results show that oil palm has a positive impact with scores ranking from 3.6 to 4 despite its overwhelming negative biophysical impact (**Table 10**).

Jobs and revenues generated by the various activities related to the establishment, maintenance and operation of a palm plantation (planting and plant maintenance, transport of FFB and oil extraction) constitute the most visible face of its socio-economic impacts able to boost the local economy if the sustainability conditions are fulfilled. Several aspects of this positive impact are to be noticed (1): the sale of FFB by farmers and elites to agro-industries; (2) the establishment of modern mills; (3) Significant induced impacts linked to a flowering of secondary processing industries in Cameroon (soap factories, cosmetics); (4) The sale of artisanally or semi-mechanically extracted oil to soap factories, on local and regional markets or at the roadside; (5) the development of income-generating activities and petty trade in these villages thanks to the oil palm cultivation; and (6) the development of cooperatives based on existing CIGs will constitute the final stage of this economic facet observed in both districts. The population perception is resumed in **Table 11**. One can observe better results in income level, Job creation and welfare while quality of social network and social infrastructure remain lukewarm.

From an economic standpoint, the benefits of oil palm cultivation are undeniable. This profitability explains the rapid development of the "red gold". Nevertheless, the contribution of the palm plantation to the local economy and to the well-being of neighboring populations does not always meet expectations. The benefits for the national economy must also be optimized. Palm oil being a source of financial evasion, it is necessary to ensure the autonomy of Cameroon in order, as much as possible, to avoid imports.

On the economic point of view, **palm oil appears as one of the most profitable land uses in the tropics because for the most producing countries, it significantly contributes to the national economies and to reduce poverty elsewhere in the producing countries** at local, regional, and national levels [30–33]. In the main producing countries like Malaysia and Indonesia, oil palm accounts for 10% of total national exports and 44% of world palm oil exports. This data is reduce by half in several smaller producing countries, such as Honduras, Papua New Guinea, Solomon Islands, and Guatemala, palm oil exports account for around 5% of total national exports [28]. This explains why oil palm generates higher incomes than rubber or cocoa and other commodities, which occupy a prominent place in exports. Cocoa, for example, represents 15% of Côte d'Ivoire's GDP and 40% of merchandise exports [34]. These laudatory results (at the economic level) are obtained mainly thanks to a cash crop organized and carried out with methods which give absolute priority to the best yields. Most of these authors show that small farmers are the most beneficial of the boom of oil palm worldwide [22, 26–28]. Despite contradictions on assessment of this economic effects of oil palm, it is evident that elaeis farming has improved incomes for rural people and supported the development of rural economies and the economies of producer countries overall.

3.3 Social impacts in the oil palm sector in Cameroon

As observed on the field, social impacts must include social protection, collective bargaining, inclusive dialogue, conflict resolution, health risk, corporate social responsibility and environmental justice. These questions variably challenge the

Component of the affected environment	Activities sources of impacts	Impacts	Characterization parameters and rafting						Final assessment
			Nature	Occurrence	Intensity	Spatial extent	Duration	Reversibility	Importance
Local economy	Harvesting bunch Packaging and sale	Development of economic activities Increase in income	▲	3	3	3	5	4	3, 6
Employment And income level	Land clearing/ deforestation Staking, hole punching Planting Plants maintenance Storage & preparation of nut Oil extraction Packaging and sale	Job creation	▲	4	4	5	4	3	4

Source: fieldwork, 2016–2020.

Table 10.
Absolute importance of impacts of the oil palm on the economy.

Area	Sanaga Maritime (n = 335)			Ngwéi (n = 290)			Ekondo-Titi (n = 260)		
Variable	**N**	**No change**	**P**	**N**	**No change**	**P**	**N**	**No change**	**P**
Income level	15	10	75	13	8	79	10	9	81
Quality of social network	25	10	65	27	5	68	15	10	75
Job creation	5	15	85	7	5	88	5	5	90
Social infrastructure	45	35	20	55	25	20	55	30	25
Welfare	10	10	80	5	5	90	4	6	90

Table 11.
Perception of livelihood economic impacts through questionnaire and landscape methods.

agro-industrial and artisanal sub-sectors. In the wake of agro-industrial activities (SOCAPALM, SAFACAM, CDC, PAMOL), more or less, there is a slight satisfaction with the social protection of employees even if controversies regularly emerge on related issues, for example at the level of wages. The fact remains that the latter are regularly paid and for the most part and benefit from some social security. Conversely, almost all of these agro-industrial companies do not adapt well to syndicate activities, especially when tackling economically sensitive issues such as salary increases, health care, paid leave, security, social benefits of family members of employees, etc. Dictatorship and dismissal are common practices without any prospects for inclusive dialogue and concerted negotiation within companies. The social situation between the owners of the elite palm plantations and the local populations is tense at Ngwéi. Because the impacts are so important (**Table 12**), it is necessary to give sustainable compensation to populations whose land has been occupied by agro-industries in the expected standards. One can add the fact that the health risk is high in agro-industries because health infrastructures are under-equipped and obsolete. The housing conditions of workers are deplorable with overpopulation, dilapidated camps, non-functional water pumps, frequency of energy power cuts, etc.

Over the 83 impacts of the table, 37.35% are positive while 62.65% are negative signifying that on social domain, oil palm can be seen as a threat. Thus, the social and economic impacts of oil palm cultivation are numerous and sometimes contradictory. It may be overshadowed by the employment and income impacts, but the social consequences of this activity remain numerous.

In the field of the artisanal sub-sector, local populations working in oil production sites take no measures to protect their health. The gloomy observations draw by such a situation are: disorganization of the sector and the market, lack of social security for smallholders, land disputes, conflicts with agro-industries (**Table 12**), lack of personal protective equipment against heat & smell, etc. The question is that of a sector that will be fully organized, where the players remain scattered and whose activities sufficiently demonstrate a collective lack of consideration of social sustainability.

The oil palm provides local communities with many material, social and cultural uses ranging from food to traditional pharmacopoeia through decoration and construction materials, contributing to their well-being and their socio-cultural development. For the traditional pharmacopoeia, red palm oil is an antidote to poisons, palm kernel oil is useful for skin care in both new-borns and adults. Lastly, palm wine appears inescapable in all traditional ceremonies and rites concerning enthronement, weddings, deaths and funerals.

Component of the affected environment	Activities sources of impacts -	Impacts	Characterization parameters						Final assessment
			Nature	Occurrence	Intensity	Spatial extent	Duration	Reversibility	Importance
Conflicts	Plant Care Storage and preparation of palm nuts Fruit bunches harvest Packaging and sales	Land dispute, Agression, violence Intimidation, threat Tense social climate	▼	4	4	1	4	2	3
Human health	Plant Care Storage and preparation of palm nuts	Degradation of workers human health of injury	▼	4	3	2	4	2	3
Insecurity	Plant Care Storage and preparation of palm nuts	Injuries Food deficit, social conflicts		3	3	2	3	3	2.8
Noise	Storage and preparation of palm nuts	Noise	▼	3	2	2	2	1	2
Odour	Storage and preparation of palm nuts Waste management	Degradation of air quality	▼	3	2	2	1	1	1.8
Cultural heritage	Craft production of palm kernel oil	Traditional Pharmacopoeia		3	3	5	4	3	3.6
Landscape aesthetics	Creation of new nurseries	Landscape embellishing	▲	2	2	1	1	1	1.4

Source: fieldwork, 2016–2020.

Table 12.
Absolute importance of impacts of the oil palm on the social environment.

Insecurity impacts can be perceived from many angles: bodily risks, lack of safety measures and injuries and accidents' risks during clearing, hole digging, cleaning and maintenance of the palm plantation; and above all, oil palm harvesting and the pruning of the palm trees. Food insecurity is caused by low consideration in subsistence or food crops for the benefit of oil palm. Food production have decreased for almost 45, 55 and 57% of the respondents in Ngwéi, Ekondo Titi and Sanaga Maritime. There are also, land tenure (97% in Ngwéi and Sanaga Maritime and 74% in Ekondo-titi) and water resources conflicts among smallholders' farmers as well as between them and hunters.

Finally, cumulative impacts (physical and human environment) affect habitat fragmentation, degradation and loss of biodiversity, deforestation coupled with the rubber and cocoa single-crop farming or the merchant crop including plantain; food insecurity; social conflicts; social protection and collective bargaining. The population perception, shown in **Table 13**, revealed relative better access to food and social infrastructure, increase in quality of housing and better access to drinking water (due to the multiplication of drilling), but significative increase in water pollution as well as insecurity and conflicts.

At the socioeconomic level, there are enormous discrepancies depending on the category of actor. The oil palm value chain seems in fact to benefit more to agro-industrial actors and operators of second and third palm oil transformations. On the contrary, smallholders, because they are not sufficiently taken into account in sectorial policies, are poorly organized, which does not allow them to take the best advantage of the still artisanal oil palm exploitation. The quantitative economic numbers therefore drown the realities.

Socially, the results above demonstrated many negative externalities, thus raising the issue of many social impacts that many authors have addressed. Is palm oil a driver of development or a driver of inequality? [7]. Because almost 70% of the elaeisfarming areas belong to Asian or European firms, Bouron [7] considered oil plantation as "the archetype of the large capitalist plantation". Indeed, the proportion of palm oil produced by smallholders has steadily increased in Cameroon from 10 to 26% today. In Indonesia and Malaysia, smallholders account for roughly 40% of the total area of planted oil palm and as much as 33% of the output, due to lower yields, on average. There is significant variation in the way that smallholder oil palm cultivation is organized [22]. It is clear that almost 50% of the oil palm land is managed by smallholders worldwide [28]. Though, it is known that oil palm is profitable for rural households and communities in terms of new

Area	Sanaga Maritime (n = 335)			Ngwéi (n = 290)			Ekondo-Titi (n = 260)		
Variable	**N**	**No change**	**P**	**N**	**No change**	**P**	**N**	**No change**	**P**
Access to food	45	20	35	50	20	30	60	15	25
Quality of Housing	12	18	70	10	15	75	20	15	65
Access to land	22	10	68	12	16	72	5	15	80
Access to social infrastructure	50	10	40	50	15	35	40	15	45
Access to drinking water	20	8	72	12	10	78	38	12	50
Water pollution	68	20	12	72	18	10	70	15	15
Insecurity and conflicts	75	10	15	83	10	7	85	5	10

Table 13.
Perception of livelihood social impacts through questionnaire and landscape methods.

employment and opportunities, farm profits, and improved rural infrastructure [28, 32, 35]. Nevertheless, this profit is not to be applied to all households and communities [36]. There are many new jobs and employment created by oil palm for landless laborers and rural households in Indonesia, in Mexico and Guatemala [37]. For some countries like Ghana and Guinea, there is a relative stable incomes and higher levels of food security [30, 38]. Migrations is another aspect underlined by [39]. Despite, employment, jobs, rural migrations, wage incomes, linked to the palm oil sector, it does not necessarily improve welfare in terms of food security, and other non-income dimensions, land conflicts, [7, 40, 41],

Some of the negative social consequences of this "oil rush" include land grabbing, large deforestation and the spoliation of indigenous peoples land rights together with unclear land property rights and laws [7, 28, 42, 43] by the large corporations and agro industries. Moreover, the educational level and financial capacities of these agro industries and corporations are also clearly higher than those of the "average" peasants, allowing them not only all the imaginable corruptive drifts (towards the administration, the traditional chiefs) but, above all, giving them an advantage in negotiation [16]. Notwithstanding efforts in developing and implementing forest protection measures, progress has been weak towards achieving this sustained goal and alleviates poverty. This has resulted in Cameroon maintaining palm oil exploitation close to protected areas. The desirability of future agricultural land to be conquered outweighs the desire to cover the forests still standing. According to data from the World Bank [44], Indonesia only granted protected area status to 12% of its vast territory, behind other comparable countries such as the DRC (13.8%) or Colombia (14.8%) and far behind Brazil (29.4%). Malaysia does better with 19.1%. In Cameroon, almost 25% of the territory is devoted to protected areas. But, the government policy can mask a great diversity of situations on the ground. Thus, the State granted 15,000 ha to Greenfil agro industry in 2014 and 50 000 ha to Camvert in 2019 near protected areas of HCV forest while de-gazetted FMU 09–025.

Such a situation shows not only the poor forest and land governance, but also, the weakness of the means of control which leads to illegal clearing, including within protected areas [45].

3.4 Institutional impacts

Outside the national framework, publics institutions seems not adapt to the local context of oil palm cultivation. Smallholder's access to land is not guarantee. This lack of good governance is a treat that can't favor sustainability of the whole sector. In addition, securing the elaeisfarming basins, prey to attacks by armed groups, is also seen as a necessity for Cameroon. An integrated and sustainable management approach in the oil palm sector takes into account all stakeholders. Governance requires having at least a national oil palm strategy still awaiting, then fighting against deforestation, approving selling prices, rationalizing production and reducing imports. Cameroon has a national strategy for sustainable development of the palm oil sector which is pending validation. This strategy identifies a set of actors and hierarchical decision-making bodies for the governance of the sector. The national steering committee is responsible for monitoring the implementation of the strategy. To this body, one can add programs and projects, professional organizations (inter-professional organizations, cooperatives and unions), consular chambers and national and regional consultation frameworks.

Overall, this analysis summarizes direct, indirect and cumulative impacts. **The direct impacts on the biophysical environment are:** air contamination, olfactory pollution, ground and surface water contamination/pollution, soil contamination and pollution, flora and fauna destruction, biodiversity degradation, deforestation and forest conversion, reduction of NTFPs and landscape aesthetics. **The direct impacts on the human environment can be summarized as:** employment and income, local economy, human health, insecurity and conflict, noise, odours, cultural heritage and waste. **Indirect impacts (physical and human)** include habitat fragmentation, degradation and loss of biodiversity, food insecurity, cultural heritage, social protection, collective bargaining and local crafts. **To end, cumulative impacts (physical and human environment)** affect habitat fragmentation, degradation and loss of biodiversity, deforestation coupled with the rubber and cocoa single-crop farming or the merchant crop including plantain; food insecurity; social conflicts; social protection and collective bargaining.

Institutional impacts are the most neglected aspects of the oil palm governance in African countries. First of all, very few countries have legislation specifically related to forest degradation and land use change and the government gives privilege on land to foreign investors and agro industries being local and not [20]. Unfortunately, without appropriate policies, smallholder production is not necessarily more rainforest-preserving, as smallholders are also significantly involved in deforestation [25, 33]. Strategies that aim at including smallholders in palm oil need to take into account: securing of land titles, access to credit, and technical support while accounting for the existing heterogeneity [46, 47]. Djouma et al. [48] propose a win-win partnerships between agro-industries and smallholders to boost the development of the national palm oil sector. Meijaard et al. [26] emphasize on certification as it is the case in Malaysia an Indonesia while African countries could not. But it is true that high carbon stock and high conservation value approaches are part of international concerns related to deforestation and oil palm environmental impacts [26, 45].

Finally, how to produce while limiting negative externalities as much as possible, one can ask? The answer can be found on several international programs launched for many agricultural crops taken individually or in groups. In the cocoa sector, for example, there are ISO 34101 standards for a sustainable and traceable cocoa bean [49]. To these initiatives must be added the certifications (like RSPO, Global Gap, Fairtrade, etc.) which give advantages on the market to producers respecting certain sustainability rules [21, 50].

4. Conclusion

The objective of this article was to assess the environmental impacts of the palm oil sector in Cameroon. The methods used gathered field observations, satellite images processing and participative survey among population through landscape perception methods. Three main production basins (Sanaga Maritime, Littoral and Southwest) were chosen.

The results revealed that the oil palm cultivation has many negative consequences on the environment such as deforestation and various forms of pollution. On deforestation Ngwéi account for 45.94% in 40 years (i.e. 1.20% per year), Sanaga Maritime, 23.61% (i.e. 0.87% yearly) and Ekondo-Titi, 22.74% (0.61% per year). The perception of rural populations confirms the results obtained on deforestation with 44.5, 33.6 and 20.39 in Sanaga Maritime, Ngwéi and Ekondo Titi respectively. The same with animal biodiversity which gave 65.81%, 47.54% and 58.53% in Sanaga Maritime, Ngwéi and Ekondo Titi respectively. Ecological impact in the matrix is 96.5% negative against 3.5 positive. The biodiversity declines and Simpson index are low in area of oil palm plantations than in other with 0.20–0.21 for Shannon index against 0.01–0.02 for Simpson index.

Economically, the sector is still dominated by small producers whose methods significantly impaired profitability. Economic impacts are 51% positive and the score varies by items with 75%, 79 and 81 for income in the Sanaga Maritime, Ngwéi and Ekondo Titi respectively; 95, 88 and 90% for Job in Sanaga Maritime, Ngwéi and Ekondo Titi respectively, also 80, 90 and 90% in Sanaga Maritime, Ngwéi and Ekondo Titi respectively for welfare.

Social impacts are diverse with 37.35% overall impacts positive against 62.65% negative. In social way, only housing (70, 75 and 65%,) access to water (72, 78 and 50) access to land (68, 72 and 80) for Sanaga Maritime, Ngwéi and Ekondo Titi respectively are 45, 50 and 60) and water pollution (68, 72 and 70) insecurity and conflicts (75, 83, 85%) are negative. At the social level, wage employment in the field is not well organized and corporate social responsibility is not applied among agro-industries and other large farmers (elites) who nevertheless deserve to be encouraged in this direction if we want to give the local riparian populations the opportunity to benefit from it.

At institutional level, governance is not well perceived apart from policies proposed to increase smallholders' areas under cultivation.

The above results revealed that the main objective of the research were fulfilled. The novelty brought by the present study lays on the effort to involve populations in the participatory assessment of their oil palm growing landscape in order to better understand the issues. Also analyzing water pollution that have not really encountered in the documents consulted. The study has equally focused on the impacts of the institutional side, little criticized in previous research in Africa, because the laws on the land are enacted by the governments which do not often hold the customary laws of the populations who are the first occupants of these territories. Indeed, already poorly organized, they are not sufficiently taken into account in sectorial policies. With regard to the environment or more specifically ecology, it is necessary to limit deforestation and the pollution induced by the palm oil sector through energetic measures, because we are witnessing a permanent granting of concessions (Greenfil SA allotted in 2014 and CAMVERT in 2019) for oil palm despite warnings and actions from environmental organizations like WWF and Rainforest. Also, it becomes necessary to respect the sustainability values, improve the agricultural yields and the livelihood, contribute to local development, and protect High Conservation Value (HCV) areas around the oil palm belt as well as preserving the environment.

Methodologically, the techniques used (transects and quadrates for biodiversity assessment, carbon assessment, remote sensing, landscape approach) without forgetting the surveys carried out with 40 students in the field made it possible to familiarize them with the impacts practices. A database has been established on the socioeconomic determinants of oil palm. In addition, the populations were made aware of how to take their landscape into account and questions of spatial justice.

The limitation aspects of this study rely on the links between climate change and oil palm plantations which have been little addressed. The same is true of the systematic census of animal species in oil palm cultivation areas. Nevertheless, the high academic contributions of the study is linked to multidisciplinary team invested (biologist, botanist, agronomist, ecologist, geographer, economist) to lead to the understanding of the socio-spatial and socio-economic and ecological changes that have occurred in Cameroon for about 30 years. The team was able to articulate questions of environmental and climatic spatial justice (subject of a current thesis) and validate the theory of the anthropocene. Practically, the impact analysis carried out reveals the need to review cultivation techniques and even agricultural policies in Cameroon, particularly the national oil palm strategy, which stills pending. A special attention is to be paid to several aspects of land management methods, availability of seeds and plant material, technical support for small growers, and awareness of the challenges of sustainable development and biodiversity conservation.

Author details

Tchindjang Mesmin[1*], Ludovic Miaro III[2], Fideline Mboringong[3], Gilles Etoga[3], Eric Voundi[1] and Emmanuel Pierre Jonathan Ngom[4]

1 Department of Geography, The University of Yaoundé 1, Yaoundé, Cameroon

2 Regional Food System & Green Growth, WWF ROA Yaoundé Hub - Panda House, Yaoundé, Cameroon

3 WWF ROA Yaoundé Hub - Panda House, Yaoundé, Cameroon

4 Ministry of Agriculture and Rural Development (MINADER), Yaoundé, Cameroon

*Address all correspondence to: mtchind@yahoo.fr; tchindjang.mesmin@gmail.com

References

[1] Bahuchet, S. & Betsch, J.-M., (2012). L'agriculture itinérante sur brûlis, une menace sur la forêt tropicale humide ? *Revue d'ethnoécologie* [En ligne], 1 | 2012, mis en ligne le 30 novembre 2012, URL: http://journals.openedition.org/ethnoecologie/768

[2] Barral S. (2017). Déforestation et palmier à huile: Diversité des productions et réception de la critique. *Savoir/Agir* 2017/1 (N° 39), pp. 110 à 115.

[3] Carlson, K. M., Curran, L. M., Asner, G. P., Pittman, A. M. D., Trigg, S. N., & Marion Adeney, J. (2013). Carbon emissions from forest conversion by Kalimantan oil palm plantations. *Nature Climate Change*, *3*(3), 283-287. https://doi.org/10.1038/nclimate1702

[4] IPCC, (2007). *Climate Change 2007: Synthesis Report. Contribution of Working Groups I, II & III to the Fourth Assessment Report of the Intergovernmental Panel on Climate Change* [Core Writing Team, Pachauri, R.K and Reisinger, A. (eds.)]. IPCC, Geneva, Switzerland, 104 p.

[5] Tchindjang, M., Saha, F., Levang, P., Voundi, E., Njombissie Petcheu, C. I. & Minka, F. (2016). Palmeraies élitistes et villageoises et développement socioéconomique dans la Sanaga Maritime: impacts, conséquences et perspectives. Revue Scientifique et Technique Forêt et Environnement du Bassin du Congo, Volume 7. pp37-52.

[6] Zhang, Q. Justice, C., Desanker, P. & Townshend, J. (2002). Impacts of simulated shilfting cultivation on deforestation and the carbon stocks of the forests of Central Africa. Agriculture, Ecosystems & Environment, vol. 90, n° , pp. 203-209.

[7] Bouron J-B., (2018). Étudier les relations environnement-sociétés à partir du cas de l'huile de palme à Bornéo. *Géoconfluences* http://geoconfluences.ens-lyon.fr/informations-scientifiques/dossiers-thematiques/changement-global/savoir-faire/huile-de-palme-et-deforestation

[8] CIRAD, (2012). Huile de palme : les idées reçues ont la peau dure. CIRAD http://www.cirad.fr/

[9] Carrere, R. (2013). *Oil palm in Africa. Past, present and future scenarios*. WRM, N° 15, 79p.

[10] Hoyle D, Levang P. (2012). *Oil palm Development in Cameroon*. An ad hoc working paper. Yaoundé, Cameroon. WWF, IRD and CIFOR.

[11] FAOSTAT (2018) http://www.fao.org/faostat/en/#search/Palm%20Oil

[12] Bogdan, V. B. (2018). *Le mythe de l'huile de palme 100% durable: Les limites des initiatives volontaires: le cas de la RSPO et de l'Alliance belge pour une huile de palme durable*, https://www.fian.be/IMG/pdf/dospalmoliefr-1217-lrnb.pdf

[13] Levang, P., Sevestre, D., Ndjogui, T.E. & Léonard, E., (2015). Accaparement des terres ou développement ? Les "élites" du palmier à huile. La Revue Foncière, (**6**), 32-38. ISSN 2417-6001.

[14] USDA, (2020) http://www.indexmundi.com/agriculture/?commodity=palmoil&graph=production

[15] Ndjogui, T. E, Nkongho, R. N., Rafflegeau, S., Feintrenie, L. & Levang, P., (2014). *Historique du secteur palmier à huile au Cameroun*. Document occasionnel 109. CIFOR, Bogor, Indonésie, 55 p.

[16] Tchindjang M. (2017). *Étude sur l'impact environnemental des palmeraies villageoises/élitistes sur la déforestation dans les paysages de la Sanaga Maritime*

et du bassin du Ndian: cas des arrondissements de Ngwéi et d'Ékondo Titi. Rapport du WWF, OPAL et CIFOR, 174p.

[17] Tchindjang, M., Levang, P., Saha, F., Voundi, E., Njombissié, Petcheu, I. C. (2015). Impact et suivi par télédétection du développement des plantations villageoises de palmiers à huile sur le couvert forestier au Cameroun : cas de la Sanaga Maritime (3°20-4°40N & 9°30-11°20E). Rapport *PALMFORCAM Principal et Rapport annexes;* Yaoundé.

[18] Ndjogui, T. E, (2018). *Elites urbaines et dynamiques socio spatiales et environnementales de l'élæiculture dans le département de la Sanaga Maritime, Région du Littoral, Cameroun*. Thèse de Doctorat PhD en Géographie, Université de Yaoundé 1, 472 p.

[19] Leduc, G. A. & Raymond, M., (2000). *L'évaluation des impacts environnementaux : Un outil d'aide à la décision.* Éditions MultiMondes, 403 p.

[20] Akmel Meless S. (2018). Impact Socio-Environnemental De L'exploitation Du Palmier À Huile (Elaeis Gineensis) En Lodjukru Dans La Région De Dabou (Côte d'Ivoire). *European Scientific Journal,* Vol.14, No.3, pp. 324-341 ISSN: 1857-7881 (Print) e - ISSN 1857-7431 Doi: 10.19044/esj.2018.v14n3p324 URL:http://dx.doi.org/10.19044/esj.2018.v14n3p324

[21] IUCN, (2018). *Oil Palm and Biodiversity – A Situation Analysis*. Gland, IUCN Oil Palm Task Force. doi: 10.2305/IUCN.CH.2018.11Kubitza, C., Krishna, V.V., Alamsyah, Z. & Qaim, M. (2018). The economics behind an ecological crisis: Livelihood effects of oil palm expansion in Sumatra, Indonesia. *Hum. Ecol.* 46:107-116

[22] Barthel, M., Jennings, S., Schreiber, W., Sheane, R., Royston, S., Fry, J, Leng Khor, Y. & McGill, J., (*2018*) *Study on the environmental impact of palm oil consumption and on existing sustainability standards.* Final Report and Appendices. European Commission, DG Environment 3Keel and LMC International.

[23] Genetet, A., (2018). Enjeux sanitaires et environnementaux de l'huile de palme. *Les notes scientifiques de l'Office Parlementaire d'évaluation des choix scientifiques et technologiques*, Notes N°7, 8p.

[24] Carlson, K. M., Heilmayr, R., Gibbs, H. K., Noojipady, P., Burns, D., Morton, D. C., et al. (2018). Effect of oil palm sustainability certification on deforestation and fire in Indonesia. Proc. Natl. Acad. Sci. U.S.A. 115, 121-126. doi: 10.1073/pnas.1704728114

[25] Ordway, E.M., Naylor, R.L., Nkongho Ndip, R. & Lambin, E.F. (2019). Oil palm expansion and deforestation in Southwest Cameroon associated with proliferation of informal mills. Nature Communications, 10:114 https://doi.org/10.1038/s41467-018-07915-2

[26] Meijaard, E., Brooks, T., Kimberly, M., Carlson, K.M., Slade, E.M., Garcia, J. et al, (2020). The environmental impacts of palm oil in context. Nature Plants, vol 6, 1418-1426.

[27] Meijaard, E. & Sheil, D. (2018). The Moral Minefield of Ethical Oil Palm and Sustainable Development. *Frontiers in Forests and Global Change* May 2019 | Volume 2 | Article 22 www.frontiersin.org

[28] Qaim, M., Kibrom, Sibhatu T., Siregar, H. & Grass, I. (2020). Environmental, Economic, and Social Consequences of the Oil Palm Boom. Annual Review of Resource Economics. 2020. 12:321-344. https://doi.org/10.1146/annurev-resource-110119-024922.

[29] Qaim et al, (2020) was already referenced at number 19

[30] Ahmed A, Dompreh E, Gasparatos A. 2019. Human wellbeing outcomes of involvement in industrial crop production: evidence from sugarcane, oil palm and jatropha sites in Ghana. PLOS ONE 14:e0215433

[31] Bou Dib, J., Krishna, V.V., Alamsyah, Z. & Qaim, M., (2018). Land-use change and livelihoods of non-farm households: the role of income from employment in oil palm and rubber in rural Indonesia. Land Use Policy 76, 828-838.

[32] Edwards, R. B. (2019). "Export agriculture and rural poverty: evidence from Indonesian palm oil," in Working Paper (Brisbane, QLD).

[33] Kubitza, C, Krishna, V.V., Urban, K., Alamsyah, Z. & Qaim, M., (2018). Land property rights, agricultural intensification, and deforestation in Indonesia. Ecol. Econ. 147:312-321

[34] World Bank (2019). Cacaoland: How to transform Côte d'Ivoire. Economic Update for Côte d'Ivoire. Report, Côte d'Ivoire.

[35] Naylor, R. L., Higgins, M. M., Edwards, R. B., and Falcon, W. P. (2019). Decentralization and the environment: assessing smallholder oil palm development in Indonesia. Ambio. 1-14. doi: 10.1007/s13280-018-1135-7

[36] Santika, T., Wilson, K.A., Budiharta, S., Law, E.A., Poh, T.M., et al. (2019). Does oil palm agriculture help alleviate poverty? A multidimensional counterfactual assessment of oil palm development in Indonesia. World Dev. 120:105-117.

[37] Abrams, J., Pischke, E.C., Mesa-Jurado, M.A., Eastmond, A., Silva C.A., Moseley, C. (2019). Between environmental change and neoliberalism: the effects of oil palm production on livelihood resilience. Soc. Nat. Resour. 32, 548-565

[38] Balde B, Diawara M, Rossignoli C, Gasparatos A. 2019. Smallholder-based oil palm and rubber production in the forest region of Guinea: an exploratory analysis of household food security outcomes. Agriculture 9 :41

[39] Green Livelihoods Alliance. (2019). Impacts and implications of oil palm on landscapes and livelihoods in Uganda's Lake Victoria Islands—an overview of recent research. Policy Brief, July. Green Livelihoods Alliance,Wageningen, Neth

[40] Castellanos-Navarrete, A., Tobar-Tomás, W.V. & López-Monzón, C.E. (2019). Development without change: oil palm labour regimes, development narratives, and disputed moral economies in Mesoamerica. J. Rural Stud. 71:169-180

[41] Hamann, S., (2018). Agro-industrialisation and food security: dietary diversity and food access of workers in Cameroon's palm oil sector. Can. J. Dev. Stud. 39, 72-88

[42] Andrianto, A., Komarudin, H., & Pacheco, P. (2019). Expansion of oil palm plantations in Indonesia's frontier: problems of externalities and the future of local and indigenous communities. Land 8:56

[43] Beyer, R.M. Durán, A.P., Rademacher, T.T., Martin, P., Sharon, C. T. Brooks, E., Coomes, D., Donald, P.F, Sanderson, F.J., (2020). The environmental impacts of palm oil and its alternatives. *bioRxiv* https://doi.org/10.1101/2020.02.16.951301 doi: preprint

[44] World Bank, (2018) World Development Indicators: Deforestation and biodiversity http://wdi.worldbank.org/table/3.4 [Retrieved on 12 June 2020].

[45] Tchindjang, M., (2021) *Identification et cartographie des aires à hautes valeurs de conservations dans certains paysages de production.* WWF, Yaoundé.

[46] Bennett, A., Ravikumar, A., McDermott, C. & Malhi, Y., (2019). Smallholder oil palm production in the Peruvian Amazon: rethinking the promise of associations and partnerships for economically sustainable livelihoods. Front. For. Glob. Change 2:14

[47] Schoneveld, G.C., Van der Haar, S., Ekowati, D., Andrianto, A., Komarudin, H., et al. (2019). Certification, good agricultural practice and smallholder heterogeneity: differentiated pathways for resolving compliance gaps in the Indonesian oil palm sector. Glob. Environ. Change 57:101933

[48] Djouma, S. H., Feintrenie, L., Levang, P., and Nji, A. (2018). Co-designing win-win partnerships between agro-industries and smallholders in the palm oil sector in Cameroon. Biotechnol. Agron. Soc. Environ. 22, 136-151. doi: 10.25518/1780-4507.16522

[49] ISO, (2019). Cacao durable et traçable-Partie 1: Exigences relatives aux systèmes de management de la durabilité du cacao. https://www.iso.org/fr/standard/64765.html

[50] Mantiaba, C-B. (2019). Engagement éthique et certification sur le marché du commerce équitable: le cas de la certification Fairtrade en Côte d'Ivoire », *Éthique publique* [En ligne], vol. 21, n° 1 | 2019, DOI: https://doi.org/10.4000/ethiquepublique.4481

Chapter 15

Oil Palm-Based Cropping Systems of the Humid Tropics: Addressing Production Sustainability, Resource Efficiency, Food Security and Livelihood Challenges

Samuel O. Agele, Friday E. Charles, Appolonia E. Obi and Ademola I. Agbona

Abstract

The oil palm (*Elaeis guineensis* Jacq), is a crop of tremendous food, nutrition and economic importance in the tropics. Weather variability and extremes profoundly impact the establishment, survival and productivity of oil palm on the field. Alleys of palm trees in plantation are cropped with arables during early years following field establishment. Studies were conducted at the Nigerian Institute for Oil Palm Research, Benin City, Nigeria. Oil palm seedlings responded to shading, irrigation and AMF inoculation via enhanced water use efficiency, growth vigor and reduced seedling mortality in the nursery during dry season. Age of oil palm and intercrops of Cassava, Maize and Pepper affected mixture productivity and competitive functions in alleys of 2 to 6 years old oil palm fields. Fertilizers (inorganic/organic) promoted agronomic and physiological efficiencies of N use by alley species. Sole crops had higher N use efficiency compared with the intercrops across the fertilizers.

Keywords: Nursery, shade, irrigation, seedling survival, intercropping, alley, fertilizer, growth resources, competition, use efficiency, rainforest

1. Introduction

The oil palm (*Elaeis guineensis Jacq) (Magnoliophyta: Arecaceae),* is a perennial monocotyledonous plant which belongs to the family *Arecaceae*. It is a crop of the humid tropics (rainforest belt). The pulp and nut that provide palm oil and kernel oil, respectively make oil palm a high yielding oil-producing crop [1, 2]. The production of the oil from oil palm fruit pulp and nut, has made the crop second to soybean oil in terms of world vegetable oil production while its demand is expected to increase in future [1, 3, 4]. The oil palm is an evergreen crop that may be used to conserve the environment while palm oil accounts for 21% and 47% of the global oil and fats production and trade [3].

IntechOpen

Palm oil is grown in many countries across Africa, South America, and Southeast Asia. The global market for palm oil and other products is dominated by Indonesia and Malaysia which collectively account for 84% of global production and other producers include Thailand, Colombia, Nigeria, Guatemala, and Ecuador. In 2018, the world produced 72 million tonnes of oil palm. Indonesia accounted for 57% of this (41 million tonnes), and Malaysia produced 27% (20 million tonnes [5]. Palm produce contributes more than 15% of the non-oil revenue of Nigeria. The industry provides extensive employment opportunities for Nigerians involved in the production, processing, marketing and distribution of both the main raw materials and the downstream products.

Palm oil is a versatile product which is used in a range of products globally. In the broad sense, it is used as **Foods**: over two-thirds (68%) is used in foods ranging from margarine to chocolate, pizzas, breads and cooking oils; **Industrial applications**: 27% is used in industrial applications and consumer products such as soaps, detergents, cosmetics and cleaning agents; and **Bioenergy**: 5% is used as biofuels for transport, electricity or heat. Palm oil and other palm products have always been used for food/pharmaceuticals/industrial and technological purposes [1, 2]. Palm oil is well suited to various food uses, particularly cooking fats and deep-frying oil, and it appears in bakery products, potato crisps and other snacks, and ice-creams and soap, margarine and cooking fat [1, 3]. Lower quality oils are used for non-edible purposes, such as soaps, resins, candles, glycerol, fatty acids, inks, polishing liquids and cosmetics. From oil palm are produced different basic products: crude palm oil (CPO), neutralized palm oil (NPO), refining bleached and deodorized palm oil (RBD), palm olein, palm stearin, palm kernel oil and palm kernel cake or meal [1, 2]. Due to the recent increase in quality and availability, and developments in technologies: refining, fractional and hydrogenation, palm oil has become highly diversified in its uses. Some of these uses are fatty acid manufacture, oleochemicals in general, additives to animal feed stuffs, potato crisp making, bread and cakes. The new compounds produced from palm oil are known collectively as oleochemicals most of these are molecules with different fatty acids attached to various simple functional groups, such as acids, amines or alcohols [1, 3], and include sulfonated methyl esters, polyols and polyurethanes. Several minor constituents of palm oil can be extracted and used separately, such as carotein, vitamin E and sterols.

The rainforest agroecologies of the humid tropics is characterized by wet and dry season transition and variability in seasonal weather conditions. In this zone, the annual rainfall ranging from 1500 to 2000 mm distributed in a bimodal pattern which results in the rainy and dry season. The dry season is a terminal drought situation characterized by inadequate rainfall, soil moisture deficit, high vapor pressure deficit and temperatures and very clear sky (high intensity of solar radiation [6]. Such unfavorable weather condition has been reported as the cause of the massive seedling and fruiting tree mortality in the dry season [7, 8]). It has been reported that high percentage of oil palm seedlings died on the field during the first and second dry season as a result of soil moisture deficit [7]. Although, fruit trees in plantations (cacao, kola, coffee, citrus species and oil palm) are seldom irrigated, despite the challenges of weather variability and extremity of the tropics. Efforts to develop sustainable production systems for oil palm would involve the evaluation of the value of agronomic practices in the amelioration of extreme growing environmental conditions especially the hydrothermal stresses of the dry season on seedling survival on the field in order to attain optimum seedling population in the field.

Climate Change (temperature and rainfall) scenarios for the deciduous and evergreen rainforest zones of West Africa including Nigeria have been variously constructed using process-based methods that rely on the General Circulation

Models (GCM) in conjunction with Simple Climate Models (SCM). The results have indicated projected decline in mean annual rainfall and increases in temperature in year 2020, 2050 and 2080 respectively. The projected climatic changes will exacerbate soil moisture and thermal conditions during the dry season (November to March) and aggravate the vulnerability of crops to such climatic conditions [9]. The changing growing environmental conditions (marginal soils and extreme weather events) impose constraints on cacao growth and productivity. In order to alleviate the constraints imposed by changing growing environmental conditions (marginal soils and extreme weather events) on fruit tree performance. It is imperative to develop climatic-stress adaptive strategies for the fruit tree-based agroforestry systems of the humid tropics in the wake of changing climate/weather conditions (climate change and weather variabilities).

2. The farming Systems of the Humid tropics

Agroforestry involves growing trees in mixtures and arable/food crops and fruit tree species simultaneously on a farm (growing arable crops and fruit tree species together). Alley cropping is an agroforestry technique in which trees are planted in hedgerows, and annuals (arable or fodder) crops are planted in the "alley ways" between the hedge row plants. Alley cropping involves growing short duration trees and shrubs that are compatible with arable or fodder crops. The trees provide other benefits such as reducing erosion, maintaining soil fertility and providing additional income to farmers crop diversity and food security in the early years of tree establishment [10, 11]. The advantages of alley cropping are attributable to improvement of soil quality, increased economic diversity, carbon sequestration, farm yield, resource use efficiencies, and environmental resilience [12, 14].

Intercropping: the simultaneous growing of two or more species in the same field for a significant period of their growth. Such crop combinations has been reported as promising to sustain soil and crop productivity. Intercropping, systems offer crucial ecosystem service that supports food supplies and other livelihood activities. Intercropping practices provide sustainable and stable yields, diversity of flora and fauna and lower risks of crop failure, and implement, sustain and enhance environmental quality, ecosystem services and livelihoods and sustainable landscapes [15, 16]. Intercropping practices are reported to optimize ecological processes including the cycling of nutrients, maintains carbon stocks (sequestration), conservation of soil water, modification of microclimate and reduce soil degradation [11, 15].

The essential features of intercropping/agroforestry systems are intensification in the use of space and time of space, light, water and nutrients. Advantages of alley cropping improves soil quality, biodiversity, carbon sequestration, farm yield, resource use efficiencies and environmental resilience (Bedou et al., 2010). Agroforestry involving alley and intercropping are important features of the farming systems of the tropics (references). Research on intercropping has shown that fruit trees can be intercropped successfully with arable crops during the early stages (1 to 5 years) of establishment [10, 17, 18].

Variable availability of growth resources with ages of trees following its establishment exist, thus the variability in the capture and use efficiencies by hedge row crops and alley crops (component species) [19]. When plants are grown together, interspecific competition may occur in relation to the use of growth resources [13] Such biophysical interactions may be positive and negative biophysical interactions exist between trees and the alley crop species. Positive and negative biophysical interactions exists between trees and alley crops in agroforestry/alley cropping systems. Competitive interactions which occur in agroforestry systems based on resource availability for use by the trees and understory plants [20, 21]. In addition, variable complimentality and compatibility between tree crops and the alley crop species in the fruit tree-based intercropping systems is reported. However, proper management of interactions ensures sustainability of the agroforestry systems [20].

Research has shown that tremendous enhancement of growth, development and yield of crops can be obtained through application of fertilizers [11]. Required, is the enhancement of uptake and utilization efficiency by crops without deleterious effect on yield and ecosystem [22]. Nutrient uptake and use efficiencies in crop production is less than 40% worldwide, improvement in the efficiencies of uptake and utilization of nutrients by crops will enhance yields and conserve the environment [23]. Research has shown that nutrient use efficiency indices are affected by cropping systems, fertilizer types and climate [6, 22, 24]. Thus, it is imperative to improve understanding of the use of fertilizers for enhancement of soil fertility, growth and yields of alley crops in the oil palm-based intercropping system of the rainforest zone of Nigeria.

Literature has reported that for most tree crops, the alleys can be intercropped successfully with arable crops during the early stages (1 to 6 years) of establishment. Species combination involving arable crops in oil palm alley and application of fertilizers may enhance competition for resources among intercropped species. Knowledge is however knowledge on the response of oil palm to the presence of cereals, root/tuber and vegetable crops and fertilizers in its alleys, performance of alley crops as sole and intercrops (Cassava, Maize and Pepper) in an oil palm-based agroforestry system, and effects of organic and inorganic fertilizers on soil, alley crops as well as uptake and use efficiencies of fertilizers in an oil palm-based agroforestry system of the rainforest agroecology of Nigeria. There is justification in giving priority to research to improve insights to the performance of cassava, maize and pepper intercrops in the alley of oil palm of different ages (2 to 6 years) as sources of food and income for oil palm farmers during the early stages of oil palm establishment.

3. Materials and methods

3.1 Experimental site and environmental conditions

Series of experiments were conducted to address the themes of the Book Chapter.

An experiment was conducted to examine the effects of shading, watering regimes and mycorrhizal inoculation on the growth and development of oil palm seedlings in the nursery in the dry season. A series of experiments were conducted at the Nursery and Field (Plantations) of the Nigerian Institute for Oil Palm Research (NIFOR), Benin City, Edo State, Nigeria between 2013 and 2018. The Nigerian Institute for Oil palm Research (NIFOR) is located in the rainforest ecological zone of Nigeria between latitude 06° 33^{1}N and longitude 05° 37^{1}E. Annual rainfall ranges between is 1500 to 2000 mm, average temperature ranges between 28° to 34° centigrade and relative humidity between 54 to 80%.

Age of oil palm and intercrops of Cassava, Maize and Pepper affected mixture productivity and competitive functions in alleys of 2 to 6 years old oil palm fields. Fertilizers (inorganic/organic) promoted agronomic and physiological efficiencies of N use by alley species. The trials were conducted in NIFOR Nursery and Field (Plantations) between 2013 and 2018.

Section A: Studies were conducted to examine the responses of oil palm seedlings to shading, irrigation and AMF inoculation with respect to growth vigor and mortality events in the nursery during dry season. Soil samples were collected from top soil under five-year fallow vegetation and were sieved to remove stones and pebbles. Black polythene bags (1400 cm^2) were filled with 6 kg of the soil media and arranged in rows 90 cm both in the open sun and in the shade. The drip lines were placed along the rows after ten days. Two months old oil palm seedlets/plantlets (*Tenera*) were obtained from NIFOR Pre-nursery and transplanted into the 6 kg pots (black polythene bags:1400 cm^2) were filled with top soil obtained from fallow vegetation (secondary forest soil) and arranged in filed plots. Shades measuring 6 x 6 m was constructed using bamboo sticks and palm fronds were used to cover the top and sides. Gravity drip irrigation system was adopted using 200 liter capacity bucket placed on a 1.5 m wooden stand; drippers were installed to apply water (2 liters per palm application). Thermometers each were installed in the shade and in the open sun for the measurement of air temperature weekly and fortnightly intervals.

4. Soil and plant measurements

Data were obtained on growth parameters, pattern of leaf production and senescence, weekly measurement of soil moisture content using a soil moisture sensor while thermometers were installed under shade and in no shade condition to measure soil and air temperatures. At the end of the experiment, 10 plants were gently uprooted and the root washed and shade dried for root measurements. Mycorrhizal Colonization of Roots and Mycorrhizal Spore Count were determined using standard methods [25].

4.1 Irrigation strategy

Oil palm seedlings were drip-irrigated weekly and fortnightly using gravity-drip irrigation system to apply 2 liters of water per plant at each irrigation via point source emitters (2 l/h discharge rate) which were installed on laterals per row of seedlings. Irrigation buckets were suspended on 1.5 m high stakes to provide the required hydraulic heads [26, 27]. There was a two-day pre-irrigation treatment (1.5litres/day) following oil palm seedling transplanting, and thereafter, the weekly and fortnight irrigation treatments were imposed.

Water requirement (WR) was determined using the relation:

$$WR = A \times B \times C \times D \times E \tag{1}$$

where: WR = Water requirement (l per day/plant) A = Open Pan evaporation (mm/day) B = Pan factor (1.0, 0.7 and 0.5), C = Spacing of plant (m^2), D = Crop factor (Crop coefficient (Kc) for oil palm seedling: initial (0.43) were obtained from Allen et al. [28].

The total water requirement (TWR) was obtained using the relation:

$$TWR = WR \times \text{No.of Plants} \tag{2}$$

Maximum allowable deficit (MAD) for oil palm was assumed as 50% of available water storage capacity of the soil (AWC).

Irrigation water requirement is determined using average season wise pan evaporation data for the area. Pan Evaporation (EPan) data used for the experiment were obtained from measurement using a Class-A Pan (121 cm in diameter and 25.5 cm in depth) from the Meteorological Station, Department of Meteorology & Climate Science, Federal University of Technology, Akure, Nigeria located near the plots.

The actual evapotranspiration (ETc) of oil palm seedlings the irrigation regimes was calculated with the water balance equation (Eq. (1)) [26].

$$ET + I + P + \Delta S – Dp - Rf \quad (3)$$

where, ET, is actual crop evapotranspiration (mm); I, the amount of irrigation water applied (mm); P the precipitation (mm); ΔSW, changes in the soil water content (mm); Dp, the deep percolation (mm); Rf, amount of runoff (mm). Since the amount of irrigation water was controlled, deep percolation and run off were assumed to be negligible.

Soil water measurements were taken throughout the growing season using the gravimetric method.

The volume of water required per plant (irrigation requirement, I_R) was estimated as:

$$I_R = ET_{peak} * area/crop/E_n \quad (4)$$

where En is emitter uniformity for drip irrigation system (0.94) and area per crop.

Peak evapotranspiration (ET_{peak}) rate for the crop under drip irrigation treatment was estimated as:

$$ET_{peak} = ETo * P/85 \quad (5)$$

where ET_{peak} is peak evapotranspiration rate for the month or period, ETo is the reference evapotranspiration, for the month/period (e.g. 5.1 mm/day), P is the proportion of total land area covered by the crop leaf area (cm) which is assumed 80% (after [26]).

Crop evapotranspiration (ETa) was also calculated using data obtained from FUTA in the formula of Doorenbos and Pruitt [29] and [28] in the form:

$$ETa = KcETo \quad (6)$$

where ETo is potential evapotranspiration and Kc is the crop coefficient.

Weather variables at site of experiment during crop growth cycle (soil and air temperatures, vapor pressure deficit (vpd), solar radiation, wind speed will be monitored from Meteorological Observatory, 500 m from site of experiment). Data collected were subjected to analysis of variance (ANOVA) while significant treatment means were separated using the Duncan Multiple Range Test (DMRT).

Section B: Experiments were conducted between 2016 and 2017 cropping seasons to examine the effects of age of oil palm in plantation on the growth, competitive interactions and mixture productivity of cassava, maize and pepper in oil palm-based strip intercropping system in the rainforest of Nigeria. The studies were conducted on 2, 3, 4 and 5 years old oil palm fields for oil palm field which were established in the fields during 2014, 2015, 2016 and 2017 using *Tenera (hybrid)*.

The experiment was 3 x 3 factorial combination of ages (2, 3, 4 and 5 years) of oil palm plantations and 3 species of arable crops arranged in a split-plot design interrow spaces (alley) between oil palm plans constitute the main plot and arable crops species as sub-plot treatment. Treatments were replicated 3 times. The planting space for oil palms was 9 x 9 m triangular with eight (8) stands per field plot. The plots were spaced by 1 m between plots and replicates (inter row) and 1 m at the borders. Strip intercropping system was adopted in the experiments. Spacing cassava, maize and pepper was 1 x1m given a plant population of 141 plants per plot.

4.2 Data collection

Oil palm data collections include number of leaves, canopy extent and number of fresh fruit bunch (FFB). These data were collected quarterly. The canopy extent was recorded in meters and canopy spread calculated using the multiplication values of the palm measured in two dimensions of 'North–South spread' and 'East–West spread' and canopy volume of oil palm trees. Oil palm canopy volume and ground coverage by Tripathy et al. [30] and canopy spread (Cs) was estimates as:

$$Cs = Ns \text{ x } EW \tag{7}$$

where: Cs = canopy spread, NS = North–South, EW = East–West.

Oil palm yield traits were sampled which include the number of fresh fruit bunches (FF) and weight of fresh fruit bunch (FBB) and total bunch yield per treatment plots. The growth and yield data were collected from ten tagged plants randomly selected from the experimental plots. Data collected include number of leaves, canopy extent and number of fresh fruit bunch (FFB). Palm yield traits: yield per palm, number of fresh fruit bunches (FF) and weight of fresh fruit bunch/palm and total palm yield/experimental plot. At harvesting, cassava tuber weight, maize seed weight and pepper fruit weight for each individual plant was according to estimate – the intra- land variability. Biomass yield, dry above ground biomass at harvest and final yield were determined from the sampled plants.

4.3 Indicators of crop mixture productivity and competitive interaction

Different measures or indices of productivity have been developed to determine the productivity of crops in crop mixtures. These indices include relative yield, relative yield total and land equivalent ratio, aggressivity. Relative yield is the biomass or yield of a species in mixture or intercropping expressed as a ratio of its yield in monoculture [31]. Relative yield total (RYT) is the sum of the relative yields of the species in mixture expressed as a ratio of its yield in monoculture [31]. Land equivalent ration (LER) is an indication of biological efficiency of intercropping in use of environmental resources compared to the sole crop [32]. The percentage land saved from intercropping was estimated using the formular described by Willey [14]. Aggressivity is a measure of competitive relationships between two crops in mixed cropping [21] and an important competition function to determine the competitive ability of a crop when grown in association with another crop [33]. This index compares the yields between intercropping and monoculture, as well as their respective land occupancy [34, 35].

Section C: Studies were conducted to examine the effects of fertilizer (NPK compound fertilizer, poultry manure and pelletized organic fertilizer) and age (3, 4 and 6 years) of oil palm on nutrient uptake and use efficiencies of strip intercropped cassava, maize and pepper in an oil palm-based intercropping system. Cassava, maize and pepper were strip-intercropped in the alleys of 3, 4 and 6 years oil palm fields. The fertilizers (NPK, Ferti plus and poultry manure) were respectively applied at the rate of 67.5, 168.75 and 337.5 g/plant) as determined by the soil test.

The indices of N uptake and agronomic and physiological efficiencies of N use were calculated using the procedures described in the literature [36–38]. Nutrient uptake refers to the ability of crop to extract or absorb nutrients from the soil. The uptake of nitrogen was calculated as the product of the measured N concentrations in shoot biomass and reproductive structures (fruit/seed/tuber) the weight of the biomass (shoot and reproductive structures) [37, 38]. The proportion of total plant N partitioned to the shoot is called the N harvest index (NHI). It is also defined as the percentage of grain N uptake to total plant N uptake [37]. Nutrient Use Efficiency (NUE): This is a term used to indicate the ratio between the amount of fertilizer N removed from the field by the crop and the amount of fertilizer applied. NUE is expressed in several ways as the efficiency of conversion of nutrient taken up by the plant into crop biomass. This ratio describes the efficiency of N fertilizer utilization in crop production.

Agronomic Efficiency (AE) is calculated as the unit of yield increase per unit nutrient applied. It reflects the direction of production impact of applied fertilizer and also relates to economic return. The calculation of AE requires knowledge of yield without nutrient input, so is only known when research plots with zero nutrient is been implemented on the farm [11]. AE is expressed as the efficiency of conversion of nutrient taken up by the plant into crop biomass.

Physiological Efficiency of N use (PE), is defined as the yield increase in relation to the increase in crop uptake of the nutrient in above ground parts of the plant (Dobermann [36]. Similar to AE, it needs a plot without application of the nutrient of interest to be implemented, and requires measurement of nutrient concentrations in crop biomass (shoot and reproductive structures).

The Apparent Recovery Efficiency (ARE) is the ratio of nutrient uptake to nutrient applied, it is also defined as the difference in nutrient uptake in above ground parts of the plant between the fertilizer treated and untreated crop relative to quantity of nutrient applied. Nutrient utilization efficiency (NUE) is calculated as the product of physiological and recovery efficiency. It is calculated based on the method described by Dobermann [36]. Internal Utilization Efficiency (IE) is defined as the yield in relation to total nutrient uptake. It varies with genotype, environment and management. It is an indication of the efficiency of internal nutrient conversion, which may be affected by other stresses (deficiencies of other nutrients, drought stress, heat stress, mineral toxicities, pest etc.) [36]. The total factor productivity (TFP) relates an index of output to a composite index of all inputs while Partial Factor Productivity measure relates output to a single input. Partial Factor Productivity (PFP) is a simple production efficiency expression, calculated in units of crop yield per unit of nutrient applied [36].

Data analysis: Data collected were subjected to analysis of variance (ANOVA) and significant treatment means were separated for 5% ($P < 0.05$) probability level.

5. Results and discussion

Section A: Response of growth and development of oil palm seedlings to shading, irrigation regimes and mycorrhizal inoculation in the dry season in the nursery.

Treatments were shading and no-shading (open sun), 7- and 14- day irrigation intervals and mycorrhizal inoculation or non-inoculation. Across irrigation treatments, un-shaded oil palm seedlings had lower biomass weights (leaf, frond and shoot) while the treatments significantly affected plant height and frond length from 2 to 20 weeks after planting (**Table 1**). The seedlings irrigated fortnightly produced longer roots compared with the unshaded and weekly irrigated (**Table 2**). Shading and weekly irrigation significantly enhanced soil moisture contents and

Treatment	Root length (cm)	Root weight (g)	Shoot weight (g)	Leaf weight (g)	Frond weight (g)	Number of roots
Shaded	39.06b	10.30a	94.17a	40.76a	35.42a	10.15a
Unshaded	45.10a	11.46b	57.83b	23.41b	20.35b	11.14a
LSD (0.05)	6.70	1.10	18.43	5.54	6.80	1.10

Table 1.
Effect of shading on growth parameters of oil palm seedlings.

Treatment	Root length (cm)	Root weight (g)	Shoot weight (g)	Leaf weight (g)	Frond weight (g)	Number of roots
Weekly	44.81b	12.61b	91.28a	39.15a	35.10a	11.37a
Fortnightly	56.72a	16.64a	65.56b	32.54b	28.40b	12.24a
LSD (0.05)	3.25	2.08	10.31	4.98	4.44	1.63

Table 2.
Effect of irrigation on growth parameters of oil palm seedlings.

seedling water use efficiency compared with the unshaded. Shaded seedlings under weekly irrigation consumed more water compared with fortnight irrigation (**Table 3**). Shading and weekly irrigated seedlings combined with mycorrhizal innoculation were more vigorous compared with the unshaded and non-inoculated (**Table 4**). Mycorrhizal inoculation enhanced oil palm seedling growth while weekly irrigation produced more vigorous seedlings. Shade conserved soil moisture whie unshaded had lower soil moisture contents across period of observation (**Figure 1**). The unshaded oil palm seedlings had significantly higher water use compared with the shaded for both weekly and fortnight irrigation (**Figure 2**). Shade combined with AMF inoculation enhanced vigor of growth across the irrigation treatments, and reduced mortality of oil palm seedlings in the nursery in the dry season (**Figure 3**).

Section B1: ***Effects of age of oil palm on the growth and yields of alley crop species.***

The performance of strip intercropped mixtures of Cassava, Maize and Pepper sown in the alleys of 2 to 6 years old oil palm fields were investigated. In the alleys of oil palm ages 2 to 6 years, sole crops of cassava, maize and pepper out-yielded the respective intercrops (**Figures 4–6**). Among oil palm ages, yields of the intercrops were higher and similar for the 3 and 4 years and lower for 6 years old fields. The ages of oil palm affected fruit yield and yield components of pepper (**Figure 4**). The weight of fruits of pepper produced was significantly influenced by ages of oil palm. The weight of pepper fruits harvested for 2 years old oil palm was higher followed by 3 and 5 old oil palm. The weight of fruits of pepper obtained for 5 years old oil palm reduced significantly. Seed yield of pepper followed the same trend for number of fruits and fruits yield. The number and weight of fruits harvested decreased as the age of oil palm increased. There were significant differences in the yield components of cassava in oil palm alleys.. Age of oil palm fields also affected number of tubers, tuber yield and harvest index (HI), significant differences were found between values of these variables for 3, 4 and 6 years oil palms (**Figure 5**). Maize cob and seed weight and number of seeds/cob were significantly different among the ages of oil palm to which the maize was alley cropped. The weight of cob was high for 2 and 3 years old oil palm. There was reduction in value for 5 years old oil palm. The seed weight of maize alley in oil palm of 2 years old recoded high compared to other ages of oil palm. The highest seed yield were obtained for 2 and

Shade	Irrigation	Plant height (cm)	Number of fronds	Frond length (cm)	Root length (cm)	Root weight (g)	Shoot weight (g)	Leaf weight (g)	Frond weight (g)	Number of roots
Shaded	Weekly	62.2	9.38	51	43	13.15	109.3	47.57	39.1	11.62
	Fortnightly	63.1	8.88	54.5	42.7	10.51	94.3	40.55	37.75	12.62
Unshaded	Weekly	43.3	9.38	34.3	40.8	12.07	61.3	25.89	22.13	12.12
	Fortnightly	40.3	9.38	33.4	55	12.61	64.8	26.13	22.65	12.25
LSD		6.42	ns	6.97	8.66	2.13	17.41	6.80	7.23	ns

Table 3.
Interaction of shade and irrigation on growth variables of oil palm seedling.

Treatment	Root length (cm)	Root weight (g)	Shoot weight (g)	Leaf weight (g)	Frond weight (g)	Number of root
Inoculated	42.35b	13.42a	89.04a	37.24a	32.73a	12.81a
Non-inoculated	48.42a	10.76a	75.79a	33.82b	28.08b	11.50a
LSD (0.05)	5.35	ns	ns	3.80	3.98	ns

Means in a column with the same letter (s) are not significantly different by DMRT (P = 0.05).

Table 4.
Effect of mycorrhizal inoculation on growth parameters of oil palm seedlings.

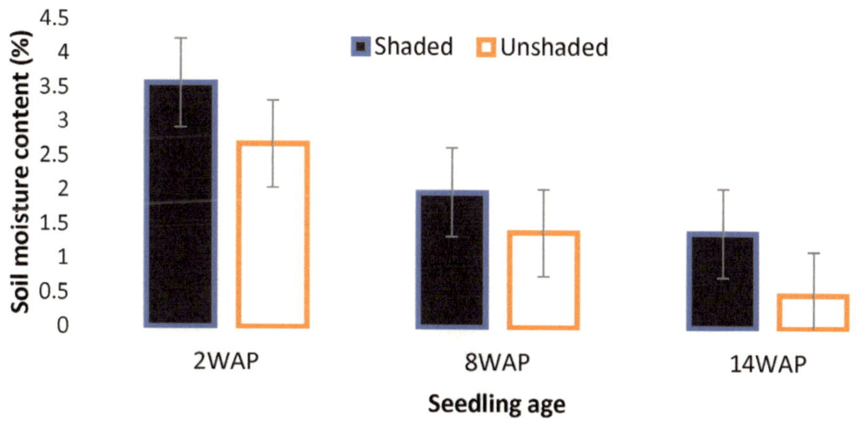

Figure 1.
Effect of shading on soil moisture contents.

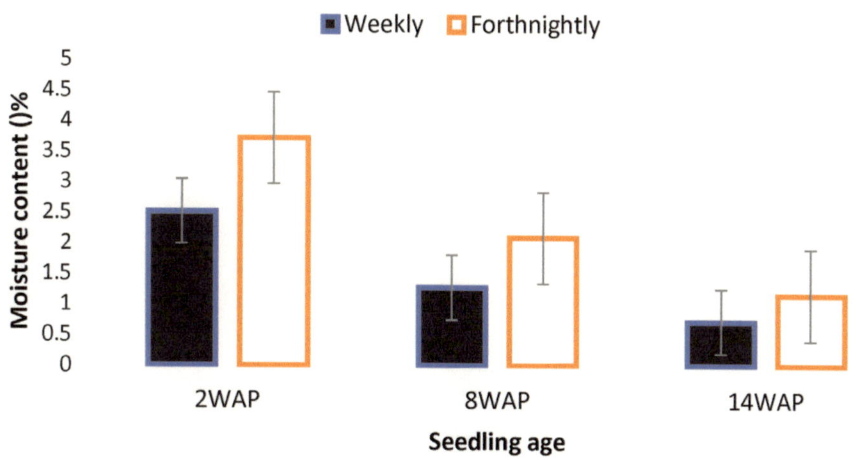

Figure 2.
Effect of irrigation on soil moisture contents.

3 years oil palm while seed yield of maize sown in the alley of 5 years old oil palm had low yield. While, the shoot biomass yield, cob weight, number of seed, seed weight, seed yield, canopy spread, air and soil temperature were significantly affected by the ages of oil palm (**Figure 6**). Ages of oil palm influenced yield and its components of the alley crop mixtures. The differences in yield of the different ages (2, 3 and 5) can be attributed to low vigorous growth as a result of low nutrient

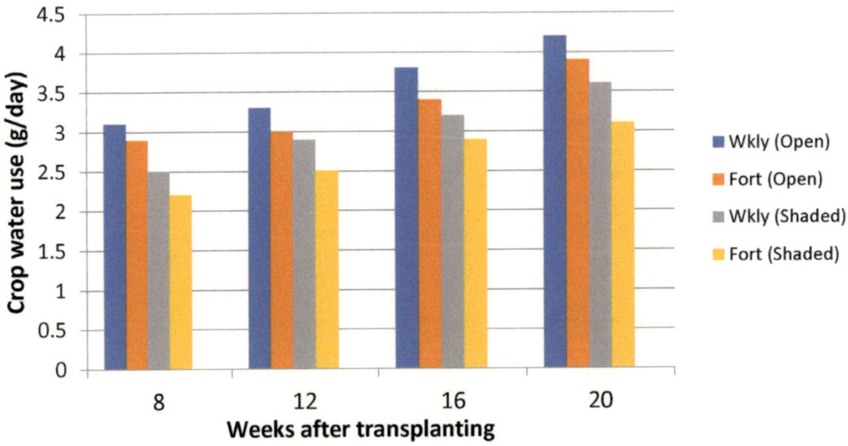

Figure 3.
Effects of irrigation and shading on water use of oil pam seedlings.

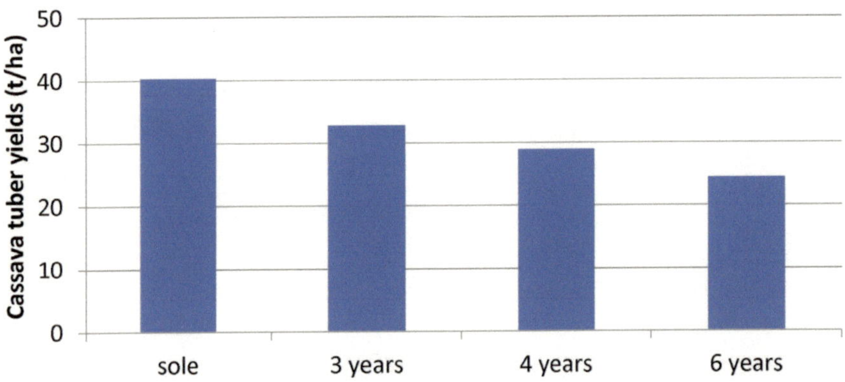

Figure 4.
Effect of age of oil palm field on cassava tuber yields.

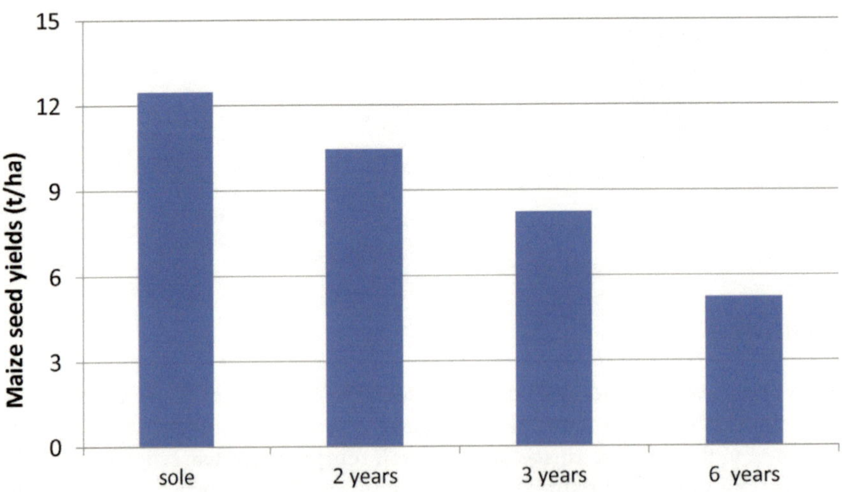

Figure 5.
Effect of age of oil palm on maize seed yields.

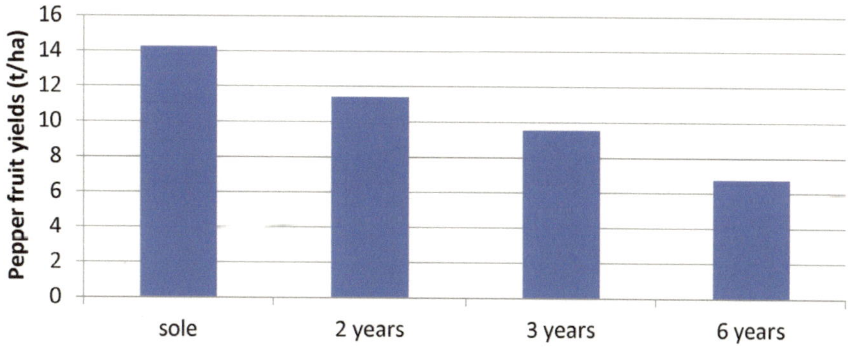

Figure 6.
Effect of age of oil palm on pepper fruit yields.

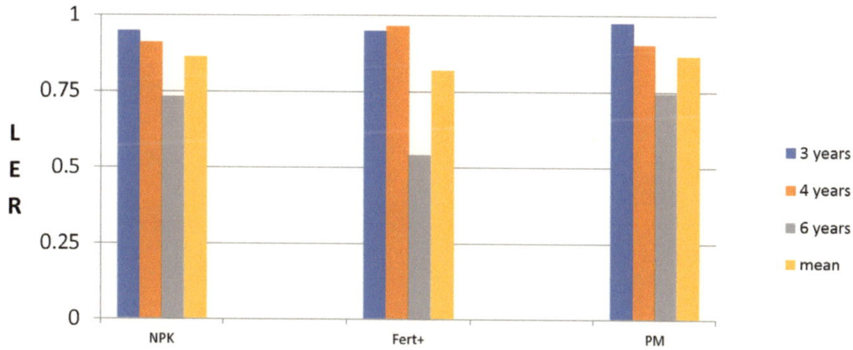

Figure 7.
Effect of age of oil palm on LER of maize.

content of the soil which led to poor vegetative growth. Contrary to the results obtained from cassava and maize, growth and fruit yields of pepper differed. Height and leaf development were not significantly different among the ages of oil palm fields despite the canopy differentials among palm ages. Although, root and shoot biomass and pepper yields of pepper were better for young palms (2 and 3 years) the differences were not significant for root and shoot biomass and pepper yields of pepper. The effect of age of palm was not significant for harvest index (HI), values were close for the different ages of oil palm.

***Section B2: Effects of age of oil palm in plantation on* competitive interactions and mixture productivity of alley intercrops of cassava, maize and pepper.**

The effects of strip intercropped species in the alleys of 2 to 6 years old oil palm fields on competitive interactions and mixture productivity were investigated. Compared with their sole crops, the intercrops had yield advantages and higher land-use efficiencies (LER >1: **Figures 7** and **9**). The land equivalent ratio (LER) expresses the magnitude/extent of yield advantage of crop mixtures over sole crops, thus represents the land required for sole crop to produce the total yield by the component crops in intercropping. The intercrop mixtures outyielded the sole crops of the component crops (cassava,maize and pepper). The magnitudes of the LER confirms larger percentage land saved by crop mixtures over the respective sole crops. This observation supported reported advantage of intercropping in terms of productivity per unit land area, use of growth resources, in addition to greater percent land saving. Age of oil palm fields significantly affected LER, lowest LER was obtained for 6 years old palm field and values were close for 3 and 4 years old fields (**Figure 7**). Relative yield (RY) is defined as the sum of relative yields of the

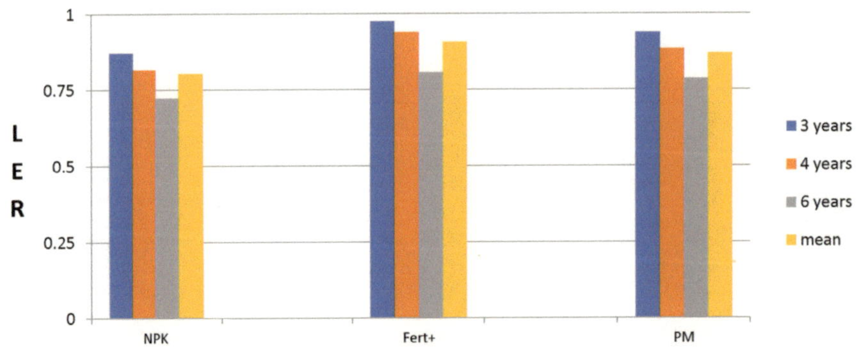

Figure 9.
Effect of age of palm and fertilizers on LER of pepper.

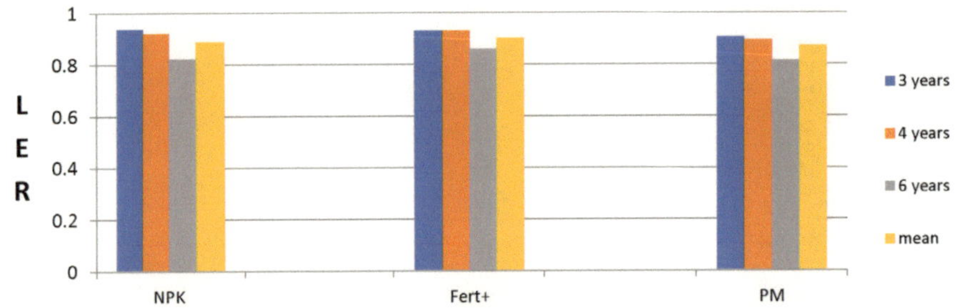

Figure 8.
Effect of age of palm on LER of cassava.

species in mixture expressed as a ratio of its yield in monoculture (**Figure 8**). The values of RY of the intercrops infer that the system is advantageous and complementary being able to utilize growth resources efficiently and were able to stimulate the growth of one another when grown in the alley of oil palm (2 to 6 years after establishment). The relative yields of component crops in the mixture over were lower compare with their counterpart sole crops. The competitive functions were computed in the form of aggressivity (Agg) while the relative species competition was therefore evaluated as competitive ratio (CR) which is a measure of the times by which the component crops are more competitive than the other. Aggressivity values were high (> 0.25) across intercropped compared to sole crops (**Figure 9**). The ages of oil palm affected competition for growth resources (space and soil nutrients) and the productivity of crop mixtures. The intensity of aggressiveness was almost similar for 3 year oil old palm field, differences were found for 4 and 6 years old. Strip intercropping of cassava, maize and pepper displayed yield advantages with high relative yield total (RYT) and land equivalence (LER), and low aggressivity in the alleys of 2 to 6 years old oil palm. These observations denote complementarity and low competitive interactions among intercrops, and the significant advantage of intercropping in terms of productivity per unit land area and greater percentage land saved under intercropping (**Tables 5–7**).

Various studies have established that cropping the alleys of oil palm with arable crops is successful for at least the first 5 years of field establishment of oil palm [2, 35]. For oil palm of 2 to 6 years old, palm leaves and canopy extent had fully developed to completely close and cover soil surface and canopy overlap over the alleys and consequent reduction in space and solar radiation interception and transmission. This is in line with [2, 35] who reported that 1, 2 and 3 years of cropping after planting produced no subsequent deleterious effects up to 16 years

Plant variables	2 years	3 years	5 years	Means	LSD (0.05)
Number of leaves	148.2	157.1	218.3	174.3	6.49
Leaf area (cm2)	18.9	23.2	23	21.7	1.53
plant height (cm)	65.3	63.2	62.9	63.8	1.21
Root biomass (g)	13.3	13.2	11.8	12.8	0.97
Shoot biomass (g)	74.2	73.5	70.5	72.7	1.46
Number of fruits/stand	66.8	65.8	63.4	65.4	1.38
Fruit yield/stand(g)	413	39.1	36.6	39.3	1.61
Fruit weight (g)	411.6	390.9	364.3	388.9	5.12
Harvest index	4.7	4.4	4.2	4.5	0.54

Table 5.
Effects of age of oil palm on growth and yield of pepper.

Plant variables	2 years	3 years	5 years	Means	LSD (0.05)
Number of leaves	9.3	9.6	7.9	8.9	1.01
Leaf area (cm2)	294.4	307.4	297.3	210	2.74
Plant height (cm)	92.6	91	86.4	90	1.88
Root biomass (g)	6.7	6.7	6.2	6.5	0.52
Shoot biomass (g)	56.7	57.2	54.9	56.2	1.15
Cob weights (g)	143.3	141.2	119.1	134.5	3.84
Seed yield (g)	500.5	510	444.4	484.9	6.25
Number of seeds/cob	87.3	90.1	76.6	84.7	2.80
Harvest index	7.9	7.1	7.0	7.6	0.73

Table 6.
Effects of age of oil palm on growth and yield of maize.

Plant variables	2 years	3 years	5 years	Means	LSD (0.05)
Number of leaves	40.3	40	38.3	39.6	1.09
Leaf area (cm2)	43.8	31.6	34.5	36.7	2.63
plant height (cm)	79.3	75.6	62.9	63.8	1.61
Shoot biomass (g)	57.2	47.8	40.4	48.5	2.53
Number of tubers/stand	4.8	4.1	3.4	4	0.51
Tuber yield/stand (kg)	5.0	4.45	4.02	4.16	0.61
Harvest index	4.7	4.4	4.2	4.5	0.54

Table 7.
Effects of age of oil palm on growth and yield of cassava.

after planting. A similar trial conducted on forest land near Benin in Nigeria including cropping for 2 years concluded that alley cropping remained possible in the early years of oil palm before complete canopy closure [2]. This early experiment evaluated maize, yams and cassava, and shade-resistant cocoyam as the only crop up until 12 years. For the first 2 or 3 years on the cropped plots, good yields from crops and growth from palms were obtained. Results were also obtained for cocoyam (*Xanthosoma sagittifolium*) in a more recent experiment near Benin City, in the fifth and sixth years after planting [2, 35]. The relative yields of component crops in the mixture were low compared with their counterpart sole crops. The strip intercrops of cassava, maize and pepper in the alleys of 2, 3 and 4 years old were characterized by relative yield (RY) values less than one (< 1). This may be attributed to inter-specific competition among oil palm of 2, 3, 4, 5 and 6 years old and the intercrop species [2, 35]. Relative yield greater than one (1) indicates that intercropping system has high competitive advantage and complementary in the utilization of resources efficiently [13]. In general, relative yield total (RYT) was best for cassava among the intercrops and lowest for maize. Apart from the yield advantage, RYT is often used to express economic feasibility of intercropping system. The low competition between intercrops implies that one component stimulated the growth of the other [13]. This is however, contrary to other reports that RYT less than one (<1) showed inter-specific competition among species. The envisage advantage of growing crops in mixture for enhanced use efficiency of growth resources and input (fertilizer) attract farmers to intercropping [10, 13, 39]. The greater yield advantage of crop mixtures was established from the land equivalent ratio (LER), land equivalence and percentage land saved (PLS)variables of competitive interaction of crop mixtures confirmed high yield advantage of cassava, maize and pepper intercrops in alleys of oil palm of 2–6 years, and also indicated high productivity of the intercropping system. This observation is consistent with those of Miller and Pallardy [40] and Agele et al. [10] on intercrop species. These results confirmed that cassava, maize and pepper can grow together in mixture without adverse effect on each other but offer yield advantage when intercropped with oil palm alley of 2 to 6 years old. Similar findings were reported by Malay et al. [13] on maize-legume intercropping systems and Agele et al. [20] on cashew-based intercrop of Sesame and bambara groundnut in the southern guinea savanna of Nigeria. Aggressivity of the intercrops was about 0.01 for sole crops of cassava, maize and pepper intercropped into 6 years old oil palm implying equal competition. However, cassava, maize and pepper had both negative and positive symbol greater than one (1). This is consistent with the report of Koohi et al. [41] who stated that crowing coefficient greater than one (>1) had yield advantage, RCC of equal to one (1) and above one (>1) has yield advantage while less than one (< 1) is disadvantageous (**Figures 10** and **11**).

Section B3: Effects of alley intercropping on the growth and yield of oil palm.

Oil palm alleys of ages 2, 3, 4 and 6 years were planted with cassava, maize and pepper in intercropping system. By this time, the oil palm plants had not completely closed canopy but expected to create competition in both above-ground and below-ground for alley crops. The alley intercrop species affected growth attributes of oil

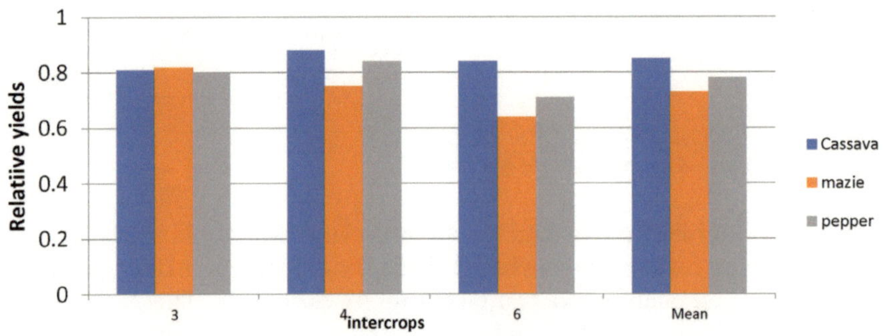

Figure 10.
Effect of age of palm on relative yields of species in the intercropping system.

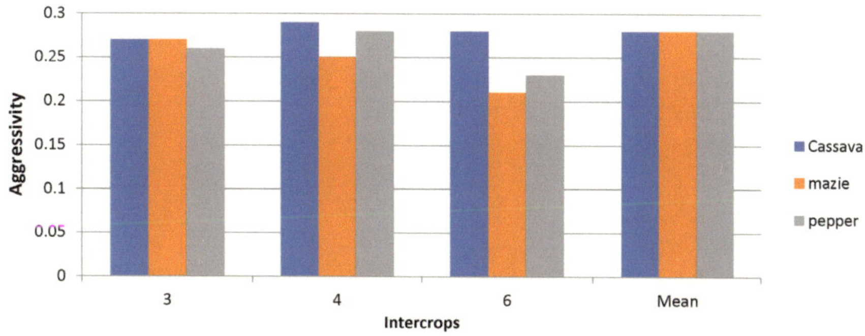

Figure 11.
Effect of age of palm on aggressivity of the intercrop species.

Plant variables	2 years	3 years	5 years	Means	LSD (0.05)
Number of leaves	16.22	17.9	35	23.1	1.13
Canopy extent (m)	1.41	1.51	2.35	1.76	0.25
Number of fresh fruit bunches/tree			9.71	9.71	
Weight of fresh fruit bunches/tree (kg)			98.68	98.68	

Table 8.
Effects of alley intercropping (cassava, maize and pepper) on the growth and yield of oil palm.

palm (**Table 8**). The growth and development of oil palm especially, canopy formation during the early years of establishment (1 to 6 years) appeared to favor sowing arable crops (such as cassava, maize and pepper) in its alleys. The alley crop species (maize, cassava and pepper) exerted no significant detrimental effects on the measured characters of oil palm (including number of leaves, canopy extent, number of fresh fruits bunch (FFB) and weigh of fresh fruits bunch (FFB) (**Table 8**). Oil palm canopy spread increased as the age of the oil palm increased, canopy extent was largest for 5 and 6 years old palm trees compared with the 2, 3 and 4 years. The number of FFB in young palms were more and small in size and lower in weight while the number of FFB in older palms were few and bigger in size. The number and weight of fresh fruit bunch (FFB) were different between the 4 years old and 5 and 6 years old palm trees. The number of fruits per palm ranged from 10 to 16, 8–16, and 10–15 and yields of FFB were on the average, 8.98, 9.28 and 31.93 $kg.m^2$ for the respective 4, 5 and 6 years old oil palm trees (**Table 8**). The number and weights of fruits were in line with NIFOR record on oil palm production [2]. The yields of FFB from 4 to 6 years old palm trees were close to NIFOR standard. The tested plant species grown in mixtures had enhanced productivity over their respective sole crops possibly by exploiting species complementarities for resource capture. The complementarity could have resulted in part from competition avoidance responses for resource capture and use for growth of the individual species [10, 13]. Individual species responses in the mixture (e.g. shade avoidance) and hence differences in resource acquisition in time and space improved the performance of the intercrop community as a whole [23, 35].

Section C1: Effects of fertilizers and age of oil palm on efficiencies of N uptake and use for biomass and yield production by alley crop species.

The fertilizers exerted significant effects on nitrogen uptake for yield production of cassava, maize and pepper strip intercropped in the alleys of 3, 4 and 6 years oil palm plot (**Table 9**). Compared with the unmanure (PuO), application of fertilizers enhanced nitrogen contents, N uptake and N yields. NPK compound fertilizer

Intercrop species		N contents	Biomass	PuN	PuO	PuN (sole)	Intercrop mixture Yield (+N)	Crop mixture Yield (unfertilized)	Sole crop Yield (N)	Sole crop Yield (unfertilized)	NHI Mixture	NHI sole crop
Cassava	NPK	1.81	1212	1775.61	1324.35	1844.3	37.8	31.2	42	35.7	0.047884	0.058013
Maize		1.68	125	164.64	113.68	178.1	11.3	10.4	14.2	11.4	0.148673	0.161538
Pepper		2.03	129	170.52	126.84	205.0	12.3	11.2	15.3	12.3	0.165041	0.18125
Cassava	Poultry manure	1.49	1108	1183.06	897.22	1247.2	33.7	27.3	37.6	35.7	0.044214	0.054579
Maize		1.38	109	111.78	87.48	187.0	10.4	8.3	13.4	11.4	0.132692	0.166265
Pepper		1.54	97	113.96	88.06	173.4	11.7	11.2	14.6	12.3	0.131624	0.1375
Cassava	Ferti Plus	1.53	1194	1320.39	957.93	1844.4	31.3	26.8	37.4	35.7	0.048882	0.05709
Maize		1.37	118	119.19	87.87	154.5	9.3	7.4	11.7	11.4	0.147312	0.185135
Pepper		1.73	123	134.94	102.96	172.4	10.8	9.3	14.1	12.3	0.160185	0.186022
	LSD	0.37	35.98	20.57	17.73	21.9	2.74	2.47	2.84	2.33	0.005	0.004
Fertilizers (Fz)		Significant	*	*	*	*	*	*	*	*	*	*
Crop types (Ct)		Significant	*	*	*	*	*	*	*	*	*	*
Fz*Ct		Significant	*	*							*	*

Table 9.
Effects of fertilizers and intercrops on N content and uptake, crop yields and nitrogen harvest index (NHI).

enhanced the contents, uptake and yields of N compared with other fertilizer treatments while poultry manure had varied effects on nitrogen contents, N uptake and N yields of the intercrops. The effects of ferti plus organo mineral fertilizer on N uptake and N yields of the intercrops was similar to that of poultry manure (**Table 9**). in addition, uptake and yields of N were higher for the respective sole crops of cassava, maize and pepper compared with the strip intercrops. While N content was highest for fruits of pepper, highest N uptake and N yields were found for cassava leaves and tubers for which in addition, shoot biomass was heaviest. The interaction of intercrop and fertilizer types were significant for most of the indicators of nutrient uptake measured (**Table 9**). The biomass yields were heaviest for cassava, therefore, its N uptake and N yields were higher compared with maize and pepper under application of poultry manure (**Table 9**). Fertilizer application using NPK, fertiplus and poultry manure enhanced the nutrient uptake in cassava planted in in the alleys 3, 4 and 6 years old oil palm. However, NPK produced slight increases in N uptake of cassava under 4 and 6 years old oil palm compared with other fertilizers (NPK, Ferti Plus and poultry manure). The age of oil palm fields influenced the effects of fertilizers on the uptake of nitrogen for biomass production in cassava. Nitrogen uptake for tuber production in cassava was better under 3 years old field by NPK and 4 years old by poultry manure. The fertilizers and age of oil palm affected nutrient uptake in the leaf and seed of maize. NPK and Ferti plus significantly enhanced leaf and seed nutrient uptake of maize compared to poultry manure across the ages of oil palm. The least values for N uptake were however recorded for poultry manure and highest for 3 years old oil palm. Among fertilizer treatments, least decline in effect of age of oil palm was found for 6 years old. NPK and Poultry manure had no differences in values for 4 years old palm hence enhanced nutrient uptake in leaf production. 6 years old oil palm had the least value. Close values were found fertilizer effects on leaf and seed nutrient uptake and for 3 and 4 years old oil palm which recorded the greatest effects (**Table 9**). Ferti plus and NPK enhanced the nutrient uptake in pepper leaves and fruits transplanted into 3 years old palm. NPK fertilizer influenced nutrient uptake compared to poultry manure for 3 and 4 years oil palm. Lowest fertilizer efficiency was found for 6 years old oil palm for both leaf and fruits of pepper. Ferti plus enhanced N nutrient uptake in leaf and fruits of pepper compared with other fertilizers (**Table 9**).

Section C2: Effects of fertilizers and age of oil palm on agronomic and physiological efficiencies of N use by alley crop species.

In general, the fertilizers following their application to the strip intercrops in the alleys of 3, 5 and 6 years old palm influenced most of the indicators of nutrients use efficiencies (**Table 10**). Agronomy Efficiency (AE) differed significantly. Apparent Recovery Efficiency (RE), Apparent recovery Efficiency by difference(RE%), Physiological Efficiency(PE), Utilization Efficiency(UE), Internal utilization Efficiency(IE) and partial factor productivity (PFP) were not significantly different among the intercrops under the fertilizers. Fertilizer treatments enhanced most of measured variables of nutrient use efficiencies compared to unfertilized plots (control) (**Tables 10** and **11**).The ages of oil palm significantly affected most of the measured variables among the intercrops except the N removed at harvest. However, apparent recovery of N differed among intercrops and ages of oil palm plots. The values of the measured parameters were highest for 5 and 6 years oil palms across the intercrops (**Table 10**). For cassava, poultry manure enhanced the utilization efficiencies while ferti plus promoted the efficiencies of N recovery from the applied fertilizers and physiological efficiency of its use. NPK enhanced both recovery and utilization efficiencies of nutrients, partial factor productivity, N removed at crop harvest and the competitive ability for uptake and use of nutrients

Intercrop species	Fertilizers	Agronomic efficiency of fertilizers	Physiological efficiency of fertilizers	N recovery efficiency	Apparent recovery of N	Utilization efficiency of N	Internal utilization efficiency	Partial factor productivity	N conversion efficiency	N removed at harvest	Relative interaction intensity	Competitive ability for nutrients
Cassava	NPK	247.61905	0.0146257	1.504200	15	0.2193857	0.0212885	0.12600	0.5524862	1771.1955	0.809406	0.2907
Maize		276.10619	0.0176609	0.169866	1.7	0.0300235	0.0686346	0.03767	0.5952381	164.26107	0.784000	0.2626
Pepper		273.17073	0.0251832	0.145600	1.4	0.0352564	0.0721323	0.04100	0.4926108	170.0972	0.651162	0.2635
Cassava	Poultry manure	4050.4451	0.0223901	1.084666	11	0.2462916	0.0284855	0.00674	0.6711409	1182.8806	0.716606	0.3032
Maize		3990.3846	0.0864198	0.004860	5	0.4320988	0.0930399	0.00208	0.7246377	111.7625	0.743119	0.2561
Pepper		4786.3248	0.019305	0.005180	6	0.1158301	0.1026676	0.00234	0.6493506	113.9423	0.762886	0.2644
Cassava	Ferti Plus	381.81818	0.0124152	0.906150	9	0.1117365	0.0237051	0.07825	0.6535948	1017.110	0.722780	0.2745
Maize		389.74359	0.0606641	0.078300	8	0.4853129	0.0780267	0.02325	0.7299270	118.9703	0.737288	0.2475
Pepper		348.93617	0.0469043	0.079950	9	0.4221388	0.0800356	0.02700	0.5780347	134.6826	0.634146	0.2423
LSD (0.05)		1991.04	0.025	0.566	4.36	0.176	0.031	0.041	0.079	62.68	0.057	0.019
Fertilizers (Fz)		*	*	*	*	*	*			*	*	*
Crop types (Ct)		*	*	*	*	*	*			*	*	*
Fz*Ct			*					*	*			*

Table 10.
Effects of fertilizers and intercrops on indices of N uptake and use efficiencies.

	N Uptake	ANR (%)	Agron NUE (kg/kg)	Physiol. NUE (kg/kg)	N yield	NHI	N removed @ harvest	N conversion efficiency
Cassava	513.65	7.02	2011.03	0.0175	21.22	0.10	691.3	0.62
Maize	337.01	8.01	2021.14	0.0211	17.14	0.11	477.7	0.65
Pepper	335.48	7.06	2031.52	0.0233	18.51	0.12	445.6	0.64
LSD	11.51	0.41	19.43	0.001	7.02	0.002	15.31	0.02
Zero	617.5	4.71	248.33	0.012	13.56	0.09	342.4	0.44
NPK	521.62	6.03	365.53	0.019	20.47	0.12	702.5	0.55
Poultry manure	357.59	7.33	427.92	0.027	17.03	0.11	470.3	0.68
Ferti Plus	382.93	8.68	373.49	0.026	18.66	0.12	524.7	0.65
LSD	40.63	0.22	28.34	0.002	0.45	0.003	23.6	0.02
Crop Type (CT)	Significant	Significant	Significant	Significant	Significant		Significant	Significant
Fertilizers (FZs)	*	*	*	*	*	*	*	*
CT x FZs	*	*	*	*	*		*	*

ANR (Apparent Recovery of N in above ground biomass); ANUE (Agronomic N use efficiency); PNUE (Physiological N use efficiency).

Table 11.
Summary of N uptake, agronomic and physiological efficiencies of N use.

from the applied fertilizers (**Table 10**). In maize, NPK enhanced recovery efficiency from applied fertilizers, partial factor productivity, N removed at harvest and competitive ability for nutrients (**Table 10**). Ferti plus enhanced physiological, utilization and conversion efficiencies of nutrients while poultry manure improved physiological and utilization efficiencies of nutrients. In pepper, ferti plus and poultry manure enhanced the utilization efficiency of nutrients while NPK significantly enhanced N removed at harvest. The fertilizers affected almost all variables measured as indicators of agronomic and physiological efficiencies of nutrient use (**Table 10**). The effects of fertilizer was significant on utilization and recovery efficiencies of N and competitive ability for nutrients, agronomic and physiological efficiency of N use and apparent recovery of N (**Table 10**). Poultry manure and ferti plus fertilizer enhanced agronomic efficiency, apparent recovery of nutrients from above ground biomass, internal utilization efficiency as well as N conversion efficiency while NPK enhanced partial factor productivity and NHI for both sole and intercrop mixtures (**Table 10**). The summary of uptake, recovery, agronomic and physiological efficiencies of N use, N removed at harvest and conversion efficiencies is presented in **Table 11**. Maize and pepper were efficient in terms of ANR, agronomic and physiological efficiencies of N use and its conversion efficiencies while cassava was outstanding for N uptake and its recovery at harvest. Among the fertilizers tested, poultry manure and Ferti plus were outstanding with respect to ANR, agronomic and physiological efficiencies of NUE and N conversion efficiencies while NPK out-performed other fertilizers for NHI and N recovery at harvest. The interactions of intercrop species and fertilizer was significant for most of the variables of agronomic and physiological efficiencies of N use evaluated (**Table 11**). The fertilizers (NPK compound fertilizer, ferti plus and poultry manure exerted significant effects on nitrogen uptake for biomass and yield production. While NPK enhanced N uptake and apparent recovery of N in above ground biomass, ferti plus and poultry manure promoted most other indicators of agronomic and physiological N use efficiencies. The efficiencies of uptake and use of N were higher for NPK and ferti plus compared with poultry manure across the ages of oil palm fields. The values of the measured indicators of N uptake and use efficiencies were highest for 4 and 6 years compared with the 3 years old fields across the intercrop species. The uptake and yields of N were higher for the respective sole crops of cassava, maize and pepper compared with the intercrops across the fertilizers compared with the control.

The age of oil palm fields and fertilizers influenced nutrient uptake and use efficiencies of the intercrop species (leaf, tuber of cassava, fruits of pepper and seeds of maize) in the alleys of 3, 4 and 6 years old oil palm trees and thus, the resultant yield improvement compared with unmanure treatments [42]. The intercrop species differed in shoot morphological and physiological attributes, and rooting patterns of rooting, biomass and nutrient accumulation and partitioning. These attributes have implications for nutrient uptake, use efficiencies and yield production among the alley crop species [22]. Crop species and fertilizer type affected the uptake and accumulation oof nutrients to the vegetative and reproductive structures [20, 23]. Information about differences among species and varieties with respect to N use efficiencies have been used to develop cultivars adapted to low fertilizer input management systems. The sole crops of the intercrop species under NPK treatment recorded the highest nutrient contents in their leaves which indicates that physiologically, nutrients uptake may depend of the degree of competition (below ground) for resources [20, 23]. The fertilizer NPK, contains high N content which is released rapidly into soil solution and promoted its availability, this must have enhanced its uptake by the plant and utilization for biomass production (improved plant growth) compared with organic fertilizers (poultry manure for example). Fertilizers enhance nutrient availability and bring about decreases in competition placed on nutrient resources by intercrop species (crop mixtures) [20]. Literature reports has indicated that differences in soil nutrient status are a major source of variation in uptake and use efficiencies and of crop yields [20, 36]. These reports also attributed crop yield enhancement by fertilizers to improvements in the efficiency of uptake and use of nutrient resources for both sole and intercropping systems. Agele et al. [6] also attributed high yield performance of sole and intercrop combinations of crop species to improvements in efficiency of nutrient utilization. Our study showed that application of poultry manure, ferti plus and NPK fertilizers to alley crops in oil palm soil affected nutrient uptake leaf and tuber/seed nutrient contents compared with the control (un-manure) treatments. The fertilizers enhanced soil nutrients (the sandy loam soil of experimental site appeared to have low fertility status especially N) while improvement in other soil (chemical, physical and biological) properties would have promoted biomass accumulation and yield production by the alley crop species. It is reported that nutrient availability depends on nutrient concentration in the soil and environment and release pattern in synchrony with the crop needs [22, 23]. However, highest N uptake values were obtained for the un-manure. Nitrogen harvest index was higher under ferti plus and poultry manure compared with NPK, this result was in line to the conclusion of Agele et al. [6, 22] that the crop yields and nutrient availability were higher in plots for which farmyard manure was applied, and to longer time availability. Manure decompose slowly and release their constituent nutrients slowly, may be to meet time dynamics of nutrient demand by growing crops [6, 43].

Highest nitrogen harvest index values for seed and leaf of intercrop species were obtained from the un-manure treated plants. The superiority of these may be attributed to more vigorous nutrient exploitation advantage [23]. Oil palm also take up nutrient from the soil for its growth, especially from sandy loam soil of experimental sites, nutrient recycling in palms is slow. However, oil palm provides nutritional elements like phosphorus [44], and nutrient P is reported to decrease species competition placed on nutrient resources [11]. For arable crops grown in the alley of oil palm (1 to 6 years old), supplementary input of nutrients especially nitrogen from fertilizers (organic and inorganic) is needed to meet the nutrient requirements of alley crops.

6. Summary and Conclusions

Based on the measured growth parameters of oil palm seedlings, shading and weekly irrigation enhanced seedling vigor compared with fortnight and open sun (unshaded) treatments. These treatments enhanced vigor, growth and reduce mortality of oil palm seedlings in the nursery in the dry season. Mycorrhizal inoculation of oil palm seedlings, shade and weekly irrigation are recommended. The findings from this study acclaimed the relevance of dry season irrigation and shade to supplement soil moisture and reduced temperature for oil palm seedling growth and development. The drip irrigation-shade strategy adopted ameliorated dry season terminal drought (hydrothermal stresses) in cacao. This is a veritable tool to scale up growth, survival, establishment and flower/pod production. The results of this study will contribute to the development of sustainable cacao production practices and development of shade and irrigation management guidelines for small holder farmers.

Oil palm age significantly affected most of the measured growth variables and indicators of competitive interactions and crop mixture productivity, and agronomic and physiological efficiencies of N use by alley intercrop species. The study established that intercrop mixtures of cassava, maize and pepper in the alleys of oil palm of 2 to 6 years old exhibited some levels of compatibility and complementarity and confirmed low competitive interactions but high growth resource use efficiencies. The results of competitive functions and crop mixture productivity indicate the yield advantage of crop mixtures over sole cropping and hence the overall biological advantage of intercropping of cassava, maize and pepper in the oil palm-based intercropping system. Strip intercropping of cassava, maize and pepper in the alleys of 2 to 6 years old oil palm in the rainforest zone had no detrimental effects on the growth and yield of oil palm. Crop performance results from the behavior of the individual plants interacting through competitiveness (vigor of growth of individual species in the mixture) and complementary which drive resource capture and utilization. The study improved understanding of compatibility and complementarity of growth resource use (space, light, nutrients and possibly water) of oil palm with some selected arable crops in oil palm-based intercropping system of the rainforest of Nigeria. Improved insight to unravel the primary drivers and dynamics of competitive and complementary growth responses of crop mixtures in cropping systems. Such knowledge is relevant to the promotion and adoption of crop mixtures (intercropping systems) for sustainable increases in crop yields at acceptable input levels. Over their respective sole crops, the strip intercrops of cassava, maize and pepper exhibited high values of land equivalence, percentage land saved (PLS) and low aggressivity in the alleys of oil palm which indicates greater yield advantage of crop mixtures over sole cropping. These observations denote the advantage of intercropping in terms of productivity per unit land area, in addition to greater percentage land saved.

The fertilizers (organic and inorganic) enhanced nutrient uptake and use efficiencies in the respective leaf, tuber, seed and fruits of cassava, maize and pepper in the alleys of 3, 4 and 6 years oil palm fields. The indicators of uptake and use efficiencies of N differed among the alley crop species and fertilizer types across the ages of oil palm plants in plantation. While NPK promoted the uptake and apparent recovery of N in above ground biomass, the organic fertilizers enhanced other indicators of agronomic and physiological N use efficiencies, and the efficiencies of uptake and use of N were higher for NPK and ferti plus compared with poultry manure. Uptake and use efficiencies of N were higher for the respective sole crops of cassava, maize and pepper compared with the intercrops across the fertilizers. N content was highest for fruits of pepper while N uptake and yields were highest for cassava tubers and seeds of maize. For arable crops grown in the alley of oil palm (1 to 6 years old), supplementary input of nutrients especially nitrogen from fertilizers (organic and inorganic) is needed to meet the nutrient requirements of alley crops.

Author details

Samuel O. Agele[1*], Friday E. Charles[2], Appolonia E. Obi[2] and Ademola I. Agbona[3]

1 Plant Physiology and Ecology Group, Department of Crop, Soil and Pest Management, Federal University of Technology, Akure, Nigeria

2 Division of Agronomy/Plantation Management, Nigerian Institute for Oil Palm Research (NIFOR), Benin City, Nigeria

3 Department of Agricultural Technology, Federal Polytechnic, Ile Oluji, Nigeria

*Address all correspondence to: soagele@futa.edu.ng

References

[1] Corley R. H. V. (1986). Oil palm: In CRC handbook of fruit set and development P. 253 – 259, CRC press Boca Raton Florida.

[2] Nigerian Institute for Oil Palm Research (NIFOR) (2010). A manual of oil palm production. The Nigerian Institute for Oil Palm Research Press. 30pp.

[3] Bateman, I. J., Fisher B., Fitzherbert E., Glew D., Naidoo R. (2010). Tigers, market and palm oil: Market potential for conservation. Oryx 44: 230 – 234.

[4] Yusof, B. (2007). Palm oil production through sustainable plantations. European Journal of Lipid Science Technology. 109: 289 – 295.

[5] Ritchie, H and Roser, M, 2020. Crop production across the world. **https://ourworldindata**.org/agricultural-production

[6] Agele, S.O., Adeyemo, A.J. and Famuwagun, I.B. (2011) Effects of Agricultural Wastes and Mineral Fertilizer on Soil and Plant Nutrient Status, Growth and Yield of Tomato. Archives of Agronomy and Soil Science, 57 (1), 91-104

[7] Agele, SO, Aiyelari, OP & Charles Friday. 2017. Effects of shading, irrigation and mycorrhizal inoculation on growth and development of oil palm *Elaeis guineensis* Jacq. (Magnoliophyta: Arecaceae) seedlings in the nursery. Brazilian Journal of Biological Sciences 4 (7): 113-126. doi.org/10.21472/bjb s.040712

[8] Gawankar, M. S. Devmore, J. P. Jamadagric, B. M. Sagvekar, V. V. Khan, H. H. (2003). Effects of water stress on growth and yield of *Tenera* oil palm. Journal of Applied Horticulture 5:39 – 40.

[9] Omorefe, A., Bonaventure, C. (2010). Identification of moisture stress tolerant oil palm genotypes. African Journal of Agricultural Research. 5: 3116 – 3121.

[10] Agele, SO, Obi, EA & Aiyelari, OP. 2019. Effects of age of oil palm on the growth, competitive interactions and mixture productivity of cassava, maize and pepper in oil palm-based strip intercropping system in the rainforest of Nigeria. London Journal Press. 19(7): 60-79. Compilation 1.0.

[11] Vanlauwe, B., Kihara, J., Chivenge, P., Pypers, P., Coe, R. and Six, J. (2011): Agronomic use efficiency of groundnut varieties as influenced by sowing dates. Journal of SAT Agricultural varieties. Research on crops, 6(1): 173 – 174.

[12] Bedoussac, L. and Justes, E. (2010). The efficiency of a durum wheat-winter pea intercrop to improve yield and wheat grain protein concentration depends on N availability during early growth. Plant Soil, 330, 19-35. http://dx.doi.org/10.1007/s11104-009-0082-2

[13] Malay, KM. Mahua, B. Hirak, B. Animesh, P. and Rajib, D. (2014). Evaluation of cereal-legume intercropping systems through productivity and competition ability. Asian Journal of Science and Technology. *5(3), 233-237*

[14] Willey, R. W. (1990). Resource use in intercropping systems. Journal of Agricultural Water Management. 17: 15-231.

[15] Agele, S. Nduka, B. & Aiyelari, P. 2018a. Agro-Waste Reuse and Effects on Growth and Yields of Sesame and Bambara Nut Alley Crops in a Cashew-Based Intercropping System of the Guinea Savanna Agroecology of Nigeria. Advances in Recycling and Waste Management 3(2): 159. DOI: 10.4172/2475-7675.1000159

[16] Ajayi, A. J., Agele, S. O. and Aiyelari, O. P. (2016). Growth, Yield

and Yield Components of Pineapple in a Pineapple-Pepper-Cowpea Intercropping System. International Journal of Horticulture. 6 (1) DOI: 10.5376/ijh.2016.06.0001

[17] Agegnehu, G., Ghizam, A. and Sinebo, W. (2008).Yield potential and land use efficiency of wheat and faba bean mixed intercropping. Agron Sustain Dev. 28:257-263.

[18] Lithourgidis A.S., Dordas C.A., Damalas C.A., and Vlachostergios D.N. (2011). Annual intercrops: an alternative pathway for sustainable agriculture (Review article). Australian journal of crop science AJCS 5(4): 396-410.

[19] Wahla, I. H., Ahmad, R., Ehsan Ullah, Ashfaq Ahmad and Jabbar, A. (2009). Competition functions of component crops in Barley-based intercropping systems. Int. J. Agric. Biol. 11:69-72.

[20] Agele, S. Nduka B., Famuwagun, B. & Adejoro, S. 2018b. Uptake and Use Efficiencies of Nutrients by Sesame and Bambara Nut Alley Crops as Influenced by Manuring in a Cashew-Based Intercropping System in the Guinea Savanna Agroecology of Nigeria. Journal of Agricultural Chemistry and Environment, 2018, 7, 153-175.

[21] Willey, R.W. (1979). Intercropping its importance and research needs. I. Competition and yield advantages. Field Crop Abst. 32: 1–10.

[22] Agele, S.O., Adeniji, I.E., Alabi, I.A. and Olabomi, A. (2008) Responses of Growth, Yeild and N use Efficiency of Selected Tomato Cultivars to Variaitons in the Hydrothermal Regimes of the Cropping Seasons in a Rainforest Zone of Nigeria. Journal of Plant Interactions, **2** (4), 273-285.

[23] Tsadik T. 2019. Nitrogen use efficiency of tef [Eragrostis tef (Zucc.) Trotter] as affected by nitrogen fertilizer under chickpea-tef rotation at Tahtay Koraro District, North Ethiopia J. Soil Sci. Environ. Manage. 10(4), pp. 58-67

[24] McCall, D. and Willumsen, J. (1998) Effects of Nitrate, Ammonium and Chloride Application on the Yield and Nitrate Content of Soil Grown lettuce. *Journal of Horticultural Science and Biotechnology*,**72** (2), 698-703.

[25] Fatima M. S., Moveira, Jerven E. Huising and David E. Bignel (2008). Tropical soil biology PP. 134 – 143.

[26] Agele, SO., Agbona, I.A., Ewulo, B. S. & Anifowose, A.Y. 2014. Drip irrigation scheduling for optimizing productivity of water use and yield of dry season pepper (*Capsicum annuum* L.) in an inland valley swamp in a humid zone of Nigeria. International Journal of Plant and Soil Science. 4 (2), 171-184

[27] International Water Management Institute (IWM). 2010. Annual Reports. IWMI Thailand. 151 pp.

[28] Allen, R. G.; Pereira, L. S.; Raes, D. & Smith, M. 1998. *Crop evapotranspiration:* guideline for computing crop water requirement. Rome: Food and Agriculture Organisation of the United Nations FAO Irrigation and Drainage Paper, 56

[29] Doorenbos, J. & Pruitt, WO. 1977. **Crop water requirements**. Rome: Food and Agricultural Organization of the United Nations FAO Irrigation and Drainage Paper, 24.

[30] Tripathy P., Sethi K., A. Patnaik K. and Mukherjee S. K. (2015). Nutrient Management in High Density Cashew Plantation under Coastal Zones of Odisha. International Journal of Bio-resource and Stress Management 2015, 6(1):093-097. DOI: 10.5958/0976-4038.2015.00025.1

[31] De Wit, C. T., and Van den Bergh, J. P. (1965) Competition between herbage plants. Netherlands Journal of Agricultural Science. 13 (2), 212–221.

[32] Mead, R., and Willey, RW. (1980) The concept of a "land equivalent ratio" and advantages in yields from intercropping. Exp. Agric. 1980; 16: 217-228.

[33] Jabbar, A. R. Ahmad, I. H. Bhatti, Z. A. Virk, W. Din and M. M. Khan (2009). Assessment of yield advantages, competitiveness and economic benefits of diversified direct seeded upland rice-based intercropping systems under strip geometry of planting. Pak. J. Agri. Sci. 46(02):96-101.

[34] Li H., Yong Y., Li, B., Jing R., Lu, C., Li, Z. (2010). Genetic analysis of tolerance to photo oxidative stress induced by high light in winter wheat (*Triticum aestivum* L) Genetics & Genomics 37(6): 399 – 412.

[35] McGilchrist, C.A. (1965) Analysis of Competition Experiments. Biometrics, 21, 975-985.http://dx.doi.org/10.2307/252825

[36] Dobermann, A. (2007). Nutrition use efficiency-measurement and management. In "IFA-International workshop: Fertilizer Best Management practices", Brussels, Belgium Pl-28

[37] Fageria N.K, Baligar V.C 2003. Methodology for Evaluation of Lowland Rice Genotypes for Nitrogen Use Efficiency. Journal of Plant Nutrition 26 (6):1315-1333

[38] Gungula, D. T. (1999). Growth and nitrogen use efficiency in maize (*Zea mays* L.) in the South Guinea Savanna of Nigeria. PhD. Thesis, University of Ibadan. 188Pp.

[39] Malézieux E, Crozat Y, Dupraz C, Laurans M, Makowski D, Ozier-Lafontaine H, Rapidel B, de Tourdonnet S, Valantin-Morison M (2009). Mixing plant species in cropping systems: concepts, tools and models:A review. Agron Sustain Dev 29:43-62.

[40] Miller, A.W., Pallardy, S.G. (2001). Resource competition across the tree-crop interface in a maize-silver maple temperate alley cropping stand in Missouri. Agroforestry Systems. 53: 247-259.

[41] Koohi, S.S. and Nasrollahzadeh, S. (2014). Evaluation of yield and advantage indices of sorghum (*Sorghum bicolor* L.) and mungbean (*Vigna radiate* L.) intercropping systems. Int. J. Adv. Biol. Biom. Res. 2(1): 151-160.

[42] Abbasi, M.K. Tahir, M.M. Rahim, N. (2013) Effect of N fertilizer source and timing on yield and N use efficiency of rainfed maize (*Zea mays* L.) in Kashmir–Pakistan. Geoderma 195/196:87–93

[43] Giller KE (2004). Emerging technologies to increase the efficiency of use of fertilizer nitrogen. In: A.R. Mosier, J. K. Syers and J.R. Freney (eds.) Agriculture and the nitrogen Cycle. Scope 65. Island Press Washington DC pp. 35-51.

[44] Tsudo, M., Walker, S., and Mukhala, E (2001). Comparisons of radiation use efficiency of mono and intercropping system with different row orientation. Field crops Res. 71: 17-29

Chapter 16

Processing of Gong Oil (*Pachymerus nucleorum*) to Obtain Biodiesel by Methyl Route

José Sebastião Cidreira Vieira, Wilame Silva Castro, Makson Rangel de Melo Rodrigues and Neemias da Silva Nascimento

Abstract

Brazil is a large world producer of vegetable oils. This condition puts it in the vanguard of the use of renewable fuels on the planet, and may constitute an excellent opportunity for scientific and technological development of the country. The most widely used method for producing biodiesel is basic homogeneoustransesterification. Its disadvantage is the use of high quality raw materials, which raises the final cost of the process by about 85%. The final cost of biodiesel production can be reduced considerably with the inclusion of raw materials of low added value containing high content of free fatty acids (FFA), the great challenge is the development of different routes and production scales capable of making the production process of this energy input viable. In addition, the valorization of different raw materials and implementation of innovative technologies to make obtaining biodiesel ecologically sustainable and more competitive in relation to fossil diesel is of fundamental importance. Among the several species with potential for biodiesel production in Brazil, the gong (Pachymerus nucleorum) stands out, which, when heated, decomposes, originating an oil equal to that extracted from the seeds of oilseed plants. This is the larval stage of a coleopteran of the family Bruchidae, a beetle, which lives inside fruits of buriti (*Mauritia flexuosa*), tucum (Bactris setosa), babaçu (*Attalea speciosa*) and carnauba (*Copernicia prunifera*) until the adult stage. This work aimed to extract the oil from the larvae of Pachymerus nucleorum and adapt it to obtain biodiesel. The physical–chemical characterization of this raw material in natura revealed high FFA content (1.63% and 1.74%). The gong oil was submitted to esterification by acid catalysis, aiming to adapt it to the basic homogeneous transesterification process. The results indicated a significant reduction of FFA (0.85% and 0.55%). The infrared spectra (FTIR) of the esterified oil and biodiesel indicated the presence of methyl esters and low moisture content. Gas chromatography (GC-FID) revealed that the composition of the biodiesel consists of saturated fatty acid methyl esters (FAME). The esterification reaction of gong oil when pretreated provided evidence that the feedstock may be suitable for biodiesel production. At present, there are no studies involving the extraction and suitability of gongo oil for the specific production of biodiesel. This issue needs to be addressed so that we can advance in the esterification-transesterification processes of feedstocks with high acidity index.

Keywords: Pachymerus nucleorum, biodiesel, sustainability

1. Introduction

Energy has established itself as an input of fundamental importance for economic growth and for raising the standard of living of modern society. Energy generation is overly dependent on petroleum derivatives. On the other hand, the gases emitted by gasoline, diesel and other derivatives have strongly contributed to environmental degradation, causing climate change, global warming, melting of the polar ice caps, rising sea levels, environmental disasters and destruction of the ozone layer. These events have negatively impacted the economy and public health policies. The limitation of oil reserves and the degradation of the environment are factors responsible for the incessant search for renewable energy sources to redeem and/or eliminate the impacts caused by fossil fuels to the environment [1, 2].

Biodiesel has become in the last decades, an alternative fuel capable of meeting the growing demand for energy. The increased demand for energy due to the world population growth has contributed to a possible depletion of fossil energy resources and logically raised the level of atmospheric pollutant emissions, causing environmental degradation [3]. In the current context, most of the energy produced in the world comes from fossil sources such as oil, coal and natural gas, which are directly associated with environmental issues and are responsible for the interest of biodiesel as a renewable fuel, capable of redeeming the emissions of greenhouse gases [4, 5].

Biodiesel is a clean-burning fuel, originating from natural and renewable sources such as vegetable oil, saturated edible oil generated from cooking and frying food, animal fat and a shortchain alcohol in the presence of a catalyst. This energy input has properties such as freedom from sulfur and aromatic compounds, high cetane number, average oxygen content, higher flash point, lower emission of hydrocarbon particles, carbon monoxide and dioxide, non-toxic and biodegradable character, which overlap in relation to the properties of petroleum derivatives [6–8].

The biodiesel production route most used today in Brazil and in the world is called transesterification. In this process, the triacylglycerides (TAG) present in the fatty raw materials, vegetable and animal oils and/or fats interact chemically with a monoalcohol (methanol or ethanol) in the presence of a basic Brönsted type catalyst (proton receptor chemical species) to be converted into a mixture of esters (methyl or ethyl) of fatty acids (biodiesel) and glycerin as a byproduct [9, 10]. **Figure 1** below shows the overall reaction process of the traditional transesterification process in the light of chemistry.

Transesterification occurs in three consecutive and reversible steps. To achieve relevant results in the course of traditional transesterification, excess short-chain alcohol is added, since the presence of water in the reaction medium (this occurs very often) even in small amounts (the reactants are hygroscopic). The basic catalysts are very sensitive by means of free fatty acids (FFA) from the fatty feedstock or formed by hydrolysis of the esters. In this system, the FFA react with the alkaline catalyst (NaOH or KOH) contributing to the formation of fatty acid salts (soaps), which in turn, at the end of the reaction, form emulsions and make it difficult to separate the product (biodiesel) from the by-product (glycerin). The use of Brönsted basic catalysts in the production of biodiesel by homogeneous catalysis requires the use of high quality grease raw materials. The acquisition of these raw materials results in a high cost and account for more than 85% of biodiesel processing expenses, as it requires the use of anhydrous alcohol and food grade oils and fats [11, 12]. Due to the basic catalysts, proton receptors (Brönsted), present high catalytic activity and are low cost and little aggressive of the equipment of the transesterification process are the most used in the biodiesel industry. The use of high purity raw materials in biodiesel production processes is one of the main

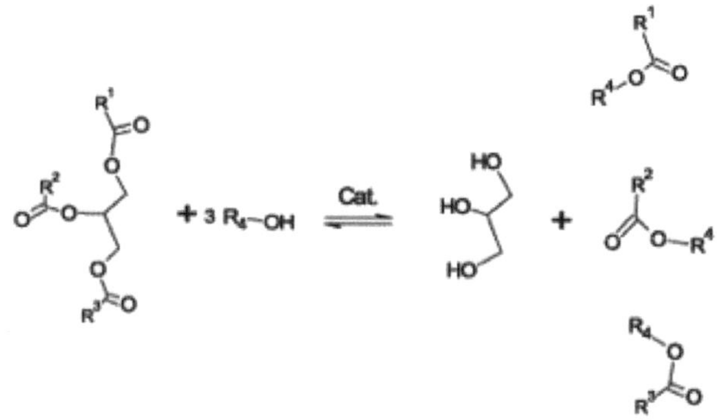

Figure 1.
Global reaction of triacylglycerides alcoholysis.

factors that make it difficult for biodiesel to be more competitive in relation to petroleum derived fuels. In view of the above, the search for alternative feedstocks capable of reducing the costs of the alkaline transesterification process has led to studies aimed at the use of materials generated from renewable resources to make the production of biodiesel ecologically sustainable and economically viable, which meets the needs of the industrial sector and puts an affordable fuel on the consumer market. Among the raw materials with potential to overcome the limitations of traditional transesterification are oils and fats from oilseed plants, animals and residual raw materials, since they can be acquired at low cost and contribute to the sustainable production of biodiesel. On the other hand, these feedstocks have high acidity index and water content, the main factors that increase the costs of the process and consequently biodiesel becomes less competitive in relation to petroleum diesel. To adapt them to basic homogeneous catalysis technology, it is necessary to previously submit them to the degumming or esterification process by acid homogeneous catalysis to reduce the intrinsic drawbacks of these feedstocks [13, 14].

Due to its territorial extension and the variety of climates and soils, Brazil performs the biodiesel processing in a decentralized way at laboratory scale level, valuing the abundant raw materials in each of its regions. Thus, new alternatives for obtaining biodiesel are constantly being tested. This means that different routes and scales of production, different raw materials and inputs should be studied, whose purpose is to evaluate the quality of biodiesel produced [3]. In this context, an attractive alternative for the production of biodiesel in the Meso Region of Alto Turi, specifically in the municipality of Zé Doca (Maranhão, Brazil), lies in the transesterification of oils extracted from numerous oleaginous plants and animal fats. This locality presents conditions to generate different biofuel production routes. Among the various species with potential for the production of biodiesel, the gong or coconut bug (Pachymerus nucleorum) stands out. When subjected to heating, it decomposes, originating an oil equal to that extracted from the seeds of oilseed plants. It is the larval stage of a coleopteran of the family Bruchidae, a beetle, which lives inside fruits of buriti (*Mauritia flexuosa*), tucum (Bactris setosa), babaçu (*Orbignya speciosa*) and carnauba (*Copernicia prunifera*) until the adult stage [3, 15, 16].

The adult female lays her eggs on the palm seeds at the time of infructescence, when the shell is forming and is less hard. The eggs hatch about 10 days later and the larvae penetrate the fruit. In this larval stage, lasting up to 90 days, the coleopteran (beetle) has a white color, black ocelli at one end of the body, along with the mouthpiece and is about 2 cm long as can be seen in **Figure 2** [16].

It is believed that the production of biodiesel from the oil extracted from the gong is an innovative idea and with relevant potential for society. In the literature,

Figure 2.
Visual aspect of the gong. Legend: a: tucum coconut, b: gong in tucum endocarp and c: gong. Source: [17].

so far, no work has been found on this energy input and its use in the production of biodiesel. The production of biodiesel in Brazil, besides being an alternative for energy self-sufficiency, can also generate employment and income opportunities and contribute to the settlement of people in the countryside.

2. Obtaining biodiesel on a laboratory scale

2.1 Sample collection

As discussed earlier, biodiesel can be obtained from low value-added fat feedstocks, i.e., with high free fatty acid (FFA) content. Such feedstocks are pretreated to suit the alkaline methanol transesterification process.

Initially, in native forest in the municipality of Zé Doca (Maranhão, Brazil), tucum (Bactris setosa) coconuts were collected, containing a small hole in its endocarp, indicative of containing the gong (Pachymerus nucleorum) inside these materials. The endocarp of the selected coconuts was broken and the gongs were removed from its interior.

2.2 Thermal extraction of coconut oil

The larvae of Pachymerus nucleorum are washed thoroughly with clean water in a 1.00 mm sieve. They are then added to an aluminum tank with a capacity of 2000 mL and heated to about 120°C, until they release all the oil contained in them. After cooling, the oil is filtered and stored in polyethylene bottles and made available for physical–chemical characterization.

The extractive performance of gong oil is evaluated as a function of the refined oil mass and the mass of oil used throughout the extraction. The yield was calculated using the following equation.

$$INCOME = \frac{OIL_{produced(g)}}{Gongo_{used(g)}} \cdot 100 \quad (1)$$

2.3 Physical-chemical characterization of the oil extracted from the gong

The oil extracted from Pachymerus nucleorum larvae was physicochemically characterized in terms of acidity index (A.I.), free fatty acids (FFA), moisture content (%H2O), density (D) and saponification index (Is) as recommended [17–19].

The acid value is defined as the number of milligrams of potassium hydroxide (mm KOH/g of sample) sufficient to neutralize the fatty acids present in a given quantity of oil or fat. This parameter is determined by the neutralization titrimetric method. In a typical assay, approximately 2.0 g of the sample is weighed into a 250 mL conical flask and 25 mL of neutralized alcohol (96°GL ethanol + ethyl ether in a 1,2 ratio) and 3 to 5 drops of the phenolphthalein indicator are added. This mixture was titrated with 0.1 Mol. L^{-1} until it turned from colorless to pink. The acidity index was expressed in mg KOH/g of sample and calculated using the equation below:

$$IA = \frac{V_{spend} \cdot [NaOH] \cdot f_c \cdot 56,1}{P_{sample}} \tag{2}$$

Where, Vgasto is the volume of NaOH spent in the titration; [NaOH] is the molar concentration of NaOH; fc is the correction factor of the NaOH solution; 56.1 milliequivalent gram of KOH and Pamsample is the weight of the sample.

The content of free fatty acids (FFA) contained in the samples is determined analogously to the acid value. The percentage of free fatty acids in terms of oleic acid (O.A.) was calculated according to the following equation.

$$AGL = \frac{V_{spend} \cdot [NaOH] \cdot f_c \cdot 28,2}{P_{sample}} \tag{3}$$

Where, 28.2 corresponds to the milliequivalent gram of oleic acid.

The determination of the water content consists of the difference between the mass of the sample considered wet (Aúmida) and the mass of the dry sample (Aseca) after being submitted to drying in an oven for a period of 3 hours at 110 ± 5°C. The percentage of water contained in the samples was determined by the following equation.

$$\%H_2O = \frac{Peso\left(A_{wet-}A_{dry}\right)}{Peso\left(A_{wet}\right)} \times 100 \tag{4}$$

The saponification index (Is) is defined as the number of milligrams of potassium hydroxide required to neutralize the fatty acids, resulting from the hydrolysis of one gram of the sample. The saponification index is determined by the Koesttstafer method, and consisted of heating a mixture containing 2.0 g of the sample and 25 mL of 4% alcoholic KOH solution, with a reflux cooler, for 30 minutes at a mild temperature (approximately 50°C), followed by titration with hydrochloric acid at 0.5 Mol. L-1. Analogously, a blank sample is carried out. The saponification index was determined by the following equation.

$$I_s = \frac{\left(V_{white} - V_{sample}\right) \times f_C \times 28}{Wheight_{sample}} = mgKOH/g_{sample} \tag{5}$$

Where, Vsample is the volume of HCl spent in the titration of the treated sample; Vblank is the volume of HCl spent in the titration of the blank sample; fc is the correction factor and SampleWeight is the weight of the sample used during the analysis.

Density is the mass per unit volume at a specified temperature. Parameter determination was performed by the pycnometry method. In a typical analysis, a 5 mL capacity pycnometer was weighed dry (m1). Vegetable oil was then added until it reached its maximum capacity. The system was weighed (m2) again and by mass difference (m2- m1) the mass of the oil and its density (D) were determined using the following equation:

$$D = \frac{mass_{oil}}{Volume_{pycnometer}} \quad (6)$$

2.4 Heterogeneous acid esterefication from gong oil

The gong oil samples containing high free fatty acid (FFA) content are previously treated by the homogeneous acid esterification method, in order to reduce the acidity content and make it suitable for homogeneous alkaline transesterification. In this step the homogeneous acid catalyst, sulfuric acid (H_2SO_4) is used.

The homogeneous acid esterification reactions are carried out in a round bottom flask of 500 mL capacity, coupled to a reflux system kept under rigorous stirring at 70°C for 3 hours.

In a typical experiment, 0.018 mol of gong oil is added 0.142 mol of methanol and 0.4% catalyst relative to the base mass of the oil, with a molar ratio of 1:8 (one to eight) between the oil and methanol. The system is kept in rigorous stirring throughout the process. The reaction time consisted of 240 minutes and temperature around 90°C. The product is recovered using the centrifugation technique, rotating at 2500 rpm, for 15 minutes. The oily part is subjected to heating at 100°C, to eliminate water and methanol residues and destined later for physicalchemical characterization, to measure the efficiency of the esterification process and also for biodiesel production by conventional alkaline transesterification methylation route.

2.5 Basic trasnesterification of esterified gong oil

The esterified gong oil containing reduced free fatty acid content is subjected to the traditional basic transesterification process to obtain biodiesel.

The biodiesel production is carried out in a flat-bottomed flask containing three mouths and 250 mL capacity, coupled to a reflux system.

In a typical experiment, 5.0 g (0.018 mol) of the starting vegetable oil is added to 4.6 g of methanol (0.144 mol) and 0.05 g of alkaline catalyst, NaOH, (1% relative to the base mass of the oil), with a molar ratio of 1:8 (one to eight) between the oil and methanol. The system is kept under strict magnetic stirring throughout the process. The reaction time consisted of 120 minutes and temperature of 90°C. The reaction mixture is added into a settling funnel for phase separation. The lower, glycerinous phase (by-product) was discarded and the upper phase (methyl ester mixture) was washed with acidulated sulfuric acid water (H_2SO_4 at 0.01 Mol. L^{-1}) and subsequently with heated water until the final product was clear. **Figure 3** shows the traditional alkaline transesterification process of obtaining biodiesel from esterified gong oil.

2.6 Characterization of gongo biodiesel

Different electroanalytical techniques are currently used to quantify and qualify biodiesel aiming the knowledge of the physical–chemical profile of this energy input and also to adapt it to the specifications required by the National Agency of Petroleum, Natural Gas and Biofuel (ANP), the Brazilian agency responsible for the quality control of petroleum products and biofuels. In this specific case, biodiesel was characterized in terms of spectroscopy in the infrared region (IR) and gas chromatography (GC) [21].

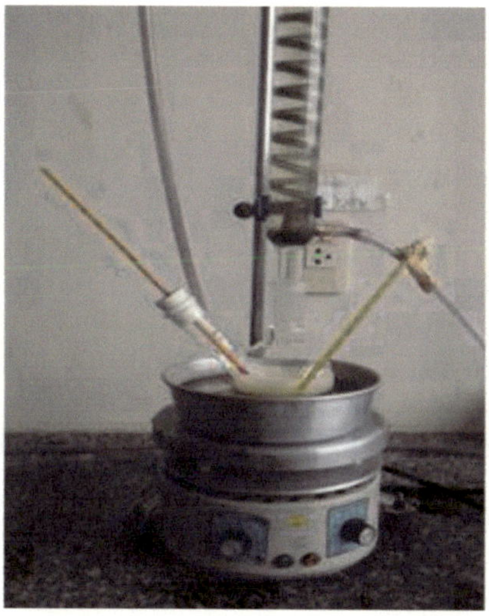

Figure 3.
Basic homogeneous transesterification process of gong oil [20].

2.6.1 Fourier transform infrared spectrocospy

Fourier transform infrared spectroscopy (FTIR) is a widely used technique in qualitative analysis, synthesis, and biochemical transformations. FTIR measures the vibrational transition when a material absorbs energy in the infrared (IR) region. Different functional groups and their bond types exhibit infrared absorption frequencies and intensities. This electroanalytical technique is a highly relevant tool for identification and structural elucidation of a chemical substance, in addition to enabling the control and monitoring of reactions [22].

The absorption spectra in the infrared region with Fourier transform (FTIR) for the esterified gong oil and biodiesel were recorded in the range of 400–4000 cm^{-1}, using a KBr tablet. The tests were performed at the Institute of Chemistry, UFRJ, in a PerkinElmer spectrophotometer, model spectrum 100 with spectral resolution of 4 cm^{-1}.

2.6.2 Gas chromatography with flame ionization detector

The chromatography technique with flame ionization detector (GC-FID) is used for the determination of the total content of fatty acid methyl or ethyl esters (FAME) from the transesterification reaction of vegetable oils and/or animal fats. The ANP specifications contained in Resolution No. 45 of August 2014 (ANP 45/2014) for analysis of the content of esters present in biodiesel were constituted based on the EN 14103 and ANBT 15342 standards. The technique of gas chromatography with flame ionization detector (GC-FID) for determination of the total ester content requires analytical standards of each FAME to confirm the results [22].

Compositional analysis of the constituent methyl esters of biodiesel produced from esterified gong oil was performed using a gas chromatograph 7890A CG series from Agilent Technologies coupled with a flame ionization detector (GC-FID). This equipment used a CPWAX 52CB capillary column 30 m long, internal diameter 0.25 mm, film thickness 0.25 μm, under the following conditions: injection volume = 0.5 μL, oven at 175°C, injector temperature at 250°C, detector temperature = 390°C, hydrogen pressure = 200 kPa, flow rate of 2 mL min^{-1} and analysis time of 20 minutes. After obtaining the chromatogram the composition was calculated from the area of each of the respective fatty acid esters. The characterization of the biodiesel fat profile of gong by GC-FID was performed at the School of Chemistry, Federal University of Rio de Janeiro.

3. Results and discussion

3.1 Thermal extraction of gong oil

The extracted gong oil presented a physical aspect, with medium viscosity, straw yellow coloration, clear and free of impurities as shown in **Figure 4**.

In the course of this work, three thermal extractions of the gong oil collected from the tucum coconut (Bactris setosa) were performed. A gravimetric yield in the range of 29 to 38% was obtained. The following equation shows the gravimetric yield of the second extraction. A total of 599 g of gong was used and 208.17 g of oil was obtained.

$$Incomeect = \frac{208,17}{599}x100 = 34,75\% \qquad (7)$$

It is believed that the production of biodiesel from the oil extracted from the gong is an innovative idea. In the literature to date no work has been found on this energy input. Comparing the gravimetric yield of some oilseed plants, such as: corn kernels with 4%, cottonseed with 15%, linseed with 34%, and soybeans with 18%. It must be agreed that the yield of thermal extraction of gong oil is of excellent quality.

3.2 Physicochemical profile of gong oil

Table 1 shows the physicochemical profile revealed for the oil extracted from the gongo in its raw or in natura form. The gongo oil was divided into two samples in order to perform its physicochemical characterization. In turn, each sample was divided into three aliquots, the analyses were performed in triplicates totaling 30 assays.

Acid value and moisture content are the main parameters affecting the sustainable production of biodiesel by the homogeneous alkaline transesterification route. A free fatty acid (FFA) content higher than 0.5% and a moisture content higher than 0.25% limit biodiesel production by basic homogeneous catalysis.

The acidity index is a quality parameter that indicates the amount of free fatty acids originating from the hydrolysis of glycerides. A high FFA content is indicative

Figure 4.
Physical aspect of gong oil in natura [20].

Samples	Quality Control Parameters				
	H_2O(%)	IA(mg KOH/g)	AGL (%A.O)	I_s(mg KOH/g)	D (Kg/m^3)
AM1	0,17 (± 0,05)	3,26 (± 0,41)	1,63 (± 0,20)	103 (± 2,16)	915 (± 0,00)
AM2	0,19 (± 0,00)	3,41 (± 0,00)	1,74 (± 0,00)	204 (± 0,00)	887 (± 0,00)

Table 1.
Physical–chemical characterization of gong oil in natura *[20].*

that the oil is undergoing breakdown in the glycerol chains, releasing its main constituents. The acidity of oils tends to increase with prolonged storage due to the oxidation of free fatty acids, which can compromise their aroma, color, and flavor due to their rancidity process [23]. In this sense, the high acidity values found in **Table 1** may be related to the storage time of the samples (6 months) and also to the rudimentary way in which the oil was extracted.

In this work we obtained AI:3.26 mg KOH/g (± 0.41) and FFA:1.63% (±0.20) for sample 1 and AI:3.41 (± 0.00) and FFA: 1.74 (±0.00) for sample 2. Obtained the following results for tucum oil extracted by mechanical pressing IA: 37.5 mg KOH/g (±0.40) and FFA:18.86% (±0.26) [24], it is noted when comparing the results that in both cases the acidity of the oil extracted from the gong contained in the tucum coconut or properly from the tucum almonds present a high acidity index indicating that the gong oil was not appropriate for biodiesel production, requiring a treatment to adapt it to the biofuel production process.

Another parameter that influences the biodiesel production is water. This substance deactivates the catalyst forming soap, hinders the separation of product phase (biodiesel) and byproduct (glycerin), besides generating effluents that contaminate the environment. **Table 1** shows that the average water content was 0.17% (± 0.05) and 0.19 (± 0.00) for samples 1 and 2, respectively. On the other hand, [25], obtained a moisture content of 0.13% (± 0.00) for tucum extracted oil. Regarding the moisture content, it can be inferred that the gong oil in natura was within the specification for biodiesel production.

The saponification index (Is) is a property that has a strong influence on the quality of an oil. The saponification reaction can indicate the degree of deterioration and stability of an oil. The gong oil revealed Is:103 mg KOH/g (± 2.16) for sample 1 and for sample 2, Is:204 (± 0.00). According to the British standard an oil that is considered first quality should have a saponification index in the range of 177 to 187 mg KOH/g of the sample. The higher the saponification index, the greater is its application for food purposes [23]. In the view of the English standard, the Is of sample 1 is not first quality. The Is of sample 2, on the other hand, [24], is ideal for human consumption and coincidentally this input is widely used for this among the countryside populations. The difference in Is between the analyzed samples may be related to regional climatic conditions, since sample 1 was collected in the dry season (summer) and sample 2 was collected in the rainy season. The rainy season contributes greatly to the rancidity of the oil in natura (crude) because with the increase in moisture content there is hydrolysis of the oil, release of fatty acids and decomposition by the action of microorganisms.

Density is a physicochemical parameter of high relevance in the quality of biodiesel. This parameter is directly related to the chemical composition of the oil used for biodiesel production. The stronger the intermolecular interactions existing in the raw materials the higher the density will be. These interactions increase with the number of carbons in the chain (single bonds) and decrease the greater the number of unsaturated bonds (double bonds) contained in the oil composition. The analysis of the chemical composition of the gong oil performed by gas chromatography (GC) revealed as the majority acid, the C12:0 (lauric acid). In **Table 1** the mean density value for gongo oil was 915 Kg/m3 (± 0.00) and 887 Kg/m3 for samples 1 and 2 respectively. On the other hand, found similar density (D:889 kg/m3 ± 0.00) [24], for the oil extracted from tucum by the physical method. When an oil suffers the influence of temperature there is a reduction in its density, in biodiesel the density is linked to the cetane number, which is an indicative property in the ignition delay time of diesel cycle engines and the calorific value, directly affecting the engine performance [23]. Fuel injection systems, thus, suffer changes in fuel density and influence engine power, in view of the addition of different mass to be injected. In addition, fuel density and viscosity affect injection pressure, fuel atomization, and engine performance.

3.3 Esterefication of gongo oil with high acidity

3.3.1 Homogeneous acid esterification of gongo oil

Table 2 presents the results of the physicochemical characterization revealed for the esterified gong oil. Regarding the water content, when compared with **Table 1** (characterization of the oil in natura), it is observed that there was a significant increase (from 0.17 to 0.8) in the moisture content of sample 1 and for sample 2 there was a decrease, around 26% of moisture (from 0.19 to 0.14). The increase in water content in sample 1 may have occurred due to the inefficiency of the dehumidification stage, since during dehumidification water is formed as a byproduct. On the other hand, it is observed that sample 2 met the specification for biodiesel production since it revealed 0.14%.

The acidity index for both sample 1 and sample 2 showed significant reduction in the range of 46% (from 3.26 to 1.5 mg KOH/g) and 32% (from 4.41 to 1.09 mg KOH/g) respectively. However, these values are above the specification of RDC270 ANVISAMS, whose optimal limit is 0.6 mg KOH/g of the sample. Oil treatment by homogeneous acid esterification significantly improved the quality of gong oil for biodiesel production. The FFA content also significantly reduced around 52% for sample1 (from 1.63 to 0.35%) and 32% for sample 2 (1.74 to 0.55%). Although the reduction is significant the treatment of the crude oil did not reach the desired specification on the order of 0.5% in sample1, however, in the light of statistics this is a small difference that does not prevent the esterified gong oil from being used in the manufacture of biodiesel.

As for the saponification index (Is), it was observed that there was some stabilization of the said quality control parameter in the range of 103–108 mg KOH/g of the sample indicating that according to English quality standards gong oil is not considered first quality.

Regarding the density, it was observed that there was a stable density in the range of 905–908 km/m3 as the majority acid in gong oil is C12:0 containing greater

Samples	Quality Control Parameters				
	H_2O(%)	IA(mg KOH/g)	AGL (%A.O)	I_s(mg KOH/g)	D (Kg/m^3)
AM1	0,8 (± 0,14)	1,5 (± 0,00)	0,85 (± 0,07)	108 (± 0,00)	908 (± 0,00)
AM2	0,14 (± 0,08)	1,09 (± 0,00)	0,55 (± 0,00)	103 (± 0,00)	905 (± 0,00)

Table 2.
Physical–chemical characterization of esterified gong oil [20].

intermolecular interaction due to its simple bonds in the carbon chain the density becomes high and tends to reduce when the raw material is subjected to treatment involving high temperature, as occurred with the gong oil in the course of the acid homogeneous esterification process.

The results obtained after the esterification of the gong oil as shown in **Table 2** led us to infer that the esterification treatment of this oil was efficient because it significantly reduced the FFA content, but not effective enough to reach the desired level (0.5% FFA). **Figure 5** illustrates the physical aspect of Pachymerus nucleorum and its oils.

3.3.2 Homogeneous transesterification of esterified oil

The production of biodiesel from fresh gong oil was carried out in two steps: homogeneous acid esterification and homogeneous basic transesterification, using 0.018 mol of oil and 0.142 mol of transesterification agent (methanol), 1% alkaline catalyst (0.05 g NaOH), in relation to the mass of oil molar ratio between oil and methanol in the order of 1:8 (one to eight), reaction temperature of 90°C and residence time of 120 minutes.

When the feedstock for biodiesel production contains a high content of free fatty acid, the yield through alkaline homogeneous transesterification method is low with soap formation and deactivation of the catalyst. A viable alternative to adapt this oil to the biodiesel process is its treatment through homogeneous acid esterification aiming to reduce its free fatty acids. **Figure 6** shows the visual aspect from the extraction of gong oil to the biodiesel obtained on a laboratory scale from this esterified oil.

3.4 Characterization of gongo oil and biodiesel

3.4.1 Fourier transform infrared spectroscopy

The elucidation of the chemical behavior of the products of the homogeneous acid esterification and homogeneous alkaline transesterification reactions of gong oil was investigated using Fourier transform infrared absorption spectroscopy (FTIR).

Figure 5.
Physical aspect of gong and its crude and esterified oils [20].

Figure 6.
Products obtained from the biodiesel processing steps. Caption: a: gongo sludge; b: gongo oil in natura; c: esterified gongo oil and d: gongo biodiesel [20].

The IR spectra of the esterified gong oil (**Figure 7a**) and biodiesel (**Figure 7b**) samples are shown in **Figure 5**. They showed similar band positions, intensity and wave number. [22, 26, 27] Characterized vegetable oils and biodiesels produced from different vegetable oils and obtained FTIR results similar to the results revealed in this work.

In the region of the functional groups that comprises the range 4000–1650 cm^{-1} few bands were observed. There, three sets of well-defined bands were revealed, two bands in the range 2853–2925 cm^{-1} referring to the CH2 stretching of alkanes and one at 1744 cm^{-1} indicative of the C=O (carbonyl) stretching that characterizes the double bond region.

The region between 1650 and 500 cm^{-1} is called the fingerprint region. In this region more bands were found in relation to the region of the functional groups, however, the bands obtained between 1171 and 1172 cm^{-1} allusive to the vibrations of the C-O bond, indicative of the stretching of the ester grouping, stand out. Between 722 and 723 cm^{-1} bands referring to asymmetric CH deformation, characteristic of long hydrocarbon chains, were revealed. Another relevant factor observed in **Figure 7a** and **b** was the absence of absorption of broad bands in the region between 2500 and 3300 cm^{-1}, showing that the gongo oil and biodiesel had low moisture content. This fact was confirmed in the course of the immediate chemical analyses shown in **Table 2**.

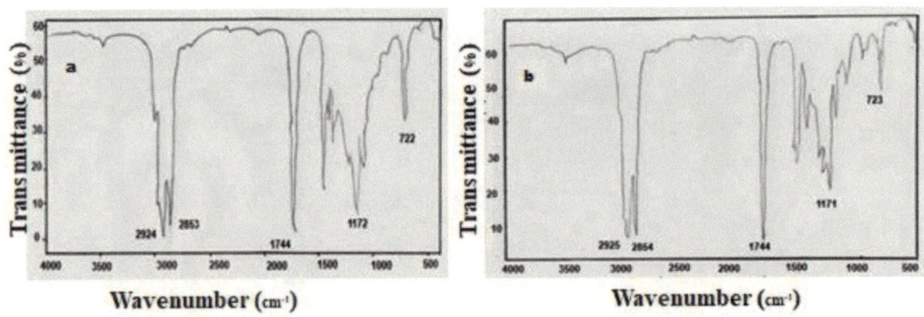

Figure 7.
FTIR spectrum. (a) Gong oil; (b) Biodiesel.

3.4.2 Fat profile of biodiesel obtained by GC-FID

In Brazil, the National Agency of Petroleum, Natural Gas and Fuel (ANP) is the body responsible for overseeing the quality parameters of biodiesel through Resolution No. 45, dated August 13, 2014. It regulates the characteristics of biofuels. The quality parameters are used to determine whether a given product is fit for consumption and not harmful to the population and the environment.

The ideal biodiesel for Resolution 045/2014-ANP should contain in its chemical constitution 100% of fatty acid methyl esters (FAME), however, in view of the low conversion of monoacylglycerols (MAG), diacylglycerols (DAG) and triacylglycerols (TAG) and competition between the secondary reactions throughout the transesterification process it becomes difficult to separate the esters formed and the reaction impurities. Consequently, at the end of the transesterification reaction, a mixture containing biodiesel and impurities is formed. Hence the importance of characterization of biodiesel and the feedstock that gave rise to it.

Among the specific quality parameters for biodiesel determined by the ANP is the ester content. The ANP has established an ester content of at least 96.5%. A low ester content directly influences the physical–chemical properties of biodiesel. Therefore, to obtain a high quality biofuel, the transesterification reaction must be complete and the purification process after the reaction must be effective. Thus, the content of remaining contaminants (FFA, methanol traces, water, etc.) should be low, so that the ester content meets the ANP specifications.

The methyl biodiesel produced from the esterified gong oil was characterized by gas chromatography with flame ionization detector (GC-FID) technique for ester conversion. The chromatographic profile in terms of fatty acid methyl esters (FAME) of the biodiesel is presented in **Figure 8**.

The FAME centesimal composition of the biodiesel is shown in **Table 3**. Therein it was observed that the majority constituent was dodecanoic acid methyl ester (methyl laurate, C 12:0) with 36.85%; followed by octadecanoic acid methyl ester (methyl stearate, C18:0) with 25.47%, tetradecanoic acid methyl ester (methyl myristate, C14:0) with 23.37% and hexadocanoic acid methyl ester (methyl palmitate, C16:0) with 14.31%.

The technique of gas chromatography with flame ionization detector (GC-FID) is suggested by standards EN 14103 and NBR 15764 for determining the content of fatty acid methyl esters (FAME) from the transesterification reaction of vegetable oils and, to confirm the results are necessary to use analytical standards for each FAME.

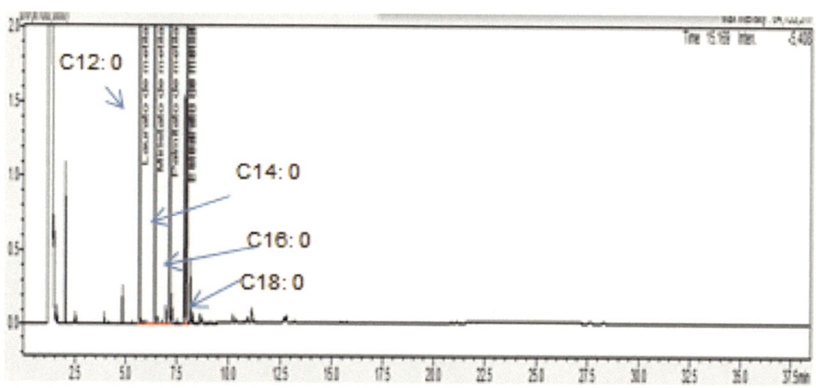

Figure 8.
Chromatographic profile of methyl esters contained in gong biodiesel. Source: the author himself.

Fatty acid methyl esters (FAME)		Amount (%)
Symbol	Name	
C 8:0	Methyl Caprylate	Nd
C 10:0	Methyl Caprolate	Nd
C 12:0	Methyl Laurate	36,85
C 14:0	Methyl Myristate	23,37
C 16:0	Methyl Palmitate	14,31
C 18:0	Methyl stearate	25,47
C 18:1	Methyl Oleate	Nd
C 18:2	Methyl Linoleate	Nd
	Total percentage	100

Legend: Nd: not detected.

Table 3.
Composition of fatty acid methyl esters from gong biodiesel [20].

According to the ANP (2014), the determination of the ester content can be done following the EN 14103 methodology. The ester content based on this standard is performed by internal standardization with methyl heptadecanoate (C17:0) and only considers the esters of chain from C14:0 to C24:1 in the calculation of the ester content. Since in the characterization of biodiesel produced from gong oil it was revealed that the majority fatty acid methyl ester was C12:0 (methyl laurate), the EN 14103 technique did not allow determining the conversion of esterified gong oil into biodiesel (methyl esters). Due to financial and technical difficulties that permeate our institution it was not possible to use other means to do the conversion. On the other hand, comparing the FTIR results for the esterified gongo oil and biodiesel with the results revealed for the fat profile of gongo biodiesel through the GC-FID technique is notable the presence of bands in the infrared absorption region allusive to the axial deformation of carbonyl (C=O) indicating the presence of fatty acid methyl esters, which were confirmed in the GC-FID tests, and the absence of bands between 2500 and 3300 cm^{-1} indicating low percentage of water in biodiesel studied. Such occurrences allow us to predict the results of some quality control parameters of biodiesel. For example, the values shown in **Table 3** (methyl ester composition) for gongo biodiesel lead us to infer as to the degree of FAME saturation that the gongo biodiesel will have:

a. **Good ignition quality** - the longer the saturated carbonic chain, the higher its cetane number (NC), which provides better combustion conditions,

b. **High oxidative stability** - the saturated chains make biodiesel more chemically stable, more resistant to oxidation if not stored or transported properly,

c. **Low cloud point** - saturated chains solidify easily at low temperatures,

d. **Good viscosity and low lubricity power** - although biodiesel naturally has good lubricity power, methyl esters of unsaturated fatty acids add greater lubricity power relative to biodiesel containing saturated FAME. Conversely, viscosity increases with increasing saturated chains.

e. **Moisture content compatible with Resolution 045/2014-ANP** - the absence of absorption bands in the infrared region, specifically in the range of

2500 to 3300 cm^{-1} allows us to predict that gong biodiesel contains low moisture content.

4. Final considerations

This work allowed the obtainment of very interesting data about the use of low value-added raw materials aiming at reducing the costs of the process of sustainable biodiesel production. The raw material used was the oil extracted from the larvae of the Pachymerus nucleorum easily found in the native forests of the mesoregion of Alto Turi (Maranhão, Brazil), which has little or almost no commercial value.

The technology for extraction of oil from the larvae of Pachymerus nucleorum is simple and to be suitable for sustainable production of biodiesel should be optimized, since it has a high acidity index, which hinders its direct use in traditional transesterification. On the other hand, the oil yield in the course of thermal extraction is very inviting, and may have a cost-effective effect, which may contribute to biodiesel being more competitive with fossil diesel.

The results revealed throughout the physical–chemical characterization of the oil (in natura and esterified) and of the biodiesel originating from the larvae of Pachymerus nucleorum indicated that this oil when previously treated can be used in the transesterification process by basic homogeneous catalysis.

Finally, Brazil in view of its geographical location is a country that emerges with a high potential for the development of innovative technologies in the area of renewable energy production, because it has tropical forests with vast biodiversity consisting of tropical forests rich in mineral resources, with exuberant fauna and flora with food, timber, steel, catalytic, medicinal, and renewable energy applications.

Acknowledgements

The authors thank the support given by IFMA-Campus Zé Doca, by the Laboratory of Reactivity of Hydrocabinetics, Biomass and Catalysis (Larhco) from UFRJ, by the Research Group in Sustainable Chemical Analysis (GPAQS) from IFMA-Campus Zé Doca. They also thank Luiz Eduardo Santos Vieira for the honorable contribution in the translation of this work written initially in Portuguese and later translated into English.

Author details

José Sebastião Cidreira Vieira*, Wilame Silva Castro,
Makson Rangel de Melo Rodrigues and Neemias da Silva Nascimento
Instituto Federal do Maranhão-Campus Zé Doca, Brazil

*Address all correspondence to: sebastiaocidreira@ifma.edu.br

References

[1] ARANSIOLA, E. F., OJUMU, T. V., OYEKOLA. O. O., MADZIMBANUTO, T. J. A review of current technology for biodiesel production: state of the art. Biomass and Bionergy, vol 61, p.276-297, 2014.

[2] DEY, S.; REANG, N. M. ; P.K. DAS, P. K.; DEB, M. A comprehensive study on prospects of economy, environment, and efficiency of palm oil biodiesel as a renewable fuel. Journal of Cleaner Production, v 286, 2021.

[3] SOUZA, B. R.; SOUSA. T. L.; ALVES, W. S.; VIEIRA, J. S. C. Resumos do 5° Congresso da Rede Brasileira de Tecnologia de Biodiesel, Salvador, Brasil, 2012.

[4] POURHOSEINI A, S. H.; NAMVAR-MAHBOUB, E. M.; HOSSEINI, B. E.; ALIMORAD, C. A. A comparative exploration of thermal, radiative and pollutant emission characteristics of oil burner flame using palm oil biodiesel-diesel blend fuel and diesel fuel. Energy, v.217, 2021.

[5] MURTA, A. L. S.; DE FREITAS, M. A. V.; FERREIRA, C. G.; PEIXOTO, M. M. C. L. The use of palm oil biodiesel blends in locomotives: An economic, social and environmental analysis. Renewable Energy, v.164, p. 521-530, 2021.

[6] POUSA, G. P. A. G.; SANTOS, A. L. F.; SUAREZ, P. A. Z. History and Policy of biodiesel in Brazil. Energy Policy, 35, 5393-5398, 2007.

[7] SUAREZ, P. A. Z.; MENEGHETTI, S. M. P.; MENEGHETTI, M. R.; WOLF, C. R. Transformação de triglicerídeos em combustíveis, materiais poliméricos e insumos químicos: algumas aplicações da catálise na oleoquímica. Quim. Nova. 2007, 30, 667.

[8] ALEGRIA, A.; ARRIBA, M. J. R.; CUELLAR, J. Biodiesel production using 4-dodecylbenzenesulfonic acid as catalyst. Appl. Catal, B.2014, 160, 743.

[9] MELERO, J. A.; IGLESIAS, J; MORALES, G. Heterogeneous acid catalysts for biodiesel produce: current status and future challenges. Green chemistry. V11, n.9, p.1285-1308,2009.

[10] ZUO, D. et al. Sulfonic acid functionalized mesoporus SBA-15 catalysts for biodiesel production. Applied Catalysis B: Environmental, V.129, p.342-350, 2013.

[11] ENCINAR, J. M.; SANCHEZ, N. G.; GARCIA, M. L. Study of biodiesel production fron animal fats with free fatty acid content. Bioresource Tecnology, v. 102, p.10907-10914. 2011.

[12] DRELINKIEWICZ, A. et al. Organosulfonic acid doped polyaniline solid acid catalyst as basic for the formation of esters bio-in reaction of esterification and transesterification. Journal Fuel, v.116, p.760-771, 2014.

[13] NATJA-NAN, B. N.; KIATKITTIPONG, W.; AIOUACHE, F.; ASSABUMRUN-GRAT, S. Process design of continuous biodiesel production by reactive distillation: comparison between homogeneous and heterogeneous catalysts. Chemical Engineering and Processing : Process Intensification, v.92, p.33-34, 2015.

[14] ALCHRIS, W. G. A,; SUTANTO, S.; ONG, L. K.; TRAN-NGUYEN, O. L.; ISMADJI, S.; JU, Y. H. Developments in in-situ (trans) esterification for biodiesel production: a critical review. Renewable and Sustainable Energy, v.60, p.284-302, 2016.

[15] OLIVEIRA, M. E. C. Produção de biodiesel de óleo de palma e seus resíduos graxos por transesterificação e esterificação. Embrapa Documentos 339. Brasília: 2008.

[16] FERREIRA, C. Empoderamento: Quem já comeu gongo de babaçu?. Disponível em: <http//saomateusdomaranhao.blogspot.com/2010/10/quem-já-comeu-gongo-de-coco-babacu.html>. acesso em 25 fev 2018.

[17] MORETO, E.; ALVES, R. F. Óleos e gorduras: processamento e análise. Florianópolis: Ed. Da UFSC, 1986.

[18] INSTITUTO ADOLFO LUTZ. Normas analíticas do Instituto Adolfo Lutz, v.1: Método químico e físicos para análise de alimentos- óleos e gorduras. 5 ed. São Paulo. IMESP, 2004.

[19] TECNOLOGIAS BIOENERGÉTICAS. Procedimentos operacional padrão da Brasil Ecodiesel-Análise do controle de Qualidade, 30p, 2008.

[20] CASTRO, W. S. Extração e preparação do óleo de gongo (Pachymerus nucleorum) para obtenção de biodiesel por rota metílica. Monografia (Graduação) – Curso de Licenciatura em Química. Instituto Federal de Educação, Ciência e Tecnologia do Maranhão – Zé Doca, 2019.

[21] Agência Nacional de Petróleo, Gás Natural e Biocombustíveis (ANP). Resolução n.45 de 25/08/2014. Disponível em: www.anp.gov.br/biocombustíveis/biodiesel. Acesso em: 15 dez 2018.

[22] RAMPAZZO, V. Avaliação térmica de óleos vegetais por termogravimetria, cromatografia gasosa e espectroscopia de infravermelho médio. 2015, 93f. Dissertação (Mestrado em Engenharia de Alimentos) – Universidade Federal do Paraná, Curitiba, 2015.

[23] VIEIRA, J. S. C. Síntese de catalisadores heterogêneos ácidos e básicos para a produção de biodiesel. 2017, 144f. Tese (Doutorado em Engenharia de Processos Químicos e Bioquímicos) – Escola de Química. Universidade Federal do Rio de Janeiro, Rio de Janeiro, 2017.

[24] BARBOSA et al. Aproveitamento do óleo das amêndoas de tucumã do amazonas na produção de biodiesel. Acta Amazônica, vol.39, pág. 371-376, 2009.

[25] COSTA, W. C. L et al. Análise físico-química de óleos provenientes de dois municípios da ilha de Marajó-PA. In: 57° CONGRESSO BRASILEIRO DE QUÍMICA, 10,2017, Gramado(RS). Anais. Gramado, 2017.

[26] ALBUQUERQUE, G. A. Obtenção e caracterização físico-química do biodiesel de canola (*Brassica napus*). 2006, 100f. Dissertação (Mestrado em Química) – Universidade Federal da Paraíba, João Pessoa, 2006.

[27] SILVA, T. A. R. Biodiesel de óleo residual: produção através da transesterificação por metanólise e etanólise básica, caracterização físico-química e otimização das condições reacionais. 2011, 151f. Tese (Doutorado em Química) – Universidade Federal de Uberlândia, Uberlândia, 2011

Chapter 17

Lignocellulosic of Oil Palm Biomass to Chemical Product via Fermentation

Farhan M. Said, Nor Farhana Hamid, Mohamad Al-Aamin Razali and Nur Fathin Shamirah Daud

Abstract

The world's largest contribution to biomass comes from lignocellulosic material. Oil palm biomass is one of the most important sources of lignocellulosic material in Asia, with biomass produced four times that of palm oil. Oil palm trunk (OPT), oil palm empty fruit bunches (OPEFB), oil palm frond (OPF), and palm oil mill effluent (POME) are examples of biomass lignocellulosic materials produced. Unfortunately, the majority of waste is disposed of in landfills, causing serious environmental issues such as global warming and the greenhouse effect. These wastes are known to contain a high concentration of cellulose and hemicellulose. Because of its high carbohydrate content, it has a promising future as a feedstock for the fermentation process, which can produce a variety of chemical products at a low cost. This chapter will describe the biochemical products produced from various oil palm biomass via various fermentation processes involving various microorganism strains.

Keywords: Oil palm, fermentation, lignocellulose, biomass, lignin

1. Introduction

The largest component of biomass available on the world is lignocellulosic material. Oil palm is one of the most important sources of lignocellulosic biomass in Asia, especially in Malaysia and Indonesia. Malaysia, after Indonesia, is the world's second-largest producer of palm oil, with a capacity of 17.32 million tonnes and a cultivated area of 5.74 million ha [1]. According to the Malaysian Palm Oil Board (MPOB) statistics from 2016, the total volume of oil palm products was 25.64 million tonnes [2]. Oil palm biomass accounts for the remaining 90% of total dry matter palms, with oil accounting for just about 10% of total dry matter palms [3]. Oil palm empty fruit bunches (OPEFB), oil palm fronds (OPF), and oil palm trunks (OPT) make up the majority of oil palm biomass in Malaysia.

Malaysia's annual production of OPEFB, OPT, and OPF is approximately 84.23 million tonnes (dry basis) (**Table 1**) [4]. This massive amount (i.e. 7 million tonnes of OPEFB, 21.4 million tonnes of OPT, and 55.8 million tonnes of OPF) suggests that oil palm biomass is a readily available feedstock for chemical products, particularly through the biological fermentation process.

Nowadays, fermentation processes are the most commonly used method of utilizing biomass because they are non-toxic, environmentally friendly, and have

Production site	Biomass type	Estimated amount (million tonnes dwb)	Remarks
Mill	Oil palm empty fruit bunch (EFB)	7.03	Based on annual fresh fruit bunch (FFB) yield
	Oil palm trunk (from replanting activity)	21.38	Based on 5% estimated oil palm planted area due for replanting
Plantation	Oil palm frond (from replanting activity)	4.16	Based on 5% estimated oil palm planted area due for replanting
	Oil palm frond (from pruning activity)	51.66	Based on 75% of oil palm planted area pruned per annum
Sources: Adapted from Bukhari et al. [4].			

Table 1.
Availability of lignocellulosic oil palm biomass in Malaysia.

a low operating cost. The current chapter outlined the role of microorganisms in the fermentation process for valorizing lignocellulosic oil palm biomass for various biochemical products.

2. Oil palm biomass

The biomass of the oil palm is a lignocellulosic biomass made up mainly of cellulose, hemicellulose, and lignin. The key component of oil palm biomass is cellulose

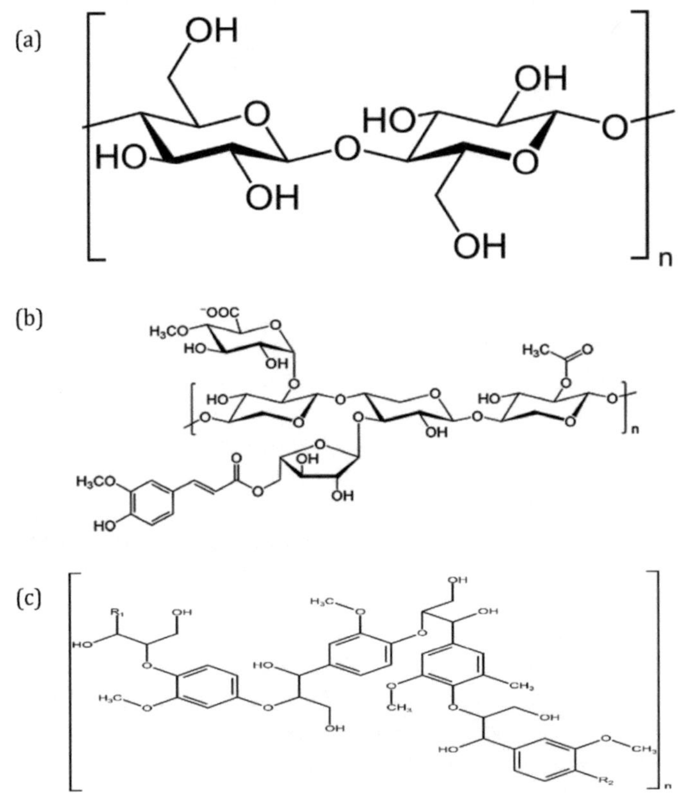

Figure 1.
Structure molecules; (a) cellulose; (b) hemicellulose, (c) lignin.

$(C_6H_{10}O_5)n$ (**Figure 1a**) which is the main component of oil palm biomass. Cellulose is a rigid, solid, and difficult-to-break linear polymer of glucose with additional hydrogen bonding [5]. By adding water to the polysaccharide, it is hydrolyzed into free sugar molecules (i.e. glucose, sucrose, and fructose) [6]. Saccharification is another name for this method.

Hemicellulose $(C_5 H_8O_4)n$ (**Figure 1b**), on the other hand, is made up of small, highly branched chains of various pentoses (such as xylose and arabinose) and hexoses and is found in secondary cell walls (i.e. mannose, galactose and glucose). Hemicellulose is simpler to hydrolyze than cellulose because of its weaker amorphous and branched structures [7].

The main non-carbohydrate component is lignin $[C_9H_{10}O_3(OCH_3)0.9–1.7]n$ (**Figure 1c**), which is a highly complex compound with a three-dimensional cross-linked polyphenolic structure [4]. Lignin is found between the cellulose cell wall and the hemicellulose cell wall, and it is responsible for the cell wall and the plant's overall strength. As a result, when used as a lignocellulosic biomass in the fermentation process, lignin presents a significant disadvantage because it is resistant to chemical and biological degradation.

3. Conversion of oil palm biomass into biochemical product

Oil palm biomass can be converted into biochemical products using a variety of methods, including fermentation, esterification, and anaerobic digestion. Fermentation is the most widely used of these because it is a non-toxic and environmentally friendly process.

Prior to the fermentation process, lignocellulosic biomass is typically pre-treated to break down the cellulose into simple sugars (i.e. maltose, glucose, fructose). The most important impact on fermentation results comes from pre-treatment [8]. Pre-treatment breaks down the biomass's recalcitrant structures, making cellulose more accessible to the organism.

The cellulose and hemicellulose content of oil palm biomass is high, which are primary sources for the fermentation process (**Table 2**). Microorganisms convert sugars extracted from lignocellulosic biomass to a variety of desired products during the fermentation phase. For example, *Saccharomyces cerevisiae* produces ethanol, *Lactobacillus sp*. produces lactic acid, and *Actinobacillus succinogenes* produces succinic acid.

There are two forms of fermentation that can be used in the fermentation method. The solid-state fermentation (SSF) method is the first. SSF creates a natural environment for filamentous fungi to grow, which has proven to be a more efficient method of producing various products [9, 10, 15–17]. Submerged fermentation (SmF) is the other kind of fermentation. Due to easier control and maintenance of fermentation factors, most cellulases and xylanase enzymes are commercially generated using SmF [18].

Oil palm biomass	Cellulose (%)	Hemicellulose (%)	Lignin (%)	Extractive (%)	Ash (%)
Oil palm empty fruit bunches (OPEFB)	38–65	17–33	13–37	2–4	1–6
Oil palm frond (OPF)	40–56	16–38	15–26	2–5	2–3
Oil palm trunk (OPT)	29–45	12–29	18–23	4–11	2–3

Sources: Adapted from references [4, 9–14].

Table 2.
Type of oil palm biomass and the chemical composition.

4. Type of oil palm biomass and the biochemical products

4.1 Oil palm trunk

The huge biomass output of the oil palm trunk (OPT) has piqued the interest of researchers. OPT has a 25 to 30 years average active life period [19]. The OPT is usually chopped into small pieces and left to rot naturally in the plantation area during the replanting period. Leaving the trunk in the plantation area may cause pollution because the OPT's high sugar and starch content will attract microflora and microfauna, raising the risk of plant diseases [20].

When compared to the other sections of oil palm trees, OPT sap contains liquid with a lower lignin percentage and a higher percentage of free fermentable sugars [20]. As a result, there is little to no need for pre-treatment (chemical or biological) to delignify or convert lignocellulose to fermentable sugar. **Table 3** lists the different biochemical products that are made with OPT as a substrate. The product's differences are mainly determined by the type of microorganisms used and the fermentation conditions (i.e. temperature, pH, oxygen level).

Organism	Fermentation system	Type of reactor	Product	References
Aspergillus fumigatus SK1	SSF	Flask	Enzymes • CMCase • Fpase • β-Glucosidase • Xylanase	[9]
Aspergillus fumigatus SK1	SSF	Flask	Enzymes • CMCase • Fpase • β-Glucosidase • Xylanase	[9]
Saccharomyces cerevisiae Kyokai no.7 (ATCC 26622)	SmF	Flask	Ethanol	[20]
A. niger	SSF	Flask	Enzymes • CMCase • FPase • β-Glucosidase • Xylanase	[21]

Table 3.
Biochemical products from different organisms using OPT as substrate.

4.2 Oil palm empty fruit bunches

One of the most significant lignocellulosic biomasses in Malaysia is oil palm empty fruit bunches (OPEFB). From palm oil production, OPEFB contributed 20% of total biomass [22]. In 2019, approximately 2.9 million tonnes of OPEFB were made, a figure that is expected to rise as global demand for oil palm grows [10]. OPEFB contains a high percentage of cellulose (38–65%), lignin (13–37%), and hemicellulose (17–33%) (**Table 2**). OPEFB can be processed into useful products, such as biochemical products, due to its abundance and high cellulose content (i.e. CMCase, Fpase, ethanol).

Organism	Fermentation system	Type of reactor	Product	References
Botryosphaeria sp.	SSF	Flask	Enzymes • CMCase • Fpase • β-Glucosidase	[9]
Aspergillus niger EFB1	SmF	Rotary drum reactor	Enzymes • CMCase • Fpase • β-Glucosidase	[23]
Yeast	SmF	Flask	Ethanol	[24]
Amauroderma rugosum SDBR-CMU-A83	SSF	Flask	Phytase	[25]
Mesophilic and thermophilic bacteria	SmF	100 ml serum vial	Biomethane	[26]
Trichoderma asperellum USM SD4	SSF	Flask	Xylanase	[27]
Pycnoporus sanguineus	SSF	Flask	Laccase	[10]

Table 4.
Biochemical production from different organisms using OPEFB as substrate.

Table 4 shows the results of various fermentation processes on biochemical products using OPEFB as the substrate. The goods are unique to each organism.

4.3 Oil palm frond

The most plentiful residue of oil palm trees is oil palm frond (OPF), which accounts for up to 70% of total palm waste [15]. According to recent studies, OPF is an excellent source of renewable carbon and lignocellulosic content for cultivating a variety of species to produce essential biochemical products such as pigments, enzymes, and succinic acid. **Table 5** summaries the biochemical product by various microorganisms using OPF as a substrate. When fermentation conditions such as temperature, pH, and oxygen level, are allowed, a variety of organisms may produce a variety of products. As shown in **Table 2**, OPF contains a high percentage of cellulose (40–56%) and hemicellulose (16–38%), making it ideal for microbial growth.

4.4 Palm oil mill effluent

The liquid waste released during the palm oil extraction process is known as palm oil mill effluent (POME). POME is one of the world's most polluting wastewaters due to its high organic matter content and it is 100 times more polluted than municipal sewage [30]. Each tonne of palm oil produces approximately 5.5–7.5 tonnes of POME [31, 32]. While, about more than 50 million m^3 of POME is generated globally each year [33].

POME is a viscous, dense brownish liquid with significant quantities of colloidal matter that is acidic (pH 3.7 to 4.5) [34]. POME also has a high chemical and biochemical oxygen demand (COD and BOD), ranging between 69,500 and 89,591 mg/L and 34,771 and 48,300 mg/L, respectively [34]. The physicochemical

Organism	Fermentation system	Type of reactor	Product	References
A. niger EFB1	SSF	Petri dish	Enzymes • Endoglucanase • β-glucosidase • Exoglucanase • Xylanase	[15]
T. asperellum UC1	SSF	Flask	Enzymes • CMCase • FPase • β-glucosidase • Xylanase	[14, 28]
R. oryzae UC2	SSF	Flask	Enzymes • CMCase • FPase • β-Glucosidase • Xylanase	[14]
Actinobacillus succinogenes 130Z	SmF	Serum vial	Succinic acid	[29]
Monascus purpureus FTC5356	SSF	Flask	Red pigment	[17]
Monascus purpureus FTC 5357	SSF	Stirred drum bioreactor	Red pigment	[16]

Table 5.
Biochemical production from different organisms using OPF as substrate.

properties of POME are shown in **Table 6**. A large amount of amino acids, inorganic nutrients (Na, K, Ca, Mg, Mn, Fe, Zn, Cu, Co, and Cd), small fibers with nitrogenous compounds, free organic acids, and carbohydrates are also found in POME [37]. Organic matter such as lignin (4700 ppm), phenolics (5800 ppm), pectin (3400 ppm), and carotene (8 ppm) are also present [34]. This suggests that POME is an appropriate source for biological treatment.

POME's physicochemical properties can vary depending on local and process factors (climate, organisms, pre-treatment, and oil extraction process, for

Parameters	Range concentration
Biochemical oxygen demand (BOD_5) mgO_2/L	34,771 – 48,300
Chemical oxygen demand (COD) mgO_2/L	69,500 - 89,591
Total dissolved solid (TDS) (mg/L)	9,310
Total suspended solid (TSS) (mg/L)	36,560 - 47,690
Total solid (TS) (mg/L)	47,050 – 62,000
pH	3.4–5.2
Reducing sugar (mg/L)	228

Sources: Adapted from references [31, 32, 35, 36].

Table 6.
Physico-chemical characteristics of POME.

Organism	Fermentation system and mode	Type of reactor	Product	References
Clostridium beijerinckii ATCC 8260	SmF batch	Hungate tube	Biohydrogen	[35]
Mixed cultures	SmF batch	Serum bottles	Biohydrogen	[31]
Mixed culture	SmF batch	Bioreactor	Biohydrogen	[38]
Mixed	SmF 2 stages operation	1. UASB -methane 2. ASBR -hydrogen	• Biohydrogen • Methane	[32]
Mixed	SmF continuous	UASB	Methane	[32]
Mixed microalgae	SmF batch	Flask	Biodiesel	[39]

Note: UASB, up flow anaerobic sludge blanket reactor; ASBR, anaerobic sequencing batch reactor.

Table 7.
Biochemical production from different organisms using POME as substrate.

example) [34]. Other biochemical products may be produced using the treatment technique. **Table 7** summarizes the different fermentation processes on various biochemical products using POME as a substrate.

5. Conclusion

Lignocellulosic material, especially from oil palm biomass, is a promising source as a feedstock for the fermentation process as it has a high content of cellulose and hemicellulose. Substrate selection is the most important factor in determining the techno-economic viability of large-scale chemical products. The substrate should be on the basis of easy availability, conversion efficiency, being toxic-free and low operational cost. Thus, the bioconversion route in chemical product production may create business opportunities to utilize the abundant agro-industrial waste that is being generated, particularly in terms of environmental pollution.

Acknowledgements

The authors would like to thank Universiti Malaysia Pahang (UMP) for the support.

Conflict of interest

The authors declare no conflict of interest.

Nomenclature

OPT	oil palm trunk
OPEFB	oil palm empty fruit bunches
OPF	oil palm frond

POME	palm oil mill effluent
ha	hectares
sp.	species
SSF	solid-state fermentation
SmF	submerged fermentation
COD	chemical oxygen demand
BOD	biochemical oxygen demand
ppm	part per million
UASB	up flow anaerobic sludge blanket reactor
ASBR	anaerobic sequencing batch reactor

Author details

Farhan M. Said*, Nor Farhana Hamid, Mohamad Al-Aamin Razali
and Nur Fathin Shamirah Daud
Faculty of Chemical and Process Engineering Technology, College of Engineering Technology, Universiti Malaysia Pahang, Gambang, Pahang, Malaysia

*Address all correspondence to: farhan@ump.edu.my; farhan_msaid@yahoo.co.uk

References

[1] N. L. Rozali, M. A. Yarmo, A. S. Idris, A. Kushairi, and U. S. Ramli, "Metabolomics differentiation of oil palm (Elaeis guineensis Jacq.) spear leaf with contrasting susceptibility to Ganoderma boninense," Plant Omics, vol. 10, no. 2, pp. 45-52, 2017, doi: 10.21475/poj.10.02.17.pne364.

[2] M. P. O. B. MPOB, "Malaysian oil palm statistics 2016," 2016.

[3] M. A. Sukiran, F. Abnisa, W. M. A. Wan Daud, N. Abu Bakar, and S. K. Loh, "A review of torrefaction of oil palm solid wastes for biofuel production," Energy Convers. Manag., vol. 149, pp. 101-120, 2017, doi: 10.1016/j.enconman.2017.07.011.

[4] N. A. Bukhari, S. K. Loh, N. A. Bakar, and J. M. Jahim, "Lignocellulose-derived sugar from oil palm biomass," *Palm Oil Eng.* Bull., no. 126, pp. 25-36, 2018.

[5] M. A. Abdel-Rahman, Y. Tashiro, and K. Sonomoto, "Lactic acid production from lignocellulose-derived sugars using lactic acid bacteria: Overview and limits," J. Biotechnol., vol. 156, no. 4, pp. 286-301, Dec. 2011, doi: 10.1016/j.jbiotec.2011.06.017.

[6] C. N. Hamelinck, G. van Hooijdonk, and A. P. Faaij, "Ethanol from lignocellulosic biomass: techno-economic performance in short-, middle- and long-term," Biomass and Bioenergy, vol. 28, no. 4, pp. 384-410, Apr. 2005, doi: 10.1016/j.biombioe.2004.09.002.

[7] M. Balat, "Production of bioethanol from lignocellulosic materials via the biochemical pathway: A review," Energy Convers. Manag., vol. 52, no. 2, pp. 858-875, Feb. 2011, doi: 10.1016/j.enconman.2010.08.013.

[8] S. Achinas and G. J. W. Euverink, "Consolidated briefing of biochemical ethanol production from lignocellulosic biomass," Electron. J. Biotechnol., vol. 23, pp. 44-53, Sep. 2016, doi: 10.1016/j.ejbt.2016.07.006.

[9] S. K. Ang, S. E.M., A. Y., S. A.A, and M. M.S, "Production of cellulases and xylanase by Aspergillus fumigatus SK1 using untreated oil palm trunk through solid state fermentation," *Process Biochem.*, vol. 48, no. 9, pp. 1293-1302, Sep. 2013, doi: 10.1016/j.procbio.2013.06.019.

[10] H. S. Hafid, A. S. Baharuddin, M. N. Mokhtar, F. N. Omar, M. A. P. Mohammed, and M. Wakisaka, "Enhanced laccase production for oil palm biomass delignification using biological pretreatment and its estimation at biorefinary scale," *Biomass and Bioenergy*, vol. 144, no. December 2020, p. 105904, 2021, doi: 10.1016/j.biombioe.2020.105904.

[11] R. Dungani *et al.*, "Biomaterial from Oil Palm Waste: Properties, Characterization and Applications. In Palm Oil," in *IntechOpen*, vol. 3, no. July, London, UK, 2018, pp. 31-51.

[12] H. P. S. Abdul Khalil, M. Siti Alwani, R. Ridzuan, H. Kamarudin, and A. Khairul, "Chemical Composition, Morphological Characteristics, and Cell Wall Structure of Malaysian Oil Palm Fibers," Polym. Plast. Technol. Eng., vol. 47, no. 3, pp. 273-280, Feb. 2008, doi: 10.1080/03602550701866840.

[13] S. Sabiha-Hanim, M. A. M. Noor, and A. Rosma, "Effect of autohydrolysis and enzymatic treatment on oil palm (Elaeis guineensis Jacq.) frond fibres for xylose and xylooligosaccharides production," Bioresour. Technol., vol. 102, no. 2, pp. 1234-1239, Jan. 2011, doi: 10.1016/j.biortech.2010.08.017.

[14] U. R. Ezeilo, R. A. Wahab, L. C. Tin, I. I. Zakaria, F. Huyop, and N. A. Mahat, "Fungal-Assisted Valorization of Raw Oil Palm Leaves for Production of Cellulase

and Xylanase in Solid State Fermentation Media," Waste and Biomass Valorization, vol. 11, no. 7, pp. 3133-3149, 2020, doi: 10.1007/s12649-019-00653-6.

[15] M. R. Mohamad Ikubar, M. Abdul Manan, M. Md. Salleh, and A. Yahya, "Solid-state fermentation of oil palm frond petiole for lignin peroxidase and xylanase-rich cocktail production," *3* Biotech, vol. 8, no. 5, p. 0, 2018, doi: 10.1007/s13205-018-1268-1.

[16] N. F. S. Daud, F. M. Said, N. H. M. Yasin, and M. A. K. M. Zahari, "Optimization of Red Pigment Production by Solid State Fermentation Using Oil Palm Frond," Mater. Sci. Forum, vol. 1025, pp. 150-156, 2021, doi: 10.4028/www.scientific.net/msf.1025.150.

[17] F. M. Said and N. F. Hamid, "Natural red colorant via solid-state fermentation of oil palm frond by Monascus purpureus FTC 5356: Effect of operating factors," J. Eng. Sci. Technol., vol. 14, no. 5, pp. 2576-2589, 2019.

[18] J. S. Tolan and B. Foody, "Cellulase from Submerged Fermentation," in *ecent Progress in Bioconversion of Ligno-cellulosics.*, T. G. T. et Al., Ed. Berlin, Heidelberg: Springer, 1999, pp. 41-67.

[19] F. N. Maluin, M. Z. Hussein, and A. S. Idris, "An overview of the oil palm industry: Challenges and some emerging opportunities for nanotechnology development," Agronomy, vol. 10, no. 3, 2020, doi: 10.3390/agronomy10030356.

[20] A. H. Norhazimah, F. M. N. Siti, M. Aida, A. B. Dilaeleyana, and M. A. Nur Shahirah, "Direct Fermentation of Oil Palm (Elaeis guineensis) Trunk Sap to Bioethanol by Saccharomyces cerevisiae," IOP Conf. Ser. Mater. Sci. Eng., vol. 943, no. 1, 2020, doi: 10.1088/1757-899X/943/1/012012.

[21] S. K. Ang, A. Yahya, S. Abd Aziz, and M. Md Salleh, "Isolation, screening, and identification of potential cellulolytic and xylanolytic producers for biodegradation of untreated oil palm trunk and its application in saccharification of lemongrass leaves," Prep. Biochem. Biotechnol., vol. 45, no. 3, pp. 279-305, 2015, doi: 10.1080/10826068.2014.923443.

[22] C. S. Goh, K. T. Tan, K. T. Lee, and S. Bhatia, "Bio-ethanol from lignocellulose: Status, perspectives and challenges in Malaysia," Bioresour. Technol., vol. 101, no. 13, pp. 4834-4841, Jul. 2010, doi: 10.1016/j.biortech.2009.08.080.

[23] K. Noratiqah, M. S. Madihah, B. S. Aisyah, M. S. Eva, A. A. Suraini, and K. Kamarulzaman, "Statistical optimization of enzymatic degradation process for oil palm empty fruit bunch (OPEFB) in rotary drum bioreactor using crude cellulase produced from Aspergillus niger EFB1," Biochem. Eng. J., vol. 75, pp. 8-20, Jun. 2013, doi: 10.1016/j.bej.2013.03.007.

[24] W. Fatriasari, R. Raniya, M. Oktaviani, and E. Hermiati, "The improvement of sugar and bioethanol production of oil palm empty fruit bunches (Elaeis guineensis Jacq) through microwave-assisted maleic acid pretreatment," BioResources, vol. 13, no. 2, pp. 4378-4403, 2018, doi: 10.15376/biores.13.2.4378-4403.

[25] K. Jatuwong, J. Kumla, N. Suwannarach, K. Matsui, and S. Lumyong, "Bioprocessing of agricultural residues as substrates and optimal conditions for phytase production of chestnut mushroom, pholiota adiposa, in solid state fermentation," J. Fungi, vol. 6, no. 4, pp. 1-21, 2020, doi: 10.3390/jof6040384.

[26] J. T. E. Lee *et al.*, "Improving methane yield of oil palm empty fruit bunches by wet oxidation pretreatment: Mesophilic and thermophilic anaerobic digestion conditions and the associated global warming potential effects," *Energy Convers. Manag.*, vol. 225, no. June, p. 113438, 2020, doi: 10.1016/j.enconman.2020.113438.

[27] K. A. Ajijolakewu, C. P. Leh, C. K. Lee, and W. A. Wan Nadiah, "Characterization of novel Trichoderma hemicellulase and its use to enhance downstream processing of lignocellulosic biomass to simple fermentable sugars," Biocatal. Agric. Biotechnol., vol. 11, pp. 166-175, Jul. 2017, doi: 10.1016/j.bcab.2017.06.005.

[28] U. R. Ezeilo, C. T. Lee, F. Huyop, I. I. Zakaria, and R. A. Wahab, "Raw oil palm frond leaves as cost-effective substrate for cellulase and xylanase productions by Trichoderma asperellum UC1 under solid-state fermentation," *J. Environ. Manage.*, vol. 243, no. May, pp. 206-217, 2019, doi: 10.1016/j.jenvman.2019.04.113.

[29] J. P. Tan, J. M. Jahim, S. Harun, T. Y. Wu, and T. Mumtaz, "Utilization of oil palm fronds as a sustainable carbon source in biorefineries," Int. J. Hydrogen Energy, vol. 41, no. 8, pp. 4896-4906, 2016, doi: 10.1016/j.ijhydene.2015.08.034.

[30] H. Kamyab, S. Chelliapan, M. F. M. Din, S. Rezania, T. Khademi, and A. Kumar, *Palm Oil Mill Effluent as an Environmental Pollutant*, vol. 32, no. July. InTech, 2018.

[31] N. Norfadilah, A. Raheem, R. Harun, and F. Ahmadun, "Bio-hydrogen production from palm oil mill effluent (POME): A preliminary study," Int. J. Hydrogen Energy, vol. 41, no. 28, pp. 11960-11964, Jul. 2016, doi: 10.1016/j.ijhydene.2016.04.096.

[32] C. Mamimin *et al.*, "Two-stage thermophilic fermentation and mesophilic methanogen process for biohythane production from palm oil mill effluent," Int. J. Hydrogen Energy, vol. 40, no. 19, pp. 6319-6328, May 2015, doi: 10.1016/j.ijhydene.2015.03.068.

[33] S. Krishnan, L. Singh, M. Sakinah, S. Thakur, Z. A. Wahid, and J. Sohaili, "Effect of organic loading rate on hydrogen (H2) and methane (CH4) production in two-stage fermentation under thermophilic conditions using palm oil mill effluent (POME)," Energy Sustain. Dev., vol. 34, pp. 130-138, 2016, doi: 10.1016/j.esd.2016.07.002.

[34] Y. Ahmed, Z. Yaakob, P. Akhtar, and K. Sopian, "Production of biogas and performance evaluation of existing treatment processes in palm oil mill effluent (POME)," Renew. Sustain. Energy Rev., vol. 42, pp. 1260-1278, 2015, doi: 10.1016/j.rser.2014.10.073.

[35] D. Rosa *et al.*, "Biological hydrogen production from palm oil mill effluent (POME) by anaerobic consortia and Clostridium beijerinckii," *J. Biotechnol.*, vol. 323, no. June, pp. 17-23, 2020, doi: 10.1016/j.jbiotec.2020.06.015.

[36] M. M. Bello, M. M. Nourouzi, L. C. Abdullah, T. S. Y. Choong, Y. S. Koay, and S. Keshani, "POME is treated for removal of color from biologically treated POME in fixed bed column: Applying wavelet neural network (WNN)," J. Hazard. Mater., vol. 262, pp. 106-113, Nov. 2013, doi: 10.1016/j.jhazmat.2013.06.053.

[37] S. J. Santosa, "Palm Oil Boom in Indonesia: From Plantation to Downstream Products and Biodiesel," CLEAN - Soil, Air, Water, vol. 36, no. 5-6, pp. 453-465, Jun. 2008, doi: 10.1002/clen.200800039.

[38] Sukaina F . A . Barghash, "Universiti Putra Malaysia Biohydrogen Production From Palm Oil Mill Effluent Using a Thermophilic Semi-Continuous Process With Recycling," 2007.

[39] H. Kamyab *et al.*, "Isolate new microalgal strain for biodiesel production and using FTIR spectroscopy for assessment of pollutant removal from Palm Oil Mill Effluent (POME)," *Chem. Eng. Trans.*, vol. 63, no. May, pp. 91-96, 2018, doi: 10.3303/CET1863016.